AF676067

# Plant Biostimulants in Sustainable Horticulture and Agriculture: Development, Function, and Applications

# Plant Biostimulants in Sustainable Horticulture and Agriculture: Development, Function, and Applications

Editors

**Mohamad Hesam Shahrajabian**
**Spyridon A. Petropoulos**

MDPI

Basel • Beijing • Wuhan • Barcelona • Belgrade • Novi Sad • Cluj • Manchester

*Editors*
Mohamad Hesam Shahrajabian
Chinese Academy of Agricultural Sciences
Beijing
China

Spyridon A. Petropoulos
University of Thessaly
Volos
Greece

*Editorial Office*
MDPI AG
Grosspeteranlage 5
4052 Basel, Switzerland

This is a reprint of articles from the Special Issue published online in the open access journal *Plants* (ISSN 2223-7747) (available at: https://www.mdpi.com/journal/plants/special_issues/44FK72777L).

For citation purposes, cite each article independently as indicated on the article page online and as indicated below:

Lastname, A.A.; Lastname, B.B. Article Title. *Journal Name* **Year**, *Volume Number*, Page Range.

**ISBN 978-3-7258-2365-9 (Hbk)**
**ISBN 978-3-7258-2366-6 (PDF)**
**doi.org/10.3390/books978-3-7258-2366-6**

© 2024 by the authors. Articles in this book are Open Access and distributed under the Creative Commons Attribution (CC BY) license. The book as a whole is distributed by MDPI under the terms and conditions of the Creative Commons Attribution-NonCommercial-NoDerivs (CC BY-NC-ND) license.

# Contents

*plants*

*Editorial*

# Editorial for the Special Issue on Plant Biostimulants in Sustainable Horticulture and Agriculture: Development, Function, and Applications

**Mohamad Hesam Shahrajabian [1,*] and Spyridon A. Petropoulos [2,*]**

[1] National Key Laboratory of Agricultural Microbiology, Biotechnology Research Institute, Chinese Academy of Agricultural Sciences, Beijing 100086, China
[2] Department of Agriculture, Crop Production and Rural Environment, University of Thessaly, 38446 Volos, Greece
* Correspondence: hesamshahrajabian@gmail.com (M.H.S.); spetropoulos@uth.gr (S.A.P.); Tel.: +86-19250156090 (M.H.S.)

The growing need for food production through sustainable cultivation practices, without reducing crop yield and producer income, is a major objective due to increased environmental pollution and the gradual degradation of cultivated soils [1,2]. Various components with bioactive characteristics can be used as biostimulants to boost plant growth and development under normal and stressful conditions [3,4]. So far, six distinct categories of biostimulants have been recognized, including microbial inoculants; humic substances, such as humic acid and fulvic acid; protein hydrolysates, and amino acids; biopolymers; inorganic compounds; and seaweed extracts, all of which are commercially available with wide applications in agriculture [5,6]. The most important biostimulant effects on crops are the acceleration of crop establishment, the improvement in nutrient uptake and nutrient use efficiency, the induction of tolerance in nutrient leaching, the improvement in root development, the removal of heavy metals from contaminated soils, the improvement in crop performance, the stimulation of the immune system of plants, the improvement in the visual quality of final products, and the induction of the biosynthesis of plant defensive biomolecules. Apart from single-product effects, there are many reports that show the combination of biostimulants results in beneficial effects on agricultural and horticultural crops, which are usually better than those of single biostimulants. Different classification approaches have been proposed so far, based either on the origin of each biostimulant, such as biological or non-biological, microbial and non-microbial, or on the mode of action which divides biostimulants into phytohormonal and non-phytohormonal ones [4–6]. Biostimulants are components which can increase plant growth, but do not qualify as essential plant nutrients, while biofertilizers are live microbes whose primary impact is to increase plant growth. It is important to consider the point that the basic difference between biofertilizers and biostimulants is that biofertilizers contain different nutrients, while biostimulants do not have the plant nutrients. Microbial biostimulants can synthesize IAA, which can promote root branching and plant growth of the plants under biotic and abiotic stresses, and they can also alleviate salt stress as they are associated with a high level of IAA and can enhance them in the re-establishing of favorable water potential gradients under water shortage conditions, as well as increasing film hydration around the roots. This Special Issue provides different examples of agricultural and horticultural plants which have been influenced by various kinds of biostimulants, while considering different mechanisms of action. Considering the above, new biostimulant products need to produced and applied; however, this should be realized under a new approach focusing on the synergistic impacts of different biostimulatory agents instead of single-product application.

**Citation:** Shahrajabian, M.H.; Petropoulos, S.A. Editorial for the Special Issue on Plant Biostimulants in Sustainable Horticulture and Agriculture: Development, Function, and Applications. *Plants* **2024**, *13*, 2342. https://doi.org/10.3390/plants13172342

Received: 14 August 2024
Accepted: 21 August 2024
Published: 23 August 2024

**Copyright:** © 2024 by the authors. Licensee MDPI, Basel, Switzerland. This article is an open access article distributed under the terms and conditions of the Creative Commons Attribution (CC BY) license (https://creativecommons.org/licenses/by/4.0/).

This Special Issue focuses on the functions and roles of various types of biostimulants in different horticultural and agricultural crops within the framework of sustainable crop management, aiming to gather critical and important information regarding their positive effects on plant growth and final yield, as well as regarding their impacts on the quality of the final products. Furthermore, the major limitations of these practices as well as the future prospects of biostimulant studies will be presented. Furthermore, we are grateful to the reviewers for dedicating their time and maintaining the standards of the publications included in this Special Issue.

**Conflicts of Interest:** The authors declare no conflicts of interest.

**List of Contributions:**

1. Lorch, M.G.; Valverde, C.; Agaras, B.C. Variability in Maize Seed Bacterization and Survival Correlating with Root Colonization by *Pseudomonas* Isolates with Plant-Probiotic Traits. *Plants* **2023**, *13*, 2130. https://doi.org/10.3390/plants13152130.
2. Ambros, E.; Kotsupiy, O.; Karpova, E.; Panova, U.; Chernonosov, A.; Trofimova, E.; Goldenberg, B. A Biostimulant Based on Silicon Chelates Enhances Growth and Modulates Physiological Responses of In-Vitro-Derived Strawberry Plants to In-Vivo Conditions. *Plants* **2023**, *12*, 4193. https://doi.org/10.3390/plants12244193.
3. Kim, Y.-N.; Choi, J.H.; Kim, S.Y.; Yoon, Y.-E.; Choe, H.; Lee, K.-A.; Kantharaj, V.; Kim, M.-J.; Lee, Y.B. Biostimulatory Effects of *Chlorella fusca* CHK0059 on Plant Growth and Fruit Quality of Strawberry. *Plants* **2023**, *12*, 4132. https://doi.org/10.3390/plants12244132.
4. Khuna, S.; Kumla, J.; Srinuanpan, S.; Lumyong, S.; Suwannarach, N. Multifarious Characterization and Efficacy of Three Phosphate-Solubilizing *Aspergillus* Species as Biostimulants in Improving Root Induction of Cassava and Sugarcane Stem Cutting. *Plants* **2023**, *12*, 3630. https://doi.org/10.3390/plants12203630.
5. Sun, W.; Shahrajabian, M.H. The Application of Arbuscular Mycorrhizal Fungi as Microbial Biostimulant, Sustainable Approaches in Modern Agriculture. *Plants* **2023**, *12*, 3101. https://doi.org/10.3390/plants12173101.
6. Sun, W.; Shahrajabian, M.H. Developing Sustainable Agriculture Systems in Medicinal and Aromatic Plant Production by Using Chitosan and Chitin-Based Biostimulants. *Plants* **2023**, *12*, 2469. https://doi.org/10.3390/plants12132469.
7. Meneguzzi, R.D.V.; Fernandez, M.; Cappellari, L.D.R.; Giordano, W.; Banchio, E. Isolation and Characterization of Plant Growth-Promoting Bacteria from the Rhizosphere of Medicinal and Aromatic Plant *Minthostachys verticillate*. *Plants* **2024**, *13*, 2062. https://doi.org/10.3390/plants13152062.
8. Kovacs, D.; Horotan, K.; Orloci, L.; Makadi, M.; Mosonyi, I.; Sutori-Dioszegi, M.; Kisvarga, S. Histological and Physiological Study of the Effects of Biostimulants and Plant Growth Stimulants in *Viburnum opulus* 'Roseum'. *Plants* **2024**, *13*, 1446. https://doi.org/10.3390/plants13111446.
9. Altuntas, O.; Dasgan, H.Y.; Akhoundnejad, Y.; Nas, Y. Unlocking the Potential of Pepper Plants under Salt Stress: Mycorrhizal Effects on Physiological Parameters Related to Plant Growth and Gas Exchange Across Tolerant and Sensitive Genotypes. *Plants* **2024**, *13*, 1380. https://doi.org/10.3390/plants13101380.
10. Rossini, A.; Ruggeri, R.; Mzid, N.; Rossini, F.; Miceli, G.D. *Codium fragile* (Suringar) Hariot as Biostimulant Agent to Alleviate Salt Stress in Durum Wheat: Preliminary Results from Germination Trials. *Plants* **2024**, *13*, 283. https://doi.org/10.3390/plants13020283.

## References


1. Shahrajabian, M.H.; Chaski, C.; Polyzos, N.; Tzortzakis, N.; Petropoulos, S.A. Sustainable Agriculture Systems in Vegetable Production Using Chitin and Chitosan as Plant Biostimulants. *Biomolecules* **2021**, *11*, 819. [CrossRef] [PubMed]
2. Shahrajabian, M.H.; Chaski, C.; Polyzos, N.; Petropoulos, S.A. Biostimulants Application: A Low Input Cropping management Tool for Sustainable Farming of Vegetables. *Biomolecules* **2021**, *11*, 698. [CrossRef] [PubMed]
3. Sun, W.; Shahrajabian, M.H.; Kuang, Y.; Wang, N. Amino Acids Biostimulants and Protein Hydrolysates in Agricultural Sciences. *Plants* **2024**, *13*, 210. [CrossRef] [PubMed]
4. Sun, W.; Shahrajabian, M.H.; Soleymani, A. The Roles of Plant-Growth-Promoting Rhizobacteria (PGPR)-Based Biostimulants for Agricultural Production Systems. *Plants* **2024**, *13*, 613. [CrossRef] [PubMed]

5. Shahrajabian, M.H.; Petropoulos, S.A.; Sun, W. Survey of the Influences of Microbial Biostimulants on Horticultural Crops: Case Studies and Successful Paradigms. *Horticulturae* **2023**, *9*, 193. [CrossRef]
6. Sun, W.; Shahrajabian, M.H. The Effectiveness of Rhizobium Bacteria on Soil Fertility and Sustainable Crop Production under Cover and Catch Crops Management and Green Manuring. *Not. Bot. Horti Agrobot. Cluj-Napoca* **2022**, *50*, 1–36. [CrossRef]

**Disclaimer/Publisher's Note:** The statements, opinions and data contained in all publications are solely those of the individual author(s) and contributor(s) and not of MDPI and/or the editor(s). MDPI and/or the editor(s) disclaim responsibility for any injury to people or property resulting from any ideas, methods, instructions or products referred to in the content.

plants

MDPI

Article

# Variability in Maize Seed Bacterization and Survival Correlating with Root Colonization by *Pseudomonas* Isolates with Plant-Probiotic Traits

**Melani G. Lorch [1,2], Claudio Valverde [1,2] and Betina C. Agaras [1,2,*]**

[1] Laboratory of Physiology and Genetics of Plant Probiotic Bacteria (LFGBBP), Centre of Biochemistry and Microbiology of Soils, National University of Quilmes, Bernal B1876BXD, Argentina; mglorch@gmail.com (M.G.L.); valverdecl@hotmail.com (C.V.)

[2] National Scientific and Technical Research Council (CONICET), Buenos Aires C1425FQB, Argentina

[*] Correspondence: betina.agaras@gmail.com or betina.agaras@unq.edu.ar; Tel.: +54-9-11-4365-7100 (ext. 5638)

**Abstract:** Seed treatment with plant growth-promoting bacteria represents the primary strategy to incorporate them into agricultural ecosystems, particularly for crops under extensive management, such as maize. In this study, we evaluated the seed bacterization levels, root colonization patterns, and root competitiveness of a collection of autochthonous *Pseudomonas* isolates that have demonstrated several plant-probiotic abilities in vitro. Our findings indicate that the seed bacterization level, both with and without the addition of various protectants, is specific to each *Pseudomonas* strain, including their response to seed pre-hydration. Bacterization kinetics revealed that while certain isolates persisted on seed surfaces for up to 4 days post-inoculation (dpi), others experienced a rapid decline in viability after 1 or 2 dpi. The observed differences in seed bacterization levels were consistent with the root colonization densities observed through confocal microscopy analysis, and with root competitiveness quantified via selective plate counts. Notably, isolates *P. protegens* RBAN4 and *P. chlororaphis* subsp. *aurantiaca* SMMP3 demonstrated effective competition with the natural microflora for colonizing the maize rhizosphere and both promoted shoot and root biomass production in maize assessed at the V3 grown stage. Conversely, *P. donghuensis* SVBP6 was detected at very low levels in the maize rhizosphere, but still exhibited a positive effect on plant parameters, suggesting a growth-stimulatory effect during the early stages of plant development. In conclusion, there is a considerable strain-specific variability in the maize seed bacterization and survival capacities of *Pseudomonas* isolates with plant-probiotic traits, with a correlation in their root competitiveness under natural conditions. This variability must be understood to optimize their adoption as inputs for the agricultural system. Our experimental approach emphasizes the critical importance of tailoring seed bacterization treatments for each inoculant candidate, including the selection and incorporation of protective substances. It should not be assumed that all bacterial cells exhibit a similar performance.

**Keywords:** bioinput; maize seed coating; inoculant; pre-inoculation; bacterial recovery

**Citation:** Lorch, M.G.; Valverde, C.; Agaras, B.C. Variability in Maize Seed Bacterization and Survival Correlating with Root Colonization by *Pseudomonas* Isolates with Plant-Probiotic Traits. *Plants* **2024**, *13*, 2130. https://doi.org/10.3390/plants13152130

Academic Editors: Spyridon A. Petropoulos and Mohamad Hesam Shahrajabian

Received: 2 July 2024
Revised: 29 July 2024
Accepted: 30 July 2024
Published: 1 August 2024

**Copyright:** © 2024 by the authors. Licensee MDPI, Basel, Switzerland. This article is an open access article distributed under the terms and conditions of the Creative Commons Attribution (CC BY) license (https://creativecommons.org/licenses/by/4.0/).

## 1. Introduction

Rhizobiomes or rhizomicrobiomes [1,2] are primarily constituted of a subset of the bulk soil microbiota, and significantly influenced by the seed microbiome [3]. In recent years, there has been a growing interest in the seed microbiomes of staple food crops, such as maize, wheat, and rice, aimed at enhancing food safety and seed quality, as well as exploiting the natural microbial diversity carried by seeds and their plant-growth promoting potential [4,5]. Most research has focused on endophytic seed microbiomes, with limited consideration of the epiphytic seed microbiome [6–8]. In all instances, crop genotype emerges as the main factor affecting microbial composition of those microbiomes, while environmental factors play a comparatively minor role [4,7,9]. Generally, a core microbiome can be described, which is rich in specific bacterial genera, such as *Pantoea*, *Enterobacter*,

*Sphingomonas*, *Pseudomonas*, *Acinetobacter*, and *Rhodococcus* sp. [3,9,10]. Regarding seed epiphytes, a recent study has indicated that maize seeds exhibit a lower natural microbial load on their surfaces than barley and wheat, with the Proteobacteria group being the most abundant [10]. This finding opens the possibility of bioaugmenting maize seeds at the planting stage through the introduction of suitable plant-growth promoting microorganisms (PGPMs) to facilitate early colonization of the seedling root.

Biological seed treatments have historically been the predominant method for ensuring high densities of introduced microbial inoculants in crop rhizospheres, mimicking the application process of agrochemical products [11–13]. This approach has been shown to improve crop production across various geographical conditions, climates, and crop types, among other factors [14,15]. It is particularly applicable to extensive crops that are directly sown into the soil, such as maize, which ranks among the top cultivated crops worldwide, with Argentina being one of the six leading maize-producing countries [16]. The concept behind seed bacterization is to position relatively high amounts of the PGPM cells in the spermosphere, thereby facilitating the establishment of the microbes within the rhizosphere and enhancing root colonization and the intimate interaction between plants and bacteria, which is critical for promoting plant growth [12,17,18]. Furthermore, there is a rising demand from seed companies and growers for access to pre-inoculated seeds that can be stored for weeks or months before sowing [11]. The technology associated with these methods has primarily been developed for legume rhizobial inoculants rather than for other PGPMs [19–21]. For rhizobial inoculants, it has been demonstrated that the initial concentration of bacterial cells in the inoculum is directly associated with the number of legume nodules formed and the resulting crop yield [22–24].

Challenges related to the survival of microorganisms on seed surfaces, compatibility with agrochemical applications, and the limited tolerance of PGPMs to prolonged exposure to elevated temperatures during the storage of pre-inoculated seeds prompt ongoing exploration for novel additives and application alternatives, such as desiccation protectants, biopriming techniques, film-type coatings, or seed pelleting [12,25]. Significant advancements have been achieved in the development of adjunct substances for formulations based on rhizobia, including sucrose, trehalose, polyethylene glycol, carboxymethyl cellulose, gum arabic, glycerol, xanthan, and other compounds that essentially protect bacteria from desiccation and improve adherence to seeds; they have been tested with varying levels of efficacy [26–28]. Trehalose, along with other non-reducing disaccharides, acts as a protectant and protein stabilizer when accumulated in bacterial cells, forming a glass-like sugar matrix that effectively replaces the water hydrogen bonds [29–31]. Therefore, to promote its accumulation under various stress conditions, microorganisms activate trehalose synthesis pathways, and/or suppress trehalose catabolism [29,32–34]. The combination of this disaccharide with polyvinylpyrrolidone (PVP) enhances the viscosity of the suspensions, potentially improving cell adherence to seeds. PVP retains significant amounts of water, maintains moisture around the cells, and slows the bacterial drying rate [35]. Additionally, the combined effects of certain protective additives, such as trehalose–PVP mixtures, have been reported for specific bacterial species, including genetically modified *Pseudomonas* [35–37].

Soybean seeds that are pre-inoculated with commercial rhizobial suspensions can be stored for up to 15 days prior to planting in a cool, well-ventilated location. It has been observed that a seed treatment performed 7 days before sowing improves yields compared to untreated controls [38]. For *Pseudomonas*—non-sporulating Gram-negative bacteria like rhizobia, that are sensitive to abrupt environmental changes such as desiccation—some of these additives have also been evaluated. Seed inoculations with suspensions supplemented with methylcellulose, carboxymethyl cellulose, trehalose, or glycerol have proven effective in wheat, cotton, tomato, and sugar beet [39–42]. Conversely, comparative studies analyzing the survival of *Pseudomonas* strains in seeds of extensive crops, with or without the addition of these substances, have not yet been conducted, unlike studies with *B. japonicum* [26]. Only in sterilized onion seeds conserved at 4 °C has it been demonstrated that *P. fluorescens* F113 survives for 70 days post-inoculation [43]. Given that, for practical reasons, liquid inoculants are generally favored by producers for seed application, and maintaining high levels of bacterial viability during pre-inoculation until sowing is ideal [13,43]; it would be pertinent to assess the capacity of candidate *Pseudomonas* isolates to remain viable in pre-bacterized seeds if new bioinputs are to be developed.

Therefore, beyond merely describing the plant-probiotic potential that can be assessed in vitro, studying the bacteria–plant interaction throughout all stages (inoculation/competition/colonization/execution of probiotic properties) is essential for optimizing the PGPM performance in agricultural settings, thereby identifying which step is the primary bottleneck for each candidate. However, there are limited studies regarding the performance of seed bacterization and/or the implication of the bacterization levels on root colonization and competitiveness for non-rhizobial microorganisms like *Pseudomonas*. For *Bacillus*, a positive correlation was observed between wheat seed bacterization density and root competitiveness in studies utilizing disinfected seeds inoculated with lyophilized bacteria and carboxymethyl cellulose as an additive [44]. Conversely, for *Stenotrophomonas maltophilia* and *Pseudomonas fluorescens* isolates, no correlation was found between bacterization levels of disinfected wheat seeds and root competitiveness in field assays using also carboxymethyl cellulose as an additive [45]. In addition to the contrasting evidence for the same crop with different bacterial isolates, there are no reports in the literature of such studies using non-disinfected seeds.

Members of the *Pseudomonas* genus are recognized as keystone components of rhizospheric microbiomes, contributing to plant growth and health [46–48]. In our laboratory, we have obtained and characterized a set of *Pseudomonas* isolates exhibiting plant-probiotic potential [49]. This collection encompasses representatives of various species, including *P. protegens*, *P. chlororaphis*, *P. soli*, and *P. donghuensis*, some of which have been recently documented in the literature, particularly concerning their biological control capabilities against phytopathogenic fungi [49–51]. Additionally, we have observed that certain isolates could improve the grain yield in wheat and maize through field assays [52]. Therefore, to optimize the development of bioformulations based on these microorganisms, the aims of this study are: (1) to analyze the bacterization level achieved through seed inoculation using non-disinfected seeds and doses recommended for agricultural producers; (2) to monitor the survival capacity of each tested strain on pre-bacterized seeds in the presence of protectant additives; (3) to characterize their root colonization pattern with confocal microscopy; (4) to evaluate their root competitiveness, quantifying their density in maize seedlings grown in soil in the presence of competing microflora.

## 2. Results

### 2.1. Chromosomal Tagging of Pseudomonas Isolates for Facilitating Enumeration in Bacterized Seeds and Visualization in Colonized Roots under Axenic Conditions

To analyze the levels of bacterization in non-disinfected seeds by plate count on selective media, and to visualize the bacterial cells on root surfaces to assess root colonization patterns of our *Pseudomonas* isolates under non-axenic conditions, we introduced antibiotic resistance markers and genes encoding fluorescent proteins. The Tn7 mutagenesis approach proved optimal for these modifications without altering the genetic background of the *Pseudomonas* strains [53–55]. Selected Tn7 derivatives maintained physiological characteristics, PGPR traits, and growth rates indistinguishable from wild type isolates (Figures S1 and S2). Growth curves with trehalose demonstrated that only SVBP6 and SVMP4 were unable to utilize this non-reducing disaccharide as a carbon source (Figure S2). Therefore, trehalose could exclusively act as an osmoprotectant for these isolates, maintaining membrane and protein integrity [56]. We also confirmed the stable integration of Tn7 cassettes in the absence of an antibiotic selection pressure (Table S1), ensuring that selective counts from bacterized seeds accurately reflected the load of viable cells after different varying incubation times.

### 2.2. Seed Bacterization Levels Were Strain-Dependent

The evaluation of seed bacterization levels at the time of inoculation (0 days post-inoculation, dpi) demonstrated that incorporating the Premax® additive in the bacterial suspensions significantly enhanced the recovery of culturable cells from the seeds. Premax® improved the bacterial recovery for all isolates compared to control treatments, albeit to varying extents (Figure 1a). The improvement ranged from 3.2× to 39.4×, indicating genuine protective and/or adhesive effects of the additive on bacterial cells. Specifically, RBAN4-*yfp* and RPAN1-*yfp* showed higher bacterization recovery values than the reference strain 1008-*cfp*, both with and without the additive (Figure S3). Conversely, SVMP4-*yfp* achieved lower bacterization levels than the reference 1008-*cfp*, both with and without the additive (Figure S3). Consequently, we tested various additive compounds with SVMP4-*yfp* to enhance its maize seed bacterization. All the tested additives improved SVMP4-*yfp* recovery, with trehalose, glycerol, and a trehalose–PVP mixture showing the highest values, increasing bacterization by 2.8×, 2.7× and 3.2×, respectively, compared to the Premax® levels (Figure 1b). Germination remained unaffected by treatments or storage conditions.

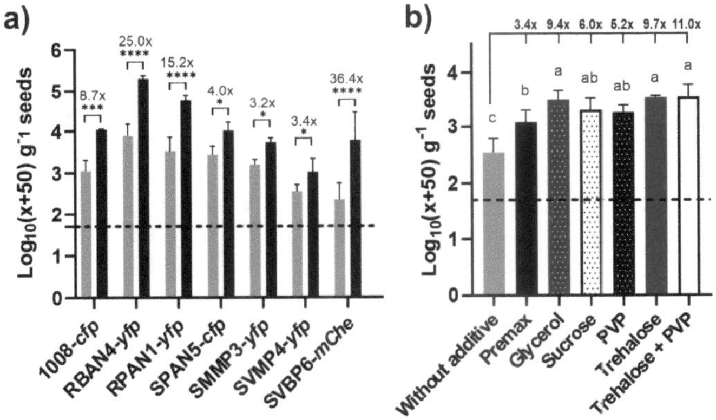


**Figure 1.** Variability in maize seed bacterization levels and protective effects within our *Pseudomonas* collection at the time of inoculation (0 dpi). The data represent transformed CFU values using the formula $\log_{10}(x + 50)$, where x is the corresponding CFU value. This transformation is pertinent for statistical comparisons of data sets including CFU null data [57]. Dashed lines indicate the $\log_{10}$ value corresponding to a transformed null CFU count. (**a**) Recovery of culturable *Pseudomonas* from maize seeds inoculated in the presence (black) or absence (grey) of the commercial bacterial protectant Premax® (Rizobacter Argentina S.A., Pergamino, Argentina). Asterisks indicate statistically significant differences between treatments with or without Premax® (ANOVA with LSD-Fisher multiple comparison test; * $p < 0.05$, *** $p < 0.001$, **** $p < 0.0001$). Numbers above treatment bars indicate fold increase in bacterial recovery for treatments with Premax® compared to the corresponding control. (**b**) Effect of different additives on recovery of SVMP4-*yfp* from maize bacterized seeds. Different letters indicate statistically significant differences (ANOVA with LSD Fisher's multiple comparison test, $p < 0.05$). Numbers above bars indicate the fold increase in bacterial recovery for treatments with additives compared to the corresponding control.

### 2.3. Pre-Hydration of Maize Seeds Improved Recovery of Culturable Bacterial Cells

Recommended seed treatment doses involve minimal liquid quantities (a few microliters per seed), incorporating both chemical and biological inputs to prevent initiation of the germination process. Consequently, bacterial cells contacting seed surfaces may undergo desiccation shock due to the rapid liquid uptake by the seeds, potentially reducing cell culturability and/or viability. Given the favorable effect of additives on seed bacterization levels, we investigated whether our isolates' performance could be impacted by seed dryness. Pre-hydrated seeds before inoculation nullified the protective benefits of the Premax® for all isolates, resulting in higher bacterization levels for almost all strains (except for RBAN4) compared to non-soaked seeds inoculated with the addition of Premax® (Figure 2). Particularly, pre-hydration significantly enhanced the recovery of culturable SVMP4-*yfp* cells compared to the Premax® treatment alone (Figure 2). The inoculant dose corresponds to 5 µL of the bacterial suspensions per gram of seeds, containing $10^4$–$10^6$ CFU. In our assays, we recovered between 0.3 and 9.1% of the bacteria incorporated in the presence of Premax®, whereas these percentages did not exceed 0.51% when no additive was used (Table S2). These recovery rates substantially improved with seed pre-hydration, reaching 7.0 to 12.0% with Premax® (Table S2). However, RBAN4-*yfp* showed consistent recovery values irrespective of pre-hydration or the Premax® addition (Figure 2), with no increase in recovery percentages. Even isolates losing the additive's effect with pre-hydration reached slightly higher recovery percentages without the additive when the seeds were pre-hydrated (Table S2). Particularly, recovery of SVMP4-*yfp* cells from pre-hydrated seeds without Premax® reached 29.47% of the inoculated bacteria, contrasting with the 1.8% recovery obtained in the presence of Premax® (Table S2).

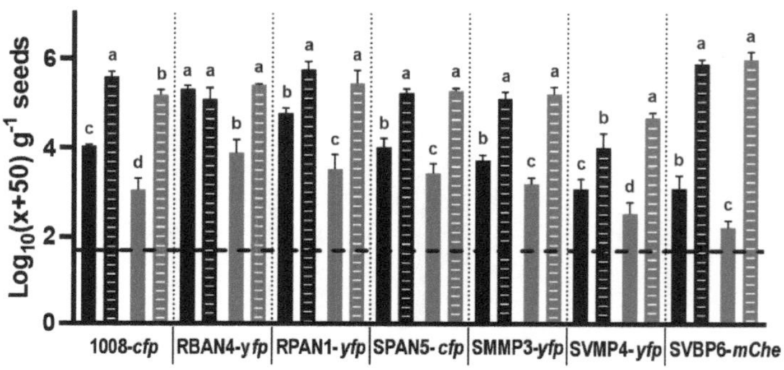


**Figure 2.** Pre-hydration of maize seeds before bacterization increases the recovery of culturable *Pseudomonas* cells at the time of inoculation (0 dpi). Treatments compared include those with (black) and without (grey) the addition of the commercial protectant Premax®, and with (dashed bars) or without (full bars) pre-hydration treatment (see main text for details). See Figure 1 legend for the reference about data transformation. Dashed line indicates the $\log_{10}$ value corresponding to a transformed null CFU count. Different letters indicate statistically significant differences between treatments for each isolate (ANOVA with LSD Fisher's multiple comparison test, $p < 0.05$).

### 2.4. *Bacterization Levels of Maize Seeds Strongly Declined up to 4 dpi*

Once we identified the optimal seed bacterization treatment for each isolate, we studied the decay kinetics of recovered culturable cells for a period of up to 4 dpi (Figure 3). Given the practical implications for agricultural producers, who often prefer seed treatments shortly before sowing, we chose a time frame comparable to a standard business week to evaluate bacterization kinetics. RBAN4-*yfp* showed the best performance, with the highest recovery values and a significant positive effect by the Premax® additive throughout the entire experiment (Figure 3). The reference strain 1008-*cfp* performed similarly, although the bacterial cell recoveries were lower (Figure 3). For the remaining isolates, the protectant effect of Premax® was observed only at 0 dpi; subsequent daily CFU counts exhibited a decaying trend that was statistically indistinguishable regardless of the use of the Premax® additive (Figure 3). This pattern was evident even for RPAN1-*yfp*, which initially exhibited higher bacterization levels at 0 dpi but decayed more rapidly than 1008-*cfp* thereafter (Figure 3). For SVMP4-*yfp* and SVBP6-*mChe*, the presence of the trehalose–PVP mixture improved the bacterial recovery, but only SVMP4-*yfp* retained recoverable cells up to 4 dpi. The bacterial cell suspensions stored under conditions identical to the bacterized seeds showed no significant viability decline up to 4 dpi (Table S3).

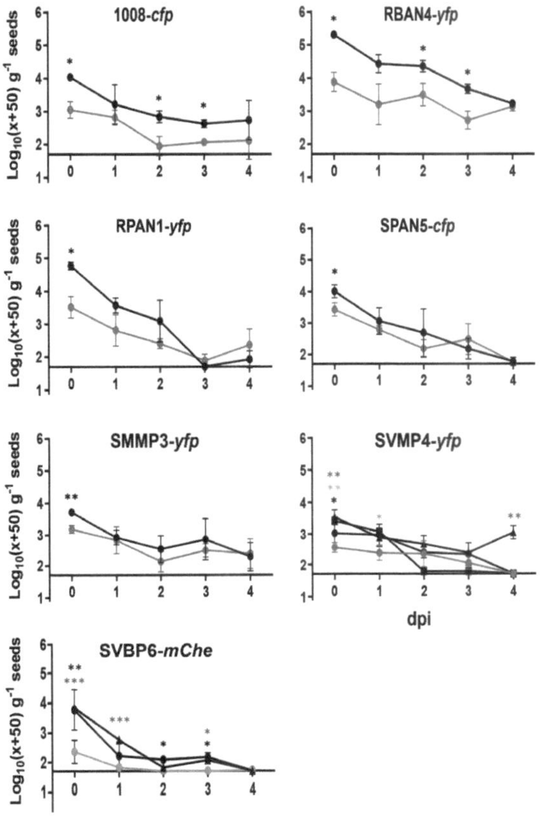


**Figure 3.** Time course recovery of *Pseudomonas* isolates from bacterized maize seeds is strain-dependent, showing a decline in the additive effect over time for some isolates. Bacterization levels (CFU) were determined in seeds sampled daily for up to 4 days after inoculation with (black) or without (grey) additives. Premax® treatment is denoted with black circles; trehalose–PVP mixture treatment is shown with triangles; and glycerol with squares. See Figure 1 legend for the reference about data transformation. Asterisks indicate statistically significant differences between additive and control treatments on the same day (Two-way ANOVA with LSD Fisher's multiple comparison test; *** $p < 0.001$; ** $p < 0.01$; * $p < 0.05$). Black asterisks denote statistically significant differences for Premax® treatment; dark grey asterisks denote differences for trehalose–PVP treatment; light grey asterisks denote differences for glycerol treatment.

### 2.5. Colonization Patterns on Maize Seedlings Were Particular and Congruent with Bacterization Levels

The observation of the maize seedlings under confocal microscopy allowed us to analyze the different colonization patterns on the root surfaces (Figure 4). We demonstrated that our isolates achieved acceptable levels of root colonization of the maize seedlings, preferentially occupying the plant cell junctions. However, we could discern three main configurations of bacterial cells on the epidermal surfaces: string-like arrangements (Figure 4e–g); micro-colonies dispersed on the surface (Figure 4b,c); or a combination of both, microcolonies and string-like arrangements (Figure 4a,d,f). RBAN4-*yfp*, which demonstrated the best seed bacterization performance (Figure 1), also displayed the highest root colonization level, with dense bacterial populations on root surfaces (Figure 4a), similar to colonization densities of the reference strain 1008 (Figure 4e). In contrast, SVMP4-*yfp* showed inconsistent patterns and low colonization levels across all replicates (Figure 4d). This observation aligns with the poor performance of this isolate during maize seed bacterization at 0 dpi (Figure 1).

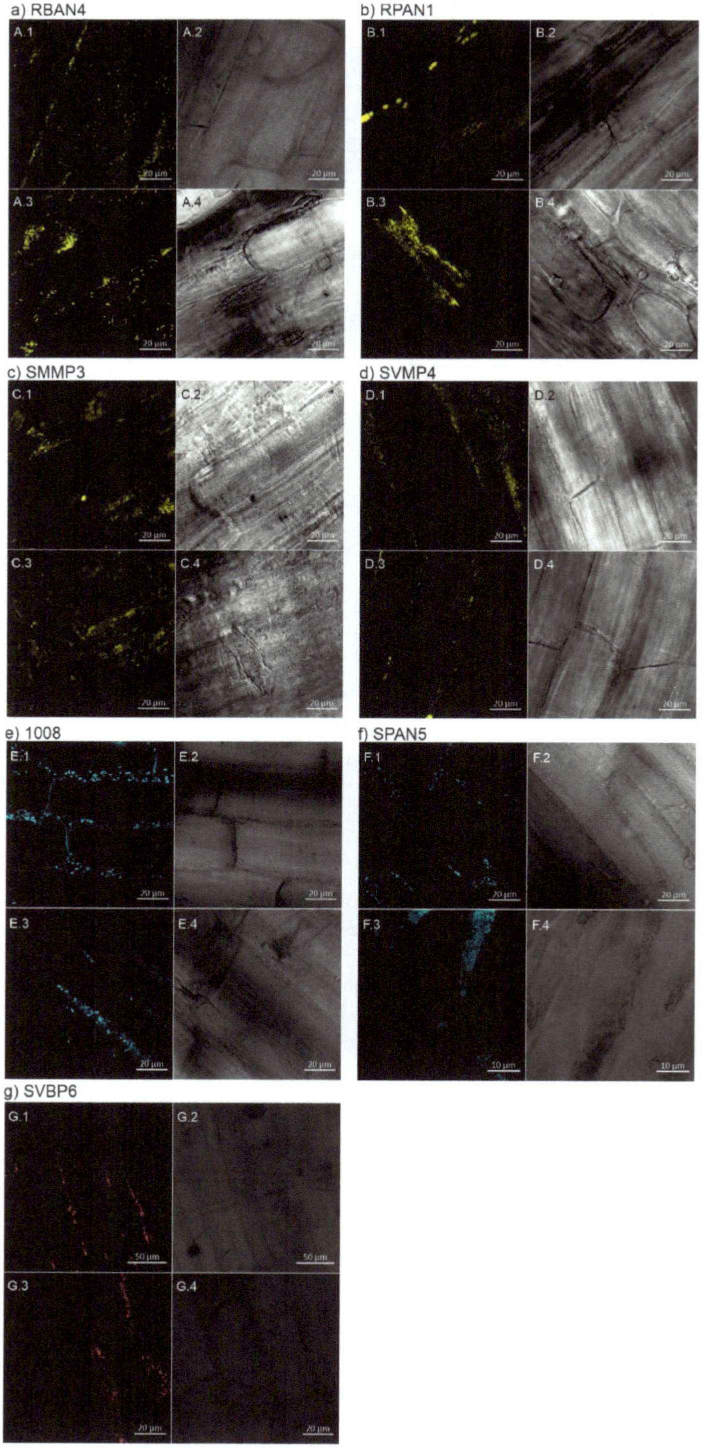


**Figure 4.** *Pseudomonas* isolates can achieve different colonization patterns on the root surface of maize seedlings after seed bacterization. Early root colonization (9 days after inoculation) was analyzed by confocal fluorescence microscopy of *Pseudomonas* isolate derivatives expressing different fluorescent

proteins: eYFP (**a–d**), eCFP (**e**,**f**) and mCherry (**g**). All images were captured at the elongation zone of the maize roots (1 cm above the root tip) using a 20× objective and 4× digital zoom, except for (**g**) (**1**,**2**) (2× zoom) and (**f**) (**3**,**4**) (40× objective and 4× zoom). In all cases, pictures from right panels (**2**,**4**) correspond to transmitted light microscopy and those from left panels (**1**,**3**) to confocal fluorescence microscopy. Images are representative of colonization patterns on all the observed plant roots (2–4 replicates per treatment). Control treatments were also observed, obtaining in all cases images without any fluorescent bacterial cells. Scales are indicated with a white bar on each image, corresponding to 20 μm, except for (**f**) (**3**,**4**) (10 μm) and (**g**) (**1**,**2**) (50 μm).

### *2.6. Maize Root Competitiveness Is Consistent with Seed Bacterization Levels, although Growth Promotion Effects Were Also Detected in Less Competitive Isolates*

After one month of plant growth in natural soil, we assessed the root competitiveness of the isolates by quantifying their abundance in a maize rhizosphere. The load of culturable, heterotrophic, mesophilic bacteria showed values between $5.7 \times 10^8$ and $3.5 \times 10^7$ CFU per gram of dry root (Table S4). We found significantly higher root colonization abundance for those strains that had shown the highest bacterization levels on the maize seeds at 0 dpi (Figure 5). Notably, SPAN5-*cfp* was the only isolate undetected in the maize rhizosphere (Figure 5) despite showing good seed bacterization levels (Figure 1). SVBP6-*mChe* and SVMP4-*yfp*, which had low seed bacterization rates (Figures 1 and 3), were also heterogeneously detected in rhizospheric samples (Figure 5). However, when we evaluated the plant-growth promoting effect of seed inoculation with these *Pseudomonas*, we observed improvements across most measured parameters (Figure 6), even for those isolates poorly detected in the rhizosphere, such as SPAN5-*cfp* and SVBP6-*mChe* (detection limit: <500 CFU $g^{-1}$ root). Particularly, the RBAN4-*yfp*, RPAN1-*yfp*, SMMP3-*yfp*, and SVBP6-*mChe* treatments significantly enhanced the shoot biomass (Figure 6a,b), even though the RBAN4-*yfp* treatment did not affect plant height (Figure 6c). Moreover, we found a positive correlation between the shoot biomass (fresh or dry), plant height, and root biomass (Figure S4). Analysis of root growth revealed increased root fresh weight by the *Pseudomonas* inoculation (Figure 6d), while root length remained similar across all the treatments (Figure 6e). Finally, the ratio of shoot to root biomass (SB/RB) indicated that SPAN5-*cfp* was the only treatment that significantly promoted shoot development without increasing root biomass, although SVBP6-*mChe* and SVMP4-*yfp* also showed similar performances (Figure 6f).

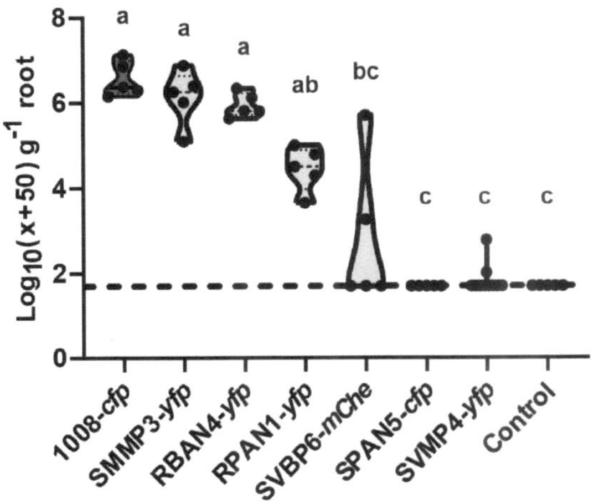


**Figure 5.** *Pseudomonas* isolates showed different root competitiveness during the early colonization of maize roots in the presence of natural soil microflora. Root colonization was quantified by selective

plate counts on S1 media supplemented with the corresponding antibiotic for every *Pseudomonas* derivative. See Figure 1 legend for reference about data transformation. Dashed line indicates the $log_{10}$ value corresponding to a transformed null CFU count. Data was corrected to express the values by the dry root weight. Different letters indicate statistically significant differences between treatments (Kruskal–Wallis non-parametric test, with the uncorrected Dunn's multiple comparison test).

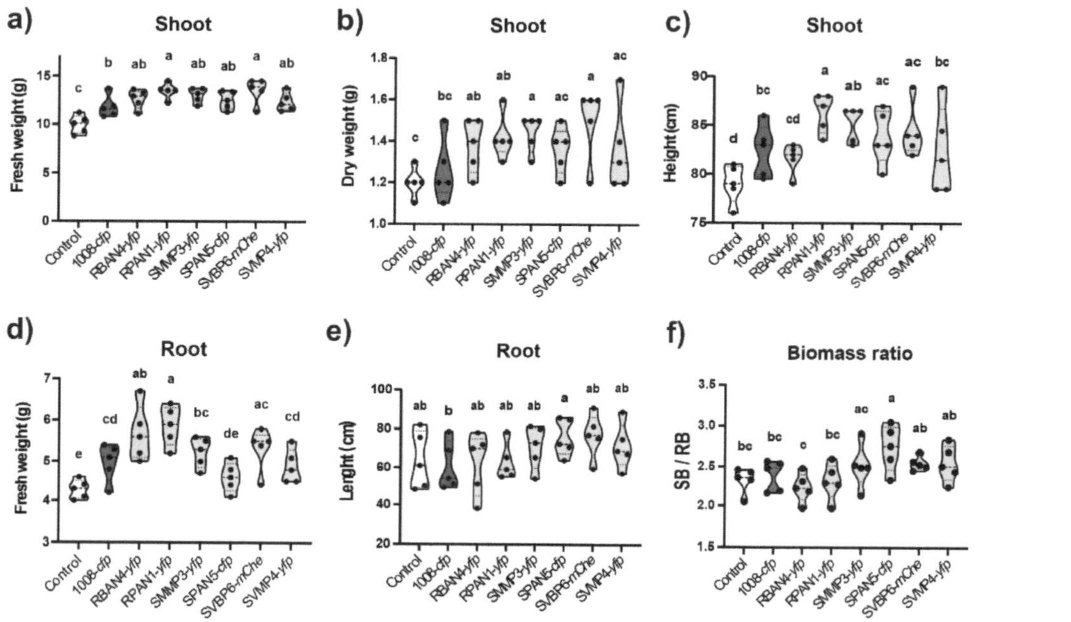


**Figure 6.** Bacterial inoculation improved several plant parameters of maize up to the V3 growth stage. After one month of incubation, we measured the shoot biomass, fresh (**a**) and dry (**b**); the shoot height (**c**); the fresh root biomass (**d**); the root length (**e**); and the shoot/root biomass ratio (SB/RB, **f**). For all the measured data, we included a non-inoculated control (white) and the reference treatment 1008 (dark grey). Different letters indicate statistically significant differences between treatments (ANOVA with LSD Fisher's multiple comparison test, $p < 0.05$).

## 3. Discussion

### *3.1. Variability in Maize Seed Bacterization among PGPMs from the Same Bacterial Genus*

It is generally assumed that different types of bacterial cells can adhere to and survive on seed surfaces with similar success [12]. Consequently, this aspect is not typically studied for inoculant candidates. In this work, we demonstrated distinct behaviors among isolates from various species within the same genus when applied to the surfaces of the same hybrid of maize seeds (Figure 1a). Comparing these isolates to a commercial *Pseudomonas* strain marketed as a bioinput for extensive crops [58], we found that some isolates achieved better, comparable, or worse bacterization levels, with CFU densities ranging from 3 to 5.5 $log_{10}$ per gram of maize seeds, depending on the isolate (Figure 1). Given an average surface area of 0.75 $\mu m^2$ per rod-shaped bacterial cell from the *Pseudomonas* genus [15], and an average maize seed surface area of 1250 $mm^2$ for the variety used in our experiments, the bacterial densities achieved imply a maximum coverage of 15–45% of the maize seed surface. Similar findings were noted across different crop seeds, suggesting an irregular spatial distribution of cells on seed surfaces [59,60]. While even higher levels of maize seed bacterization (over $10^6$ CFU per g) have been reported for a *Pseudomonas* isolate like the strain *P. putida* mt-2, our results generally align with other studies testing various *Pseudomonas* isolates using different inoculation techniques, highlighting limitations in achieving high bacterial densities on maize seed surfaces [61,62]. In previous studies, the adherence of *Pseudomonas* to seeds has been linked to the synthesis of cell surface proteins, like LapA and LapF, as well as flagella-mediated processes [63–67]. Consequently, differential biochemical and/or expression properties of these surface macromolecules in the tested isolates may impact their ability to adhere to seeds. Consistent with our seed bacterization results, the evaluation of adhesion capabilities to abiotic surfaces (e.g., polystyrene) has shown that RBAN4, RPAN1, SPAN5, and SMMP3 adhered more efficiently to such surfaces compared to SVBP6 and SVMP4 [49].

### 3.2. Enhancement of Seed Bacterization with a Commercial Bacterial Protectant Additive Was Also Strain-Dependent

Seed bacterization success can be significantly enhanced by incorporating a cell protectant and/or adhesive compounds into inoculant formulations [68–70]. Our study demonstrates that different protectants indeed improve bacterization performance of the tested isolates (Figure 1). The efficacy of Premax® in enhancing the bacterial recovery from the seeds was evident at 0 dpi (Figure 1a), and up to 3 dpi (Figure 3) for RBAN4-*yfp*, a member of the *P. protegens* subgroup (Table 1). However, at 4 dpi this enhancement was no longer observed, with approximately $10^3$ CFU per gram of seeds recovered regardless of the presence of the additive. This behavior, unique to this isolate among those tested, suggests a physiological feature enabling it to maintain viability and culturability throughout our evaluation period. Such persistence could facilitate pre-inoculation strategies for bioinputs formulated with RBAN4. Given that members of the *P. protegens* species are characterized by their multiple plant-growth promotion abilities, mainly associated with biocontrol [71], preserving viable cells at sowing becomes crucial for further colonizing the rhizosphere. Notably, the *P. protegens* strain CHA0 has been reported to enter a viable but non-culturable (VBNC) state under stressful conditions and in particular soil microbial habitats [72,73]. Therefore, RBAN4 might similarly enter this state under the potentially stressful conditions offered by seed surfaces. Future investigations employing VBNC detection techniques [74] could elucidate if RBAN4 cells persist longer with higher viable cell densities on maize seed surfaces.

### 3.3. The Incorporation of Trehalose and PVP Ameliorates Seed Bacterization for Isolates Experiencing Higher Dehydration Stress Upon Seed Inoculation

For all tested strains, pre-hydration (imbibition) of the seeds enhanced the bacterization levels at 0 dpi more effectively than the addition of protectants (Figure 2). This suggests that the *Pseudomonas* isolates studied herein undergo dehydration stress on the seed surfaces, leading to decreased viability in the absence of protectants. This effect was particularly pronounced for SVMP4-*yfp*, which exhibited the lowest bacterization levels (Figures 1 and 2). Previous studies have shown that the environmental model microorganism *P. putida* KT2440, a plasmid-cured derivative of the strain mt-2, is highly sensitive to desiccation stress, inducing a VBNC state under such conditions [75,76]. However, this sensitivity can be mitigated by introducing the *otsBA* operon from *E. coli*, which is involved in trehalose synthesis, enabling *P. putida* KT2440 to accumulate this osmolyte in its cytoplasm [37,77]. From our pseudomonads collection, strains SVBP6 and SVMP4 lack the ability to metabolize trehalose (Figure S2), and these isolates showed the lowest bacterization recovery and a pronounced decline in their survival on the seeds up to 4 dpi (Figures 1 and 3). Furthermore, supplementing NYB media with 20 mM trehalose did not alter their bacterization performances. Supporting these experimental results, genomic analyses of SVBP6 revealed the absence of homologous genes associated with trehalose metabolism (*treP*, *treA* and *treR*). Although we have not sequenced the genome of SVMP4, these trehalose-related genes are not present in the genomes of other *P. soli* isolates. However, with a similar approach, we found homologous genes related to trehalose synthesis, *treS* and *treY/Z*. Therefore, we suggest that both isolates are not capable of internalizing trehalose molecules, and that under the stressful experimental conditions of seed bacterization, these isolates might not activate trehalose synthesis or that the amount of trehalose synthetized is not sufficient to support their viability, as it occurs in *P. putida* KT2440 [37]. Nevertheless, we observed a better seed bacterization performance of SVMP4-*yfp* with the addition of a trehalose–PVP mixture compared to Premax® (Figure 1b), whereas SVBP6-*mChe* performed similarly with both additive treatments (Figure 3). This suggests that the efficacy of this mixture for these two isolates relates to modifying external conditions that favor their adherence and reduce dehydration shock. Previous reports have demonstrated the beneficial effects of externally added trehalose in non-trehalose-producing cells under stress [37,56,78–80]. Additionally, the protective role of the trehalose–PVP mixtures has been reported for various bacteria [35–37]. Given that both SVBP6 and SVMP4 belong to the extensive *P. putida* subgroup [81], our results suggest that these isolates might also enter the VBNC state after the inoculation process in response to the desiccation shock on the seed surfaces, potentially explaining the decline in culturable counts. Further investigation using VBNC approaches would contribute to explore this possibility [76,82].

### *3.4. The Beneficial Effects of Plant-Growth Promoting Microorganisms Might Not Solely Stem from Root Colonization and Competitiveness*

It is conventionally assumed that root colonization is essential for the effective promotion of plant growth by microorganisms, as direct contact with plant tissues enables PGPMs to exert their beneficial effects on plants while utilizing root exudates as nutrient sources [18]. Cell motility and biofilm formation are associated with a superior colonization performance [83–85], and we have previously demonstrated that all our isolates display these abilities in vitro [49]. As microorganisms can exhibit distinct spatial arrangements along the root, occupying specific niches [67,86,87], we used confocal microscopy to analyze root colonization patterns of our *Pseudomonas* isolates (Figure 4). The presence of bacterial cells on the rhizoplane confirmed their ability to exit seed coats and colonize the emerging roots. We observed a differential localization of bacterial cells on maize root epidermal surfaces (Figure 4). Typically, microorganisms show a preference for cell junctions of the root epidermis, a region proposed as highly permeable for root exudates through the apoplastic pathway [88,89]. Notably, the commercial reference strain 1008-*cfp* and our derivative SVBP6-*mChe* primarily occupied cell junctions, leaving outer surfaces of epidermal cells largely unoccupied (Figure 4e,g). In contrast, SMMP3-*yfp* and RPAN1-*yfp* predominantly occupied outer epidermal surfaces (Figure 4b,c). Other isolate derivatives (RBAN4-*yfp*, SVMP4-*yfp*, and SPAN5-*cfp*) displayed a mixed arrangement (Figure 4a,d,f). These differential colonization patterns likely influence their ability to effectively exert biocontrol activity, as competition for niches and nutrients is one of the strategies involved in plant protection by *Pseudomonas* [87,90]. From our collection of autochthonous *Pseudomonas* isolates with plant-probiotic traits [49,52], RBAN4-*yfp*, RPAN1-*yfp*, and SMMP3-*yfp* exhibited the highest colonization densities on maize radicles (Figure 4). These isolates have shown superior bacterization levels, with RBAN4-*yfp* and RPAN1-*yfp* significantly outperforming the control treatment 1008-*cfp* (Figure S3). Thus, root colonization densities appear correlated with seed bacterization levels, as previously shown for other *Pseudomonas* isolates on alfalfa [91].

The confocal microscopy findings agree with our root competitiveness assay, where these three isolates were the most abundant in rhizospheric samples after one month of maize growth, i.e., up to the V3 stage, despite competition from soil microflora (Figure 5). However, the plant-growth promotion observed post-seed inoculation extended to all the *Pseudomonas* isolates, regardless of their root competitiveness (Figure 6). In previous results under axenic conditions, we also observed that SMMP3 significantly improved the early growth of maize [49]. Moreover, this isolate increased maize grain yield in experimental field assays when inoculated alone or in combination with a *Trichoderma* fungal isolate [52]. Therefore, SMMP3's robust root surface colonization likely contributes to its positive effects on plant growth and crop yield. Conversely, we also noted an enhanced shoot length, shoot biomass, and root biomass in plants treated with SVBP6-*mChe* (Figure 6b–d), despite its heterogeneous and low detection in rhizospheric samples

(Figure 5) and microscopy-based colonization (Figure 4g). Overall, confocal microscopy revealed reasonable bacterial colonization densities in the maize root elongation zone for all tested strains, indicating an efficient root colonization performance by these isolates through the strategy of seed bacterization before sowing. However, intriguingly, after one month of competition with the natural microflora, certain isolates such as SVMP4, SVBP6, and SPAN5, were unable to maintain detectable colonization levels in the maize rhizosphere (Figure 5). The improvement in seed bacterization achieved by SVMP4 with the addition of the trehalose–PVP mixture (Figure 1b) seemed insufficient to enable this isolate to compete with the natural soil microflora (Figure 5). Although we did not detect these isolates in the rhizospheric samples of maize plants at the V3 stage, our results suggest that the interaction between bacteria and plant organs during early plant development stages stimulates plant growth, even after the decline in the inoculated strain density on the root. In this regard, a recent study has shown that the initial contact between a plant-growth promoting *Bacillus subtilis* strain and melon seeds modifies the seedling development process, leading to a positive effect on adult plants [92]. Furthermore, it demonstrated that some *Bacillus* derivative strains, despite not colonizing the melon radicle, maintain their promoting effect on adult plants due to the production of specific metabolites such as the amyloid protein TasA and the lipopeptide fengycin, which alter plant metabolism [92]. Therefore, we propose that some of our *Pseudomonas* isolates may similarly impact maize growth through the production of specific metabolites. However, further experiments are necessary to challenge this hypothesis thoroughly.

## 4. Materials and Methods

### 4.1. Bacterial Strains, Growth Condition and Plant Species

Six bacterial isolates from our *Pseudomonas* collection [49] and their respective Tn7 tagged strains were utilized in this study (Table 1). *P. pergaminensis* 1008, the active principle of the commercial inoculant Rizofos® (Rizobacter Argentina s.a., RASA, Pergamino, Argentina), recommended for maize inoculation in Argentina [58], served as a reference. Bacterial strains were stored in 20% *w*/*v* glycerol at −80 °C, and routinely cultured on nutrient agar (NA, tryptone soy agar 40 g $L^{-1}$; yeast extract 5 g $L^{-1}$, Biokar, Cedex, France) or nutrient yeast broth (NYB, nutrient broth 20 g $L^{-1}$; yeast extract 5 g $L^{-1}$, Biokar Cedex, France). Incubation was carried out at 28 °C (except for *Escherichia coli* strains, which were grown at 37 °C), with agitation at 200 rpm for liquid cultures. The selective medium Gould's S1 [93] was used to specifically quantify the endogenous CFU load of *Pseudomonas* isolates of non-disinfected, untreated seeds. Media were supplemented with antibiotics as needed (Gm, gentamycin 20 μg $mL^{-1}$; Km, kanamycin 25 μg $mL^{-1}$; Ap, ampicillin 100 μg $mL^{-1}$; Sm, streptomycin 100 μg $mL^{-1}$; all from Sigma Aldrich, Darmstadt, Alemania). The commercial target crop was *Zea mays* (maize) and seeds belonged to the variety KM 8701 VIP3 or KM 87 VIP3 (KWS Argentina s.a., Balcarce, Argentina). Due to their larger size compared to typical maize seeds, 2–3 seeds per gram were used.

### 4.2. Chromosomal Tagging of Pseudomonas Isolates

Isolates were tagged with fluorescent proteins and an antibiotic resistance gene using a Tn7 transposition-based system, which inserts a single copy of the antibiotic resistance gene along with a gene encoding a fluorescent protein (eYfp, yellow; eCfp, cyan; mCherry, red) at a neutral chromosomal site [53,94]. Briefly, equal volumes from overnight cultures of *E. coli* SM10λpir/pUX-BF13, *E. coli* HB101/pME497, the corresponding *E. coli* carrying the eyfp, or ecfp delivery vector (MT102) or the mCherry delivery vector (DH5α), and each target *Pseudomonas* isolate (Table 1) were combined in a tetraparental mating mix on the edge of a NA plate and incubated for 4 h at 37 °C. This conjugation mixture was resuspended in 1 ml of NYB and 100 μL were spread onto M9 minimal medium agar plates [95] supplemented with citrate 0.2% as a carbon source and the corresponding antibiotic from the Tn7 cassette [53,96]. After incubation at 28 °C for 48 h, 3 or 4 colonies of each *Pseudomonas* isolate were streaked onto new Gould's S1 agar plates containing the appropriate antibiotic. Colony PCR was conducted to confirm the correct Tn7 insertion using specific oligonucleotides Tn7R109 and Tn7-*glmS* [53]. To verify that the tagged strains retained the same phenotype as the wild type, the PGPM properties demonstrated in vitro by the wild type isolates [49] were re-evaluated, and growth curves on M9-glucose (20 mM) and M9-trehalose (10 mM) were performed and compared between the variants and wild type strains. The growth curve with trehalose aimed to assess if this disaccharide could be used as an osmoprotectant if not metabolized by either strain [41]. Stability of the genomic insertion was also analyzed by subculturing the variants for 5 days without the selective pressure of the antibiotic, and then plating onto agar plates with or without the antibiotic.

**Table 1.** Bacterial isolates and plasmids used in this work.

| ***Pseudomonas* Isolates (Wild Type)** | | | |
|---|---|---|---|
| **Name** | **Taxonomical Affiliation [a]** | **Origin** | **Reference** |
| RBAN4 | *P. protegens* [b] | Pasture rhizosphere, Natural Environment, Bengolea, Córdoba, Argentina | [49] |
| SMMP3 | *P. chlororaphis* subsp. *aurantiaca* | Bulk soil from soybean plot, Monte Buey, Córdoba, Argentina | [49] |
| SPAN5 | *P. chlororaphis* | Bulk soil from pasture plot, Pergamino, Buenos Aires, Argentina | [49,97] |
| SVMP4 | *P. soli* | Bulk soil from pasture plot, Viale, Entre Ríos, Argentina | [49,98] |
| SVBP6 | *P. donghuensis* | Bulk soil from soybean plot, Viale, Entre Ríos, Argentina | [49,50] |
| RPAN1 | *P. chlororaphis* subsp. *piscium* | Pasture rhizosphere, Natural Environment, Pergamino, Buenos Aires, Argentina | [49] |
| 1008 | *P. pergaminensis* | Wheat rhizosphere from a productive plot, Pergamino, Buenos Aires, Argentina | [58] |
| ***Pseudomonas* derivative strains** | | | |
| | **Name** | **Relevant genetic and/or phenotypic features** | **Reference** |
| | RBAN4-*yfp* | RBAN4::*att*Tn7-*eyfp*; $Km^r$ $Sm^r$ | This study |
| | SMMP3-*yfp* | SMMP3::*att*Tn7-*eyfp*; $Km^r$ $Sm^r$ | This study |
| | SPAN5-*cfp* | SPAN5::*att*Tn7-*ecfp*; $Km^r$ $Sm^r$ | This study |
| | SVMP4-*yfp* | SVMP4::*att*Tn7-*eyfp*; $Km^r$ $Sm^r$ | This study |
| | SVBP6-*mChe* | SVBP6::*att*Tn7-*mCherry*; $Gm^r$ | This study |
| | RPAN1-*yfp* | RPAN1::*att*Tn7-*eyfp*; $Gm^r$ | This study |
| | 1008-*cfp* | 1008::*att*Tn7-*ecfp*; $Km^r$ $Sm^r$ | This study |
| ***Escherichia coli* strains** | | | |
| | **Strains** | **Relevant genetic and/or phenotypic features** | **Reference** |
| | SM10 | *thi-1, thr, leu, tonA, lacY, supE, recA*::RP4-2-Tc::Mu, λ*pir*. $Km^r$ | [99] |
| | HB101 | Laboratory strain. K12 derivative. $F^-$ $Pro^-$ $Gal^-$ $Rec^-$ $Sm^r$ | [100] |
| | DH5α | Laboratory strain. K12 derivative. $F^-$ *recA1*$^-$ *endA1*$^-$ *lacZ*ΔM15 | [101] |
| | MT102 | Laboratory strain. *ara*D139 D*(ara-leu)7697* Δ*lac thi hsdR* derivate of *E. coli* K-12 substrain MC1000. $Sm^r$ $Rif^r$ $Azide^r$ | [102] |
| **Plasmids** | | | |
| | pUX-BF13 | Helper plasmid for Tn7-based transposon mutagenesis containing the transposition functions; R6K-replicon; $Ap^r$ | [103] |
| | pME497 | Mobilizing plasmid; IncP-1, Tra; RepA(Ts); $Ap^r$ | [104] |
| | pME9407 | Delivery plasmid for mini-Tn7-*mCherry*; pME3280a carrying *mCherry* placed under Plac control; $Ap^r$ $Gm^r$ | [94] |

**Table 1.** *Cont.*

| | | |
|---|---|---|
| miniTn7(Km, Sm)$P_{A1/04/03}$–*ecfp*-a | Delivery plasmid for mini-Tn7-*ecfp*; pUC19 derivative carrying *ecfp* under a constitutive promoter (Plac derivative). Not replicative in *Pseudomonas*. $Ap^r$ $Km^r$ $Sm^r$ | [53] |
| miniTn7(Km, Sm)$P_{A1/04/03}$–*eyfp*-a | Delivery plasmid for mini-Tn7-*eyfp*; pUC19 derivative carrying *eyfp* under a constitutive promoter (Plac derivative). Not replicative in *Pseudomonas*. $Ap^r$ $Km^r$ $Sm^r$ | [53] |

[a] For SVBP6, RBAN4, SMMP3, and 1008, the taxonomy affiliation was performed by phylogenomic analyses. For SVMP4, SPAN5, and RPAN1, the affiliation was performed with taxonomical markers (16S rDNA, oprF and rpoB genes, see [49]; [b] The taxonomy affiliation was recently corrected with an ANI performed with the sequenced genome.

### 4.3. Seed Inoculation and Recuperation of Bacteria from Seeds

To prepare the inoculation mix, *Pseudomonas* tagged strains were grown in Erlenmeyer flasks with rotary agitation with a 5:1 flask/culture volume ratio, for 16 h at 28 °C. Bacterial cells were centrifuged at 5000 rpm and 4 °C for 10 min. The pellet was washed twice with saline solution (NaCl 0.85% *w*/*v*) and normalized to an $OD_{600}$ of 1.0 for seed inoculation. Viable cell counts were determined by drop plate count onto NA supplemented with the corresponding antibiotic (Table S1). Non-disinfected maize seeds were inoculated following the recommended dose of Rizofos® inoculant for maize at the start of the experiment (7 mL $kg^{-1}$). Initially, we assessed bacterization in the presence or absence of the commercial additive Premax® (RASA, Pergamino, Argentina). For this purpose, the inoculation mix was prepared with 2 mL $kg^{-1}$ of the additive and 5 mL $kg^{-1}$ of the bacterial suspension. Control treatments replaced additive volume with saline solution. When Premax® did not enhance seed bacterization levels, we evaluated alternative additives: (i) trehalose 1M [37]; (ii) polyvinylpyrrolidone (PVP) 1.5% *w*/*v* [40]; (iii) a mixture of both, trehalose and PVP; (iv) glycerol 20% *w*/*v* [40]; and (v) sucrose 20% *w*/*v* [105]. The ratio between bacterial suspension and additive was consistent across all cases (5 mL $kg^{-1}$ for the bacterial suspension and 2 mL $kg^{-1}$ for the additive). The total volume of each inoculation mix was added to seeds in a sterile plastic container and vigorously shaken by hand for 1.5 min.

Immediately after inoculation, bacteria were recovered from seeds and quantified on selective solid medium (day 0). For this, 5 g of seeds were suspended in 15 mL of saline solution in 50 mL screw-cap tubes. After vortexing for 10 s and immersing the tubes in a sonication bath (40 kHz, 160 W, Testlab TB04, Bernal, Argentina) for 10 min to promote detachment of bacteria from seed surface and dispersion of bacterial aggregates. Tubes were centrifuged at 50× *g* at room temperature for 1 min to remove coarse material. The supernatant containing bacteria was transferred to sterile 50 mL tubes, and serial dilutions were plated in triplicate onto Gould's S1 plates with the corresponding antibiotic from the Tn7 cassette for quantification of culturable pseudomonads recovered from seeds using the drop count method [106].

Bacterization decay kinetics were assessed for up to 4 days post inoculation (dpi), with daily recovery and enumeration of bacteria. During this period, seeds were incubated in a plant growth chamber (Ingelab Argentina, I-501 PF, Llavallol, Argentina) in darkness with temperature cycling from 24 °C to 13 °C (16/8 h, respectively). As a control for the intrinsic cell survival, daily viable cell counts were conducted in the bacterization suspensions with $OD_{600}$ = 1.0, maintained under the same storage condition as the bacterized seeds. Additionally, the effect of seed bacterization on germination was assessed by sowing seeds on 1% (*w*/*v*) water agar plates and determining the proportion of germinated seeds out of the total sown.

### 4.4. Bacterization of Pre-Hydrated Seeds

To determine if seed water content influenced bacterization levels, seeds were first soaked in sterilized, deionized water for 16 h at 120 rpm, dried with absorbent paper to prevent dripping; and then inoculated following the same procedure described in

Section 4.3, both with and without the addition of Premax®. Bacteria were recovered from seeds immediately after inoculation (0 dpi) and quantified by drop plate count on selective medium, as previously described.

### 4.5. Visualization of Bacterial Root-Colonization Patterns by Confocal Laser Scanning Microscopy (CLSM)

Maize seeds were surface disinfected with 70% *v/v* ethanol for 1 min, followed by treatment with 1.1% *w*/*v* NaClO for 10 min [107]. Then, seeds were washed 5 times with sterile distilled water for 2 min each and finally dried on a paper towel for 15 min in a laminar flow. Seed inoculation was carried out as described above, using the optimal additive for each strain. Control treatment involved a mixture of saline solution and the respective additive at the specific usage proportion. After inoculation, seeds were placed in sterile cotton-stoppered glass tubes containing 30 mL of semi-solid (0.5% *w*/*v* agar) Jensen's mineral solution [108] with 50 mM ok $KNO_3$ and incubated for 9 days in a plant growth chamber in darkness, under the conditions mentioned above. Roots segments from seedlings were cut into small pieces (around 1 cm) from 1 cm of root tips and visualized by scanning different focal planes of the surface using a laser scanning confocal microscope (Zeiss LSM 880 with Airyscan, Leloir Institute, Buenos Aires, Argentina) with either a Plan-Apochromat 20×/0.8 or C-Apochromat 40×/1.2 objectives, and the electronic zoom to improve visualization. Excitation wavelengths were between 405 and 458 nm for isolates tagged with eCFP; 488 nm for those tagged with eYFP; or 543 nm for mCherry. Emission wavelengths were between 463 nm and 498 nm for eCFP; between 515 nm and 574 nm for eYFP; or between 583 nm and 675 nm for mCherry (Table 1).

### 4.6. Root Competitiveness and Plant-Growth Promotion of Maize Grown from Bacterized Seeds

To analyze if the *Pseudomonas* isolates can occupy the rhizosphere of maize from bacterized seeds competing with the natural soil microflora, we designed a greenhouse assay. We used 1 L pots filled with sterilized perlite and natural soil mixed in 1:1 volume ratio, with one plant per pot and five plants per treatment (n = 5). The soil was obtained from a pristine environment located in Llavallol, Buenos Aires Province, Argentine (34°47′12.8″ S, 58°26′28.6″ W), and sieved by a 2 mm mesh prior to use. Non-disinfected maize seeds (variety KM 87 VIP3) were inoculated with the optimal additive for each isolate (Premax® for RBAN4-*yfp*, SMMP3-*yfp*, RPAN1-*yfp*, SPAN5-*cfp*, and the reference 1008-*cfp*; trehalose–PVP mixture for SVMP4-*yfp* and SVBP6-*mChe*) or treated with saline solution for the negative control. Pots were incubated under greenhouse conditions with temperature and humidity ranging from 20 to 24 °C and from 60 to 80%, respectively; and a day–night cycle of 16–8 h (supplemented with artificial light to mitigate seasonal variations in the natural photoperiod). Substrate humidity was maintained at field capacity with sterilized distilled water applied 3 times a week. After 1 month of incubation (V3-V4 maize stage), plants were harvested and the root systems were gently shaken to remove large, adhered soil particles and obtain the "rhizospheric sample" (i.e., the complete root system with the small soil particles tightly adhered to its surface). Rhizospheric samples were collected and kept at 4 °C until processing as previously described [57] to quantify bacterial colonization using the drop plate assay [106]. Total heterotrophic, mesophilic, and culturable bacteria were quantified by plating onto diluted tryptone soy agar (TSA 10%, Biokar, Cedex, France) supplemented with cycloheximide 100 μg $mL^{-1}$ (Anedra, Buenos Aires, Argentina) to inhibit fungal growth [57]. Specific counts for each inoculated isolate were performed on S1 agar plates supplemented with cycloheximide 100 μg $mL^{-1}$ and the corresponding antibiotic for each isolate. For isolates resistant to both Km and Sm (Table 1), plates were supplemented with both antibiotics to prevent growth of soil microflora with natural Km resistance. Colony counts were conducted after 48 h of incubation at 28 °C. At the time of plant harvest, we measured the height of the aerial part, the length of the primary root, and the fresh and dry weight of the aerial part, as well as the fresh weight of the whole root system after removing large and loosely adhered soil particles. A subsample of the root system was dried to determine bacterial colonization density based on dry root biomass.

### 4.7. Statistical Analyses

To compare the daily recovery of bacteria from seeds, three replicate samples were used (n = 3). *In planta* assay under greenhouse conditions was performed with 5 biological replicates. CFU counts were performed on Gould's S1 in triplicates for each replicate. CFU values were transformed using the formula $\log_{10}(x + 50)$ prior to statistical analyses, to account for null values [57]. Analysis of variance (ANOVA) or the Kruskal–Wallis non-parametric test were performed as appropriate, followed by the "Least Significant Difference" (Fisher's LSD) or "Uncorrected Dunn's" multiple comparison tests, respectively, using GraphPad Prism V. 8.00 for Windows (GraphPad Software, La Jolla, CA, USA, www.graphpad.com) to evaluate significant differences between values. Statistical significance was set at $p < 0.05$.

Due to unsynchronized seed germination, the number of replicate samples for visualizing bacterial root-colonization patterns varied among strains. The number of replicate samples ranged from 2 to 4 seeds per treatment.

## 5. Conclusions

In our study, we characterized the distinct performances of a group of autochthonous plant-probiotic *Pseudomonas* strains concerning their capacities to remain viable and culturable for up to 4 days post-inoculation on the surface of inoculated maize seeds. Our observations revealed significant variability in the ability of isolates to adhere to and persist in the culturable state on maize seeds, with a correlation between root colonization density, root competitiveness, and bacterization levels. Additionally, we demonstrated that bacterial protectants added to the inoculant suspensions promoted seed bacterization, with a differential impact depending on the isolate. Notably, the performance of three isolates from our *Pseudomonas* collection was comparable to that of 1008-*cfp*, a Tn7 derivative from a strain commercialized as a maize bioferstilizer with demonstrated benefits [58]. This suggests that maize seed bacterization with these isolates could be implemented similarly to achieve crop yield improvements in the field. Finally, we observed that although some isolates did not achieve high root colonization densities in adult plants under natural soil conditions, they still exerted a plant growth promotion effect one month after seed bacterization and sowing. These findings underscore the importance of thoroughly examining these factors when implementing a seed inoculation strategy to introduce PGPMs into agricultural ecosystems. Our results highlight the need to carefully consider seed adhesion and persistence characteristics to optimize the effectiveness of PGPMs in promoting plant growth. Ongoing analyses are also underway to extend our investigation to wheat and soybean seeds, aiming to broaden our understanding of the interactions between *Pseudomonas* strains and diverse plant species within agricultural settings.

**Supplementary Materials:** The following supporting information can be downloaded at: https://www.mdpi.com/article/10.3390/plants13152130/s1, Figure S1: Phenotypic check for the neutrality of mini-Tn7 cassette integration into the chromosome of wild type *Pseudomonas* isolates; Figure S2: Growth curves of wild type isolates (black symbols) and the corresponding Tn7 tagged variants (grey symbols) in M9 minimal medium supplemented with 20 mM glucose (triangles) or 10 mM trehalose (circles); Figure S3: Impact of the commercial bacterial protectant Premax® on maize bacterization levels of the *Pseudomonas* isolates at the day of inoculation (0 dpi); Figure S4: Pearson correlation analysis between the plant parameters that were evaluated in the greenhouse experiment; Table S1: Transposon stability under the absence of antibiotic selection pressure; Table S2: Soil properties; Table S3: Cell density of the seed bacterizing suspensions and percentage of culturable bacterial cells recovered from bacterized seeds; Table S4: Strain survival (CFU $mL^{-1}$) in seed bacterizing suspensions stored in parallel and under the same conditions as maize bacterized seeds; Table S5: Rhizospheric abundance of culturable, heterotrophic, mesophilic bacteria.

**Author Contributions:** Conceptualization, B.C.A.; methodology, B.C.A. and C.V.; software, B.C.A and M.G.L.; formal analysis, M.G.L. and B.C.A.; investigation, M.G.L.; resources, B.C.A. and C.V. data curation, M.G.L.; writing—original draft preparation, B.C.A.; writing—review and editing, M.G.L., C.V., and B.C.A.; supervision, B.C.A.; project administration, B.C.A.; funding acquisition B.C.A. and C.V. All authors have read and agreed to the published version of the manuscript.

**Funding:** This research was funded by the Agencia I + D+i, grant numbers PICT 2016 N°4362 and PICT 2019 N° 3617 (Argentina); and by the Universidad Nacional de Quilmes, grant numbers PUNQ 1306/19 and 591 2295/22 (Argentina); and by CONICET, grant numbers PIP 11220150100388CO and PIP 11220200101442CO (Argentina).

**Data Availability Statement:** The raw data supporting the conclusions of this article will be made available by the authors on request.

**Acknowledgments:** We gratefully acknowledge Andrés Hugo Rossi, Carla Pascuale and Esteban Alberto Miglietta from the Imaging and Microscopy Service of the Leloir Institute Foundation for their kind assistance and support in obtaining the confocal microscopy images. We also extend our gratitude to Rizobacter Argentina S.A., particularly Marisa Díaz, for providing us with the maize seeds; and to Mariana Marrero, from the National University of Lomas de Zamora, for providing us the access to the Natural Reserve Santa Catalina where we obtained the soil samples.

**Conflicts of Interest:** The authors declare no conflicts of interest.

## References

1. Venturi, V.; Keel, C. Signaling in the Rhizosphere. *Trends Plant Sci.* **2016**, *21*, 187–198. [CrossRef] [PubMed]
2. Olanrewaju, O.S.; Ayangbenro, A.S.; Glick, B.R.; Babalola, O.O. Plant Health: Feedback Effect of Root Exudates-Rhizobiome Interactions. *Appl. Microbiol. Biotechnol.* **2019**, *103*, 1155–1166. [CrossRef] [PubMed]
3. Johnston-Monje, D.; Gutiérrez, J.P.; Lopez-Lavalle, L. Seed-Transmitted Bacteria and Fungi Dominate Juvenile Plant Microbiomes. *Front. Microbiol.* **2021**, *12*, 737616. [CrossRef] [PubMed]
4. Sun, Z.; Adeleke, B.S.; Shi, Y.; Li, C. The Seed Microbiomes of Staple Food Crops. *Microb. Biotechnol.* **2023**, *16*, 2236–2249. [CrossRef] [PubMed]
5. Wassermann, B.; Adam, E.; Cernava, T.; Berg, G. Understanding the Indigenous Seed Microbiota to Design Bacterial Seed Treatments. In *Seed Endophytes*; Springer International Publishing: Cham, Switzerland, 2019; pp. 83–99.
6. Wallace, J.G. Maize Seed Endophytes. *Mol. Plant Pathol.* **2023**, *24*, 801–810. [CrossRef] [PubMed]
7. Özkurt, E.; Hassani, M.A.; Sesiz, U.; Künzel, S.; Dagan, T.; Özkan, H.; Stukenbrock, E.H. Seed-Derived Microbial Colonization of Wild Emmer and Domesticated Bread Wheat (*Triticum dicoccoides* and *T. aestivum*) Seedlings Shows Pronounced Differences in Overall Diversity and Composition. *mBio* **2020**, *11*, 10–128. [CrossRef] [PubMed]
8. Zhang, J.; Zhang, C.; Yang, J.; Zhang, R.; Gao, J.; Zhao, X.; Zhao, J.; Zhao, D.; Zhang, X. Insights into Endophytic Bacterial Community Structures of Seeds Among Various *Oryza sativa* L. Rice Genotypes. *J. Plant Growth Regul.* **2019**, *38*, 93–102. [CrossRef]
9. Morales Moreira, Z.P.; Helgason, B.L.; Germida, J.J. Crop, Genotype, and Field Environmental Conditions Shape Bacterial and Fungal Seed Epiphytic Microbiomes. *Can. J. Microbiol.* **2021**, *67*, 161–173. [CrossRef] [PubMed]
10. Tkalec, V.; Mahnic, A.; Gselman, P.; Rupnik, M. Analysis of Seed-Associated Bacteria and Fungi on Staple Crops Using the Cultivation and Metagenomic Approaches. *Folia Microbiol.* **2022**, *67*, 351–361. [CrossRef]
11. Catroux, G.; Hartmann, A.; Revellin, C. Trends in Rhizobial Inoculant Production and Use. *Plant Soil* **2001**, *230*, 21–30. [CrossRef]
12. O'Callaghan, M. Microbial Inoculation of Seed for Improved Crop Performance: Issues and Opportunities. *Appl. Microbiol. Biotechnol.* **2016**, *100*, 5729–5746. [CrossRef] [PubMed]
13. Idígoras, G. Mejoramiento de Cultivos y Producción de Semillas; Buenos Aires. 2013. Available online: https://www.argentina.gob.ar/sites/default/files/mejoramiento_cultivos_y_produccion_semillas-doc.pdf (accessed on 20 June 2024).
14. Lamichhane, J.R.; Corrales, D.C.; Soltani, E. Biological Seed Treatments Promote Crop Establishment and Yield: A Global Meta-Analysis. *Agron. Sustain. Dev.* **2022**, *42*, 45. [CrossRef]
15. Ali, M.A.; Ilyas, F.; Arshad, M.; Hussain, S.; Iqbal, M.; Ahmad, S.; Saboor, A.; Mustafa, G.; Ahmed, N. Microbial Inoculation of Seeds for Better Plant Growth and Productivity. In *Priming and Pretreatment of Seeds and Seedlings*; Springer Singapore: Singapore, 2019; pp. 523–550.
16. Global Market Analysis, Foreign Agricultural Service. *World Agricultural Production*; Foreign Agricultural Service: Washington, DC, USA, 2024.
17. Kumar, B.; Dube, H. Seed Bacterization with a Fluorescent *Pseudomonas* for Enhanced Plant Growth, Yield and Disease Control. *Soil Biol. Biochem.* **1992**, *24*, 539–542. [CrossRef]
18. Philippot, L.; Raaijmakers, J.M.; Lemanceau, P.; van der Putten, W.H. Going Back to the Roots: The Microbial Ecology of the Rhizosphere. *Nat. Rev. Microbiol.* **2013**, *11*, 789–799. [CrossRef] [PubMed]
19. Gopalakrishnan, S.; Sathya, A.; Vijayabharathi, R.; Varshney, R.K.; Gowda, C.L.L.; Krishnamurthy, L. Plant Growth Promoting Rhizobia: Challenges and Opportunities. *Biotech* **2014**, *5*, 355–377. [CrossRef] [PubMed]

20. Deaker, R.; Roughley, R.; Kennedy, I. Legume Seed Inoculation Technology—A Review. *Soil Biol. Biochem.* **2004**, *36*, 1275–1288. [CrossRef]
21. Silva Araujo, R.; Purin da Cruz, S.; Souchie, E.; Martin, T.; Shigueyoshi Nakatani, A.; Nogueira, M.; Hungria, M. Preinoculation of Soybean Seeds Treated with Agrichemicals up to 30 Days before Sowing: Technological Innovation for Large-Scale Agriculture. *Int. J. Microbiol.* **2017**, *2017*, 5914786. [CrossRef]
22. Roughley, R.J.; Gemell, L.G.; Thompson, J.A.; Brockwell, J. The Number of *Bradyrhizobium* sp. (Lupinus) Applied to Seed and Its Effect on Rhizosphere Colonization, Nodulation and Yield of Lupin. *Soil Biol. Biochem.* **1993**, *25*, 1453–1458. [CrossRef]
23. Hume, D.J.; Blair, D.H. Effect of Numbers of *Bradyrhizobium japonicum* Applied in Commercial Inoculants on Soybean Seed Yield in Ontario. *Can. J. Microbiol.* **1992**, *38*, 588–593. [CrossRef]
24. Elegba, M.; Rennie, R. Effect of Different Inoculant Adhesive Agents on Rhizobial Survival, Nodulation, and Nitrogenase (Acetylene-Reducing) Activity of Soybeans (*Glycine max* (L.) Merrill). *Can. J. Soil Sci.* **1984**, *64*, 631–636. [CrossRef]
25. Mahmood, A.; Turgay, O.; Farooq, M.; Hayat, R. Seed Biopriming with Plant Growth Promoting Rhizobacteria: A Review. *FEMS Microbiol. Ecol.* **2016**, *98*, fiw112. [CrossRef] [PubMed]
26. Temprano, F.J.; Albareda, M.; Camacho, M.; Daza, A.; Santamaría, C.; Rodríguez-Navarro, D.N.; Rodríguez-Navarro, N.D.; Rodríguez-Navarro, D.N. Survival of Several *Rhizobium/Bradyrhizobium* Strains on Different Inoculant Formulations and Inoculated Seeds. *Int. Microbiol.* **2002**, *5*, 81–86. [CrossRef]
27. Herrmann, L.; Lesueur, D. Challenges of Formulation and Quality of Biofertilizers for Successful Inoculation. *Appl. Microbiol. Biotechnol.* **2013**, *97*, 8859–8873. [CrossRef]
28. Streeter, J.G. Effect of Trehalose on Survival of *Bradyrhizobium japonicum* during Desiccation. *J. Appl. Microbiol.* **2003**, *95*, 484–491. [CrossRef]
29. Ruhal, R.; Kataria, R.; Choudhury, B. Trends in Bacterial Trehalose Metabolism and Significant Nodes of Metabolic Pathway in the Direction of Trehalose Accumulation. *Microb. Biotechnol.* **2013**, *6*, 493–502. [CrossRef] [PubMed]
30. Kaushik, J.K.; Bhat, R. Why Is Trehalose an Exceptional Protein Stabilizer? *J. Biol. Chem.* **2003**, *278*, 26458–26465. [CrossRef]
31. Craig, K.; Johnson, B.R.; Grunden, A. Leveraging Pseudomonas Stress Response Mechanisms for Industrial Applications. *Front. Microbiol.* **2021**, *12*, 660134. [CrossRef]
32. Hengge-Aronis, R.; Klein, W.; Lange, R.; Rimmele, M.; Boos, W. Trehalose Synthesis Genes Are Controlled by the Putative Sigma Factor Encoded by RpoS and Are Involved in Stationary-Phase Thermotolerance in *Escherichia coli*. *J. Bacteriol.* **1991**, *173*, 7918–7924. [CrossRef] [PubMed]
33. De Virgilio, C.; Hottiger, T.; Dominguez, J.; Boller, T.; Wiemken, A. The Role of Trehalose Synthesis for the Acquisition of Thermotolerance in Yeast. *Eur. J. Biochem.* **1994**, *219*, 179–186. [CrossRef]
34. Reina-Bueno, M.; Argandoña, M.; Nieto, J.J.; Hidalgo-García, A.; Iglesias-Guerra, F.; Delgado, M.J.; Vargas, C. Role of Trehalose in Heat and Desiccation Tolerance in the Soil Bacterium *Rhizobium etli*. *BMC Microbiol.* **2012**, *12*, 207. [CrossRef]
35. Singleton, P.; Keyser, H.; Sande, E. Development and Evaluation of Liquid Inoculants. In *Proceedings of the Inoculants and Nitrogen Fixation of Legumes in Vietnam*; Herridge, D., Ed.; ACIAR: Canberra, Australian, 2002; pp. 52–66.
36. Kuppardt, A.; Chatzinotas, A.; Breuer, U.; van der Meer, J.R.; Harms, H. Optimization of Preservation Conditions of As (III) Bioreporter Bacteria. *Appl. Microbiol. Biotechnol.* **2009**, *82*, 785–792. [CrossRef] [PubMed]
37. Manzanera, M.; García de Castro, A.; Tøndervik, A.; Rayner-Brandes, M.; Strøm, A.R.; Tunnacliffe, A. Hydroxyectoine Is Superior to Trehalose for Anhydrobiotic Engineering of *Pseudomonas putida* KT2440. *Appl. Environ. Microbiol.* **2002**, *68*, 4328–4333. [CrossRef] [PubMed]
38. Frontera, G. Estrategias de inoculación en soja. In *Producción Agroindustrial*. Tucumán, November–December 2010; pp. 60–61. Available online: https://www.produccion.com.ar/ver_nota.php?edicion=Nov_Dic2016&numero=187&id=772 (accessed on 20 June 2024).
39. Heydari, A.; Naraghi, L. Application of Antagonistic Bacteria for the Promotion of Cotton Seedlings Growth Characteristics. *Int. J. Agric. Crop Sci.* **2014**, *7*, 1267–1273.
40. Manikandan, R.; Saravanakumar, D.; Rajendran, L.; Raguchander, T.; Samiyappan, R. Standardization of Liquid Formulation of *Pseudomonas fluorescens* Pf1 for Its Efficacy against *Fusarium* Wilt of Tomato. *Biol. Control* **2010**, *54*, 83–89. [CrossRef]
41. Seong, K.-Y.; Ryu, O.-R.; Choi, W.-Y. Effect of Trehalose on the Viability of Fluorescent *Pseudomonas*, Strain SSL3. *Korean J. Soil Sci. Fertil.* **2000**, *33*, 292–301.
42. Shah-Smith, D.A.; Burns, R.G. Shelf-Life of a Biocontrol *Pseudomonas putida* Applied to Sugar Beet Seeds Using Commercial Coatings. *Biocontrol Sci. Technol.* **1997**, *7*, 65–74. [CrossRef]
43. O'Callaghan, M.; Swaminathan, J.; Lottmann, J.; Wright, D.A.; Jackson, T.A. Seed Coating with Biocontrol Strain *Pseudomonas Fluorescens* F113. *N. Z. Plant Prot.* **2006**, *59*, 80–85. [CrossRef]
44. Kuz'mina, L.; Melent'ev, A. The Effect of Seed Bacterization by *Bacillus* Cohn Bacteria on Their Colonization of the Spring Wheat Rhizosphere. *Mikrobiologiya* **2003**, *72*, 268–274.
45. Juhnke, M.E.; Mathre, D.E.; Sands, D.C. Relationship between Bacterial Seed Inoculum Density and Rhizosphere Colonization of Spring Wheat. *Soil Biol. Biochem.* **1989**, *21*, 591–595. [CrossRef]
46. Wang, N.; Wang, T.; Chen, Y.; Wang, M.; Lu, Q.; Wang, K.; Dou, Z.; Chi, Z.; Qiu, W.; Dai, J.; et al. Microbiome Convergence Enables Siderophore-Secreting-Rhizobacteria to Improve Iron Nutrition and Yield of Peanut Intercropped with Maize. *Nat. Commun.* **2024**, *15*, 839. [CrossRef]

47. Mendes, R.; Kruijt, M.; de Bruijn, I.; Dekkers, E.; van der Voort, M.; Schneider, J.H.M.; Piceno, Y.M.; DeSantis, T.Z.; Andersen, G.L.; Bakker, P.A.H.M.; et al. Deciphering the Rhizosphere Microbiome for Disease-Suppressive Bacteria. *Science* **2011**, *332*, 1097–1100. [CrossRef]
48. Zheng, Y.; Han, X.; Zhao, D.; Wei, K.; Yuan, Y.; Li, Y.; Liu, M.; Zhang, C.-S. Exploring Biocontrol Agents from Microbial Keystone Taxa Associated to Suppressive Soil: A New Attempt for a Biocontrol Strategy. *Front. Plant Sci.* **2021**, *12*, 655673. [CrossRef] [PubMed]
49. Agaras, B.; Scandiani, M.; Luque, A.; Fernández, L.; Farina, F.; Carmona, M.; Gally, M.; Romero, A.; Wall, L.; Valverde, C. Quantification of the Potential Biocontrol and Direct Plant Growth Promotion Abilities Based on Multiple Biological Traits Distinguish Different Groups of *Pseudomonas* spp. Isolates. *Biol. Control* **2015**, *90*, 173–186. [CrossRef]
50. Agaras, B.C.; Iriarte, A.; Valverde, C.F. Genomic Insights into the Broad Antifungal Activity, Plant-Probiotic Properties, and Their Regulation, in *Pseudomonas donghuensis* Strain SVBP6. *PLoS ONE* **2018**, *13*, e0194088. [CrossRef]
51. Muzio, F.M.; Agaras, B.C.; Masi, M.; Tuzi, A.; Evidente, A.; Valverde, C. 7-Hydroxytropolone Is the Main Metabolite Responsible for the Fungal Antagonism of *Pseudomonas donghuensis* Strain SVBP6. *Environ. Microbiol.* **2020**, *22*, 2550–2563. [CrossRef] [PubMed]
52. Agaras, B.C.; Noguera, F.; González Anta, G.; Wall, L.; Valverde, C. Biocontrol Potential Index of Pseudomonads, Instead of Their Direct-Growth Promotion Traits, Is a Predictor of Seed Inoculation Effect on Crop Productivity under Field Conditions. *Biol. Control* **2020**, *143*, 104209. [CrossRef]
53. Lambertsen, L.; Sternberg, C.; Molin, S. Mini-Tn7 Transposons for Site-Specific Tagging of Bacteria with Fluorescent Proteins. *Environ. Microbiol.* **2004**, *6*, 726–732. [CrossRef]
54. Choi, K.-H.H.; Schweizer, H.P. Mini-Tn7 Insertion in Bacteria with Single AttTn7 Sites: Example *Pseudomonas aeruginosa*. *Nat. Protoc.* **2006**, *1*, 153–161. [CrossRef] [PubMed]
55. Koch, B.; Jensen, L.E.; Nybroe, O. A Panel of Tn7-Based Vectors for Insertion of the Gfp Marker Gene or for Delivery of Cloned DNA into Gram-Negative Bacteria at a Neutral Chromosomal Site. *J. Microbiol. Methods* **2001**, *45*, 187–195. [CrossRef]
56. Leslie, S.B.; Israeli, E.; Lighthart, B.; Crowe, J.H.; Crowe, L.M. Trehalose and Sucrose Protect Both Membranes and Proteins in Intact Bacteria during Drying. *Appl. Environ. Microbiol.* **1995**, *61*, 3592–3597. [CrossRef]
57. Agaras, B.C.; Wall, L.G.; Valverde, C. Influence of Agricultural Practices and Seasons on the Abundance and Community Structure of Culturable Pseudomonads in Soils under No-till Management in Argentina. *Plant Soil* **2014**, *382*, 117–131. [CrossRef]
58. Díaz, M.; Bach, T.; González Anta, G.; Agaras, B.; Wibberg, D.; Noguera, F.; Canciani, W.; Valverde, C. Agronomic Efficiency and Genome Mining Analysis of the Wheat-Biostimulant Rhizospheric Bacterium *Pseudomonas pergaminensis* sp. nov. Strain 1008T. *Front. Plant Sci.* **2022**, *13*, 894985. [CrossRef] [PubMed]
59. Fukui, R.; Poinar, E.; Bauer, P.; Schroth, M.; Hendson, M.; Wang, X.; Hancock, J. Spatial Colonization Patterns and Interaction of Bacteria on Inoculated Sugar Beet Seed. *Phytopathology* **1994**, *84*, 1338–1345. [CrossRef]
60. Tombolini, R.; van der Gaag, D.J.; Gerhardson, B.; Jansson, J.K. Colonization Pattern of the Biocontrol Strain *Pseudomonas chlororaphis* MA 342 on Barley Seeds Visualized by Using Green Fluorescent Protein. *Appl. Environ. Microbiol.* **1999**, *65*, 3674–3680. [CrossRef] [PubMed]
61. Molina, L.; Ramos, C.; Duque, E.; Ronchel, M.C.; García, J.; Wyke, L.; Ramos, J. Survival of *Pseudomonas putida* KT2440 in Soil and in the Rhizosphere of Plants under Greenhouse and Environmental Conditions. *Soil Biol. Biochem.* **2000**, *32*, 315–321. [CrossRef]
62. Rojas-Solís, D.; Hernández-Pacheco, C.E.; Santoyo, G. Evaluation of *Bacillus* and *Pseudomonas* to Colonize the Rhizosphere and Their Effect on Growth Promotion in Tomato (Physalis Ixocarpa Brot. Ex Horm.). *Rev. Chapingo Ser. Hortic.* **2016**, *XXII*, 45–57. [CrossRef]
63. Martínez-Gil, M.; Yousef-Coronado, F.; Espinosa-Urgel, M. LapF, the Second Largest *Pseudomonas putida* Protein, Contributes to Plant Root Colonization and Determines Biofilm Architecture. *Mol. Microbiol.* **2010**, *77*, 549–561. [CrossRef]
64. Espinosa-Urgel, M.; Salido, A.; Ramos, J.-L. Genetic Analysis of Functions Involved in Adhesion of *Pseudomonas putida* to Seeds. *J. Bacteriol.* **2000**, *182*, 2363–2369. [CrossRef] [PubMed]
65. DeFlaun, M.; Marshall, B.; Kulle, E.; Levy, S. Tn5 Insertion Mutants of *Pseudomonas fluorescens* Defective in Adhesion to Soil and Seeds. *Appl. Environ. Microbiol.* **1994**, *60*, 2637–2642. [CrossRef]
66. Hinsa, S.M.; Espinosa-Urgel, M.; Ramos, J.L.; O'Toole, G.A. Transition from Reversible to Irreversible Attachment during Biofilm Formation by *Pseudomonas fluorescens* WCS365 Requires an ABC Transporter and a Large Secreted Protein. *Mol. Microbiol.* **2003**, *49*, 905–918. [CrossRef]
67. Yousef-Coronado, F.; Travieso, M.L.; Espinosa-Urgel, M. Different, Overlapping Mechanisms for Colonization of Abiotic and Plant Surfaces by *Pseudomonas putida*. *FEMS Microbiol. Lett.* **2008**, *288*, 118–124. [CrossRef]
68. Fipke, G.M.; Martin, T.N.; Müller, T.M.; Cunha, V.d.S.; Munareto, J.D.; Schönell, A.T.; Grando, L.F.T.; Rossato, A. da C. Osmoprotectant in Soybean Seeds Can Increase the Inoculation and Co-Inoculation Time in Pre-Sowing. *Aust. J. Crop Sci.* **2020**, *14*, 905–912. [CrossRef]
69. Arora, N.K.; Mishra, J. Prospecting the Roles of Metabolites and Additives in Future Bioformulations for Sustainable Agriculture. *Appl. Soil Ecol.* **2016**, *107*, 405–407. [CrossRef]
70. Tittabutr, P.; Payakapong, W.; Teaumroong, N.; Singleton, P.W.; Boonkerd, N. Growth, Survival and Field Performance of Bradyrhizobial Liquid Inoculant Formulations with Polymeric Additives. *ScienceAsia* **2007**, *33*, 69–77. [CrossRef]

71. Ramette, A.; Frapolli, M.; Fischer-Le Saux, M.; Gruffaz, C.; Meyer, J.-M.; Défago, G.; Sutra, L.; Moënne-Loccoz, Y. *Pseudomonas protegens* Sp. Nov., Widespread Plant-Protecting Bacteria Producing the Biocontrol Compounds 2,4-Diacetylphloroglucinol and Pyoluteorin. *Syst. Appl. Microbiol.* **2011**, *34*, 180–188. [CrossRef] [PubMed]
72. Troxler, J.; Svercel, M.; Natsch, A.; Zala, M.; Keel, C.; Moënne-Loccoz, Y.; Défago, G. Persistence of a Biocontrol *Pseudomonas* Inoculant as High Populations of Culturable and Non-Culturable Cells in 200-Cm-Deep Soil Profiles. *Soil Biol. Biochem.* **2012**, *44*, 122–129. [CrossRef]
73. Arana, I.; Muela, A.; Orruño, M.; Seco, C.; Garaizabal, I.; Barcina, I. Effect of Temperature and Starvation upon Survival Strategies of *Pseudomonas fluorescens* CHA0: Comparison with *Escherichia coli*. *FEMS Microbiol. Ecol.* **2010**, *74*, 500–509. [CrossRef] [PubMed]
74. Pazos-Rojas, L.A.; Cuellar-Sánchez, A.; Romero-Cerón, A.L.; Rivera-Urbalejo, A.; Van Dillewijn, P.; Luna-Vital, D.A.; Muñoz-Rojas, J.; Morales-García, Y.E.; Bustillos-Cristales, M. del R. The Viable but Non-Culturable (VBNC) State, a Poorly Explored Aspect of Beneficial Bacteria. *Microorganisms* **2023**, *12*, 39. [CrossRef] [PubMed]
75. García de Castro, A.; Bredholt, H.; Strøm, A.R.; Tunnacliffe, A. Anhydrobiotic Engineering of Gram-Negative Bacteria. *Appl. Environ. Microbiol.* **2000**, *66*, 4142–4144. [CrossRef] [PubMed]
76. Pazos-Rojas, L.A.; Muñoz-Arenas, L.C.; Rodríguez-Andrade, O.; López-Cruz, L.E.; López-Ortega, O.; Lopes-Olivares, F.; Luna-Suarez, S.; Baez, A.; Morales-García, Y.E.; Quintero-Hernández, V.; et al. Desiccation-Induced Viable but Nonculturable State in *Pseudomonas putida* KT2440, a Survival Strategy. *PLoS ONE* **2019**, *14*, e0219554. [CrossRef]
77. Vílchez, J.; García-Fontana, C.; Román-Naranjo, D.; González-López, J.; Manzanera, M. Plant Drought Tolerance Enhancement by Trehalose Production of Desiccation-Tolerant Microorganisms. *Front. Microbiol.* **2016**, *7*, 01577. [CrossRef] [PubMed]
78. Matuszewska, E.; Leszczyńska, D.; Kuczyńska-Wiśnik, D.; Algara, M.M.; Stojowska, K.; Augustynowicz, M.; Laskowska, E. Lack of Intracellular Trehalose Affects Formation of *Escherichia coli* Persister Cells. *Microbiology* **2015**, *161*, 786–796. [CrossRef]
79. Lim, B.R.; Choi, H.J.; Kwon, G.-S.; Joo, W.H. Enhancement of Solvent Tolerance in *Pseudomonas* sp. BCNU 106 with Trehalose. *Lett. Appl. Microbiol.* **2015**, *61*, 607–612. [CrossRef] [PubMed]
80. Kumaresan, G.; Reetha, D. Survival of Azospirillum Brasilense in Liquid Formulation Amended with Different Chemical Additives. *J. Phytol.* **2011**, *3*, 48–51.
81. Lalucat, J.; Gomila, M.; Mulet, M.; Zaruma, A.; García-Valdés, E. Past, Present and Future of the Boundaries of the *Pseudomonas* Genus: Proposal of Stutzerimonas Gen. Nov. *Syst. Appl. Microbiol.* **2022**, *45*, 126289. [CrossRef] [PubMed]
82. Zhang, J.; Khan, S.; Chousalkar, K.K. Development of PMAxxTM-Based QPCR for the Quantification of Viable and Non-Viable Load of *Salmonella* From Poultry Environment. *Front. Microbiol.* **2020**, *11*, 581201. [CrossRef] [PubMed]
83. Martínez-Gil, M.; Romero, D.; Kolter, R.; Espinosa-Urgel, M. Calcium Causes Multimerization of the Large Adhesin LapF and Modulates Biofilm Formation by *Pseudomonas putida*. *J. Bacteriol.* **2012**, *194*, 6782–6789. [CrossRef]
84. Achouak, W.; Conrod, S.; Cohen, V.; Heulin, T. Phenotypic Variation of *Pseudomonas brassicacearum* as a Plant Root-Colonization Strategy. *Mol. Plant-Microbe Interact.* **2004**, *17*, 872–879. [CrossRef] [PubMed]
85. Sang, M.K.; Kim, K.D. Biocontrol Activity and Root Colonization by *Pseudomonas corrugata* Strains CCR04 and CCR80 against *Phytophthora* Blight of Pepper. *BioControl* **2014**, *59*, 437–448. [CrossRef]
86. Chin-A-Woeng, T.F.C.; de Priester, W.; van der Bij, A.J.; Lugtenberg, B.J.J. Description of the Colonization of a Gnotobiotic Tomato Rhizosphere by *Pseudomonas fluorescens* Biocontrol Strain WCS365, Using Scanning Electron Microscopy. *MPMI* **1997**, *10*, 79–86. [CrossRef]
87. Pliego, C.; de Weert, S.; Lamers, G.; de Vicente, A.; Bloemberg, G.; Cazorla, F.M.; Ramos, C. Two Similar Enhanced Root-Colonizing *Pseudomonas* Strains Differ Largely in Their Colonization Strategies of Avocado Roots and Rosellinia Necatrix Hyphae. *Environ. Microbiol.* **2008**, *10*, 3295–3304. [CrossRef] [PubMed]
88. Campbell, R.; Greaves, M.P. Anatomy and Community Structure of the Rhizosphere. In *The Rhizosphere*; Lynch, J.M., Ed.; John Wiley and Sons Ltd.: Hoboken, NJ, USA, 1990; pp. 11–34, ISBN 0-471-92548-9.
89. Canarini, A.; Kaiser, C.; Merchant, A.; Richter, A.; Wanek, W. Root Exudation of Primary Metabolites: Mechanisms and Their Roles in Plant Responses to Environmental Stimuli. *Front. Plant Sci.* **2019**, *10*, 157. [CrossRef] [PubMed]
90. Validov, S.; Kamilova, F.; Qi, S.; Stephan, D.; Wang, J.J.; Makarova, N.; Lugtenberg, B. Selection of Bacteria Able to Control *Fusarium oxysporum* f. sp. radicis-lycopersici in Stonewool Substrate. *J. Appl. Microbiol.* **2007**, *102*, 461–471. [CrossRef] [PubMed]
91. Hatzinger, P.B.; Alexander, M. Relationship between the Number of Bacteria Added to Soil or Seeds and Their Abundance and Distribution in the Rhizosphere of Alfalfa. *Plant Soil* **1994**, *158*, 211–222. [CrossRef]
92. Berlanga-Clavero, M.V.; Molina-Santiago, C.; Caraballo-Rodríguez, A.M.; Petras, D.; Díaz-Martínez, L.; Pérez-García, A.; de Vicente, A.; Carrión, V.J.; Dorrestein, P.C.; Romero, D. *Bacillus subtilis* Biofilm Matrix Components Target Seed Oil Bodies to Promote Growth and Anti-Fungal Resistance in Melon. *Nat. Microbiol.* **2022**, *7*, 1001–1015. [CrossRef] [PubMed]
93. Gould, W.D.; Hagedorn, C.; Bardinelli, T.R.; Zablotowicz, R.M. New Selective Media for Enumeration and Recovery of Fluorescent Pseudomonads from Various Habitats. *Appl. Environ. Microbiol.* **1985**, *49*, 28–32. [CrossRef] [PubMed]
94. Rochat, L.; Péchy-tarr, M.; Baehler, E.; Maurhofer, M.; Keel, C. Combination of Fluorescent Reporters for Simultaneous Monitoring of Root Colonization and Antifungal Gene Expression by a Biocontrol Pseudomonad on Cereals with Flow Cytometry. *MPMI* **2010**, *23*, 949–961. [CrossRef] [PubMed]
95. Green, M.R.; Sambrook, J. *Molecular Cloning: A Laboratory Manual*, 4th ed.; Cold Spring Harbor Laboratory: Cold Spring Harbor, NY, USA, 2012.

96. Martínez-García, E.; Aparicio, T.; de Lorenzo, V.; Nikel, P.I. New Transposon Tools Tailored for Metabolic Engineering of Gram-Negative Microbial Cell Factories. *Front. Bioeng. Biotechnol.* **2014**, *2*, 46. [CrossRef] [PubMed]
97. Agaras, B.; Valverde, C. A Mutation in the *gltA* Gene from a Native Isolate of the *Pseudomonas chlororaphis* Subgroup Induces a Phenotypic Change Associated with Phenazine Production. *Arch. Phytopathol. Plant Prot.* **2019**, *52*, 601–624. [CrossRef]
98. Garavaglia, M.; Muzlera, A.; Valverde, C. Comparative Genomics and Informational Content Analysis Uncovered Internal Regions of the Core Genes *rpoD, pepN* and *gltX* for an MLSA with Genome-Level Resolving Power within the Genus *Pseudomonas Mol. Phylogenet. Evol.* **2023**, *179*, 107663. [CrossRef]
99. Simon, R.; Priefer, U.; Pühler, A. A Broad Host Range Mobilization System for In Vivo Genetic Engineering: Transposon Mutagenesis in Gram Negative Bacteria. *Bio/Technol.* **1983**, *1*, 784–791. [CrossRef]
100. Boyer, H.W.; Roulland-Dussoix, D. A Complementation Analysis of the Restriction and Modification of DNA in *Escherichia coli*. *J. Mol. Biol.* **1969**, *41*, 459–472. [CrossRef] [PubMed]
101. Woodcock, D.M.; Crowther, P.J.; Doherty, J.; Jefferson, S.; DeCruz, E.; Noyer-Weidner, M.; Smith, S.S.; Michael, M.Z.; Graham, M.W. Quantitative Evaluation of *Escherichia coli* Host Strains for Tolerance to Cytosine Methylation in Plasmid and Phage Recombinants. *Nucleic Acids Res.* **1989**, *17*, 3469–3478. [CrossRef] [PubMed]
102. Valcek, A.; Overballe-Petersen, S.; Hansen, F.; Dolejska, M.; Hasman, H. Complete Genome Sequence of *Escherichia coli* MT102, a Plasmid-Free Recipient Resistant to Rifampin, Azide, and Streptomycin, Used in Conjugation Experiments. *Microbiol. Resour. Announc.* **2019**, *8*, e00383-19. [CrossRef] [PubMed]
103. Bao, Y.; Lies, D.P.; Fu, H.; Roberts, G.P. An Improved Tn7-Based System for the Single-Copy Insertion of Cloned Genes into Chromosomes of Gram-Negative Bacteria. *Gene* **1991**, *109*, 167–168. [CrossRef] [PubMed]
104. Voisard, C.; Rella, M.; Haas, D. Conjugative Transfer of Plasmid RP1 to Soil Isolates of *Pseudomonas Fluorescens* Is Facilitated by Certain Large RP1 Deletions. *FEMS Microbiol. Lett.* **1988**, *55*, 9–13. [CrossRef]
105. Taurian, T.; Anzuay, M.S.; Angelini, J.G.; Tonelli, M.L.; Ludueña, L.; Pena, D.; Ibáñez, F.; Fabra, A. Phosphate-Solubilizing Peanut Associated Bacteria: Screening for Plant Growth-Promoting Activities. *Plant Soil* **2010**, *329*, 421–431. [CrossRef]
106. Naghili, H.; Tajik, H.; Mardani, K.; Razavi Rouhani, S.M.; Ehsani, A.; Zare, P. Validation of Drop Plate Technique for Bacterial Enumeration by Parametric and Nonparametric Tests. *Vet. Res. Forum* **2013**, *4*, 179–183. [PubMed]
107. Marrero, M.A.; Agaras, B.C.; Wall, L.G.; Valverde, C. Enriquecimiento Diferencial de *Pseudomonas* Spp. En El Rizoplano de Distintas Especies Cultivadas. *Rev. Argent Microbiol.* **2015**, *47*, 132–137. [CrossRef]
108. Vincent, J. A Manual for Practical Study of Root-Nodule Bacteria. In *IBP Handbook*; Backwell Scientific Publications: Oxford, UK, 1970; p. 164.

**Disclaimer/Publisher's Note:** The statements, opinions and data contained in all publications are solely those of the individual author(s) and contributor(s) and not of MDPI and/or the editor(s). MDPI and/or the editor(s) disclaim responsibility for any injury to people or property resulting from any ideas, methods, instructions or products referred to in the content.

plants

MDPI

Article

# Isolation and Characterization of Plant Growth-Promoting Bacteria from the Rhizosphere of Medicinal and Aromatic Plant *Minthostachys verticillata*

**Romina del Valle Meneguzzi, Marilina Fernandez, Lorena del Rosario Cappellari, Walter Giordano and Erika Banchio ***

INBIAS Instituto de Biotecnología Ambiental y Salud, CONICET-Universidad Nacional de Río Cuarto, Río Cuarto 5800, Argentina; rmeneguzzi@exa.unrc.edu.ar (R.d.V.M.); mfernandez@exa.unrc.edu.ar (M.F.); lcappellari@exa.unrc.edu.ar (L.d.R.C.); wgiordano@exa.unrc.edu.ar (W.G.)
* Correspondence: ebanchio@exa.unrc.edu.ar

**Abstract:** This study aimed to isolate and characterize *Pseudomonas* native strains from the rhizospheric soil of *Minthostachys verticillata* plants to evaluate their potential as plant growth-promoting rhizobacteria (PGPR). A total of 22 bacterial isolates were obtained and subjected to various biochemical tests, as well as assessments of plant growth-promoting traits such as phosphate solubilization, hydrogen cyanide production, biocontrol properties through antibiosis, and indole acetic production. Genotypic analysis via 16S rRNA gene sequencing and phylogenetic tree construction identified the strains, with one particular strain named SM 33 showing significant growth-promoting effects on *M. verticillata* seedlings. This strain, SM 33, showed high similarity to *Stutzerimonas stutzeri* based on 16S rRNA gene sequencing and notably increased both shoot fresh weight and root dry weight of the plants. These findings underscore the potential application of native *Pseudomonas* strains in enhancing plant growth and health, offering promising avenues for sustainable agricultural practices.

**Keywords:** native rhizobacteria; PGPR; genotyping; peperina; *Stutzerimonas stutzeri*; *Pseudomonas*

**Citation:** Meneguzzi, R.d.V.; Fernandez, M.; Cappellari, L.d.R.; Giordano, W.; Banchio, E. Isolation and Characterization of Plant Growth-Promoting Bacteria from the Rhizosphere of Medicinal and Aromatic Plant *Minthostachys verticillata*. *Plants* **2024**, *13*, 2062. https://doi.org/10.3390/plants13152062

Academic Editors: Spyridon A. Petropoulos and Mohamad Hesam Shahrajabian

Received: 4 July 2024
Revised: 22 July 2024
Accepted: 23 July 2024
Published: 26 July 2024

**Copyright:** © 2024 by the authors. Licensee MDPI, Basel, Switzerland. This article is an open access article distributed under the terms and conditions of the Creative Commons Attribution (CC BY) license (https://creativecommons.org/licenses/by/4.0/).

## 1. Introduction

*Minthostachys* (Benth.) Spach (Lamiaceae) comprises 17 species native to South America, found across the Andes from Venezuela to Argentina. It is a significant genus in Andean traditional medicine and commerce [1]. *Minthostachys verticillata* (Griseb.) Epling, also known as 'peperina', is a perennial, gynodioecious semishrub unique to Argentina [1,2]. Peperina is the most commonly utilized species in traditional medicine by local communities [3,4]. Additionally, it is commercially used in the production of liqueurs, bitter drinks, and aromatic mate (a traditional South American infusion) [3–7].

Peperina is a species of high conservation priority in central Argentina [3,8] because it is one of the species under the greatest collection pressure. This is evidenced by the decline, weakening in terms of number and density, or disappearance of populations in numerous localities [3,6]. This situation is due to urbanization, collection practices that involve uprooting or not allowing the plants to seed, grazing in these areas, and fires that devastate the vegetation [3,6]. This also leads to a loss of genetic diversity in medicinal and aromatic plants [7]. Productive crops involving peperina are currently scarce, and wild populations of this plant are becoming depleted, putting the species at risk. Meanwhile, companies that demand this aromatic plant—mostly yerba mate producers—require increasingly larger quantities of high-quality raw material [2].

On the other hand, it is important to highlight that for several years, various properties of this aromatic plant have been studied, such as its antibacterial, antiviral, and antiseptic capabilities through its essential oils (EOs), obtaining encouraging results [9–13]. It has been proven that peperina EOs can be used safely and are free from any toxic risk [14]. The capability of these EOs was evaluated against aflatoxin $B_1$-induced damage in rats [13] and the antimicrobial power of the EO [15]. It has been demonstrated that peperina EO and one of its pure compounds (limonene) have antimicrobial activity, inhibiting the growth and biofilm formation of pathogens causing bovine mastitis [16].

Soils are complex and constantly changing environments with a delicate balance of biological, organic, and mineral components. This reflects intricate connections involving nourishment, coexistence, decay, decomposition, and recycling. The rhizosphere is the part of the soil influenced by the presence of living plant roots [17,18]. Plant roots release various substances such as sugars, amino acids, organic acids, phenolic compounds, vitamins, proteins, and polysaccharides through processes like secretion, exudation, and autolysis of old roots. This increases local nutrient levels, leading to the concentration and assembly of microorganisms in the rhizosphere through chemotaxis and biofilm formation [19,20]. Microorganisms and roots form distinct communities where interaction and communication processes are tightly regulated, controlling or modifying the activities of the entire community [21–24]. The interacting organisms produce species-specific molecular signals that alter their defense mechanisms and tend to maintain positive relationships among the organisms.

Plant growth-promoting rhizobacteria (PGPR) refers to a group of bacteria that live in the rhizosphere of plants and aid in plant development through various direct and indirect mechanisms [25–30]. These rhizobacteria produce growth-regulating substances such as gibberellins, cytokinins, and auxins, which promote the growth of root hairs, increasing the number, length, and surface area of plant roots. This enhances the plant's ability to absorb water and nutrients, making it more resilient and productive and better able to withstand adverse weather conditions [26,31]. Additionally, these microorganisms help in the solubilization of less mobile soil nutrients, such as phosphorus, which is the second most important nutrient for crop growth after nitrogen. They also contribute to the natural control of plant diseases by boosting the plant's defense system, thus enhancing its resistance to diseases [28,31].

Among these organisms, *Pseudomonas* is thought to be the most promising group of rhizobacteria that promote plant growth and help control plant diseases [32]. Various *Pseudomonas* strains have several ecologically beneficial traits, such as forming biofilms, producing antifungal substances, mediating quorum sensing, interacting synergistically with plant roots, mediating chemotaxis, and utilizing various plant secretions. *Pseudomonas* species are generally known for promoting plant growth (both roots and shoots) at the molecular level and providing a range of benefits to plants. Multiple strains of *Pseudomonas* contribute to plant health by enhancing growth and biologically suppressing diseases, such as *Fusarium* wilts in wheat and root rot in tobacco [32–34]. Particularly, the genus *Stutzerimonas,* previously classified within the genus *Pseudomonas,* is known for its significant plant growth-promoting effects [35]. *Stutzerimonas stutzeri,* in particular, exhibits capabilities such as phosphate solubilization, production of indole-3-acetic acid (IAA), and biocontrol of pathogens [36].

The cultivation of native plants for commercial use is currently not common. To successfully grow these plants, it is essential to understand the soil and climate conditions they require [5]. It is also important to improve our understanding of the rhizobacteria that are associated with the plant's root systems. The rhizosphere plays a significant role in promoting plant health and growth by aiding in nutrient absorption, tolerating environmental stress, and protecting against plant diseases [28]. Therefore, rhizobacteria can be a valuable resource for supporting the cultivation of these plant species. Additionally, identifying and studying rhizobacteria from mountainous soils that have characteristics promoting plant growth from untapped sources can have many practical applications [27,37,38].

The objectives of this study were to isolate, identify, and characterize PGPR strains from the rhizosphere of *M. verticillata.* The study also aimed to evaluate the beneficial effects of these strains on plant growth-promoting properties. Additionally, the study analyzed the phylogenetic relationships based on 16S rRNA sequences of the isolated strains and compared the rhizosphere of *M. verticillata* with the rhizosphere of another aromatic plant: *Mentha piperita*. This comparison was chosen to understand the microbial diversity and potential plant growth-promoting activities in the rhizosphere of these aromatic plants, which are known for their medicinal and aromatic properties.

## 2. Results and Discussion

### 2.1. Physico-Chemical Soil Characteristics

The analysis reveals that the soil possesses a loamy-clay texture with a high clay content (sand: 50–500 μm, silt: 2–50 μm, clay: 2 μm). Table 1 shows a significant amount of organic matter (5.48%) compared to other agricultural soils in the region, indicating excellent soil fertility. Organic matter is crucial as it improves soil structure, water retention, and nutrient supply. The soil also has high levels of phosphorus and nitrogen (23.30 ppm N-$NO_3$ and 103.22 ppm $NO_3$), which are essential for plant growth. High phosphorus levels promote root development and flowering, while elevated nitrogen levels support vigorous vegetative growth. The excellent water retention capacity (32.10%) is beneficial for crops, particularly in regions with variable rainfall, ensuring that plants have a steady water supply. The near-neutral pH of 7.50 is ideal for most crops, as it optimizes the availability of essential nutrients and minimizes the risk of toxicities associated with very acidic or alkaline soils. The soil's balanced base ratio and high Cation Exchange Capacity (CEC) of 27.83 cmol/kg, resulting from its substantial clay and organic matter content, enhance its ability to retain and supply essential cations like calcium (16.00 cmol/kg), magnesium (6.50 cmol/kg), sodium (0.90 cmol/kg), and potassium (2.31 cmol/kg). High CEC values indicate that the soil can hold and exchange a large amount of nutrients, which is beneficial for long-term fertility and reduces the risk of nutrient leaching. Additionally, the calcium and magnesium levels are adequate to support plant structure and enzyme function, while the potassium content is sufficient to enhance water uptake and resistance to diseases. Sodium levels are low, which is advantageous as high sodium can lead to soil dispersion and poor structure. Overall, these findings suggest that the soil has significant nutrient and water retention capabilities, making it highly suitable for crop growth and development [39].

**Table 1.** Chemical characteristics of rhizosphere soil from *M. verticillatta* plants.

| Chemical Characteristics | |
|---|---|
| pH | 7.50 |
| OM (%) | 5.48 |
| H° (%) | 32.10 |
| P (ppm) | 57.00 |
| N-$NO_3$ (ppm) | 23.30 |
| $NO_3$ (ppm) | 103.22 |
| $Ca^{+2}$ (cmol/Kg) | 16.00 |
| $Mg^{+2}$ (cmol/Kg) | 6.50 |
| $Na^{+}$ (cmol/Kg) | 0.90 |
| $K^{+}$ (cmol/Kg) | 2.31 |
| CEC (cmol/Kg) | 27.83 |

### 2.2. Strains Isolatated form *M. verticillata* Rhizosphere

The isolation of native strains involved collecting rhizospheric soil from spontaneously grown *M. verticillata*, with a total bacteria count of $2.7 \times 10^6$ CFU/g. Among the microorganisms that inhabit the soil, bacteria, including actinomycetes, are the most abundant, as they are found in a range of approximately $10^6$ to $10^8$ cells per gram of soil [40]. This finding aligns with previous studies examining rhizospheric soil from *Mentha piperita* commercial crops grown in the San Javier department, Córdoba province, Argentina, where a total bacteria count of $3.05 \times 10^7$ CFU/g was reported [41]. Similar bacterial counts have been observed in other aromatic plants, such as *Ocimum basilicum* ($0.75 \times 10^6$ CFU/g), *Melissa officinalis* ($10.00 \times 10^6$ CFU/g), and *Origanum syriacum* ($20.0 \times 10^6$ CFU/g), all grown in loam sandy soil in Egypt [42].

The microbial populations in the rhizospheric soils of different medicinal and aromatic plant species seem to have similar levels of bacterial abundance despite potential differences in environmental conditions and management practices. This consistency highlights the importance of rhizospheric microbiota in aromatic medicinal plants and emphasizes their potential role in plant health and productivity. It is important to note that the number of bacteria in the rhizosphere depends on various factors such as season, soil type, vegetation, moisture content, tillage, and fertilization [43]. However, plants clearly have a significant effect on microbial communities. The diverse microbial communities found in different types of soil used for crop production are influenced by the introduction of crop plants, which can either enhance or suppress the growth of native bacteria [44,45]. There is an ongoing debate about whether soil properties or the introduction of plants primarily determine the structure of microbial communities. It is evident that plants have a significant impact on microbial communities through the composition of their root exudates, which is closely linked to their nutritional status and growth stage [44].

### 2.3. Phenotypic Characterization

Forty-four colonies were randomly isolated from the highest dilution plate counts, representing the most abundant bacterial members of each rhizospheric community. From these, 22 bacterial isolates from the M. verticillata rhizosphere were selected for testing. These isolates were chosen for their Gram-negative characteristics with the goal of identifying native strains of the genus *Pseudomonas*. *Pseudomonas* is the most extensively researched of the known PGPR genera. It has been widely studied due to its wide distribution in various environments and its ease of cultivation under laboratory conditions [46,47]. The collected isolates strains displayed various biochemical activities. Out of the 22 *Pseudomonas* isolates, 8 (36.36%) showed the ability to hydrolyze casein, 2 (9.09%) hydrolyzed lecithin, 2 (9.09%) hydrolyzed starch, and 3 (13.63%) hydrolyzed lipids. Additionally, 8 (36.36%) strains were able to tolerate saline concentrations of 5%, and 10 (45.45%) could tolerate low temperatures (Table 2).

**Table 2.** Phenotypic characterization of native *Pseudomonas* strains, based on morphological, physiological, and biochemical evaluation of native strains isolated from the *M. verticillata* rhizosphere. '+' indicates presence, '−' indicates absence, and '+/−' indicates variable presence.

| Strain | Gram | Catalase | Oxidase | Pigmentation | Growth | | | | Hydrolysis | | | |
|---|---|---|---|---|---|---|---|---|---|---|---|---|
| | | | | King B | 4 °C | 42 °C | NaCl 5% | NaCl 6.5% | Lipids | Starch | Casein | Lecithinase |
| SM 2 | − | + | + | green | − | − | +/− | − | | − | − | − |
| SM 6 | − | + | + | orange | − | − | − | − | + | − | + | − |
| SM 8 | − | + | + | green | − | − | +/− | − | − | − | − | − |
| SM 12 | − | + | + | orange | − | − | − | − | | − | − | − |
| SM 17 | − | + | − | | − | − | − | − | + | − | + | − |
| SM 18 | − | + | − | orange | − | − | +/− | − | − | − | − | − |
| SM 19 | − | + | − | orange | − | − | +/− | − | − | − | − | − |
| SM 20 | − | + | + | green | + | − | − | − | − | − | + | − |
| SM 21 | − | + | + | green | + | − | − | − | − | − | + | − |
| SM 24 | − | − | + | | − | − | − | − | − | − | − | − |
| SM 25 | − | + | − | | − | + | − | − | + | − | − | + |
| SM 27 | − | + | + | green | − | − | − | − | − | − | − | − |
| SM 29 | − | + | + | | + | − | − | − | − | − | − | − |
| SM 33 | − | + | + | | + | − | +/− | − | − | − | − | − |
| SM 35 | − | + | + | green | − | + | − | − | − | − | − | − |
| SM 40 | − | + | + | | + | + | + | − | + | + | + | + |
| SM 41 | − | + | + | | + | − | + | − | − | − | + | − |
| SM 42 | − | + | + | | + | − | − | − | − | − | + | − |
| SM 46 | − | + | + | green | − | − | +/− | − | − | − | − | − |
| SM 47 | − | + | + | green | + | − | − | − | − | − | − | − |
| SM 49 | − | + | + | | + | − | − | − | − | − | + | − |
| SM 51 | − | + | + | | + | + | − | − | + | + | − | − |

A multivariate cluster analysis was performed to categorize the strains isolated from the *M. verticillata* rhizosphere based on their similar biochemical characteristics. Figure 1 illustrates the relationships between the strains, providing a better understanding of the variability present in the *M. verticillata* rhizosphere. The analysis grouped the strains into seven categories with similar biochemical profiles. One group consisted of strains SM 20, 21, 29, 33, 41, 42, 47, and 49, characterized by their ability to grow at low temperatures, with 5 of them capable of hydrolyzing casein. The second group, comprising strains SM 2, 8, 18, 19, and 46, was characterized by their growth in 5% saline concentrations. The third group included strains 12, 24, 27, and 35, identified as not exhibiting hydrolytic activity. The fourth group contained strains SM 6 and 17, which were able to hydrolyze casein and lipids. The fifth and sixth groups consisted of strains 25, 40, and 51, known for their growth at high temperatures, with strains SM 40 and 51 also capable of hydrolyzing starch and lipids.

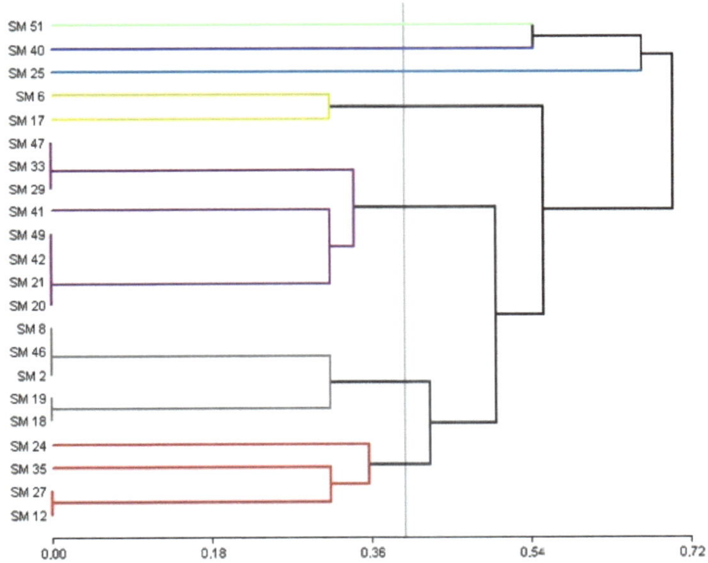


**Figure 1.** Cluster analysis constructed with variables resulting from the biochemical characterization of *Pseudomonas* strains isolated from the *M. verticillata* rhizosphere.

## 2.4. In Vitro Characterization of PGPR Strains

Based on the obtained results, potential mechanisms for enhancing plant growth were identified in the native strains isolated from the *M. verticillata* rhizosphere. As shown in Table 3, out of the 22 isolated strains, 60% exhibited the ability to solubilize phosphates and 50% to produce IAA. The production of IAA is a crucial trait of bacteria that promotes plant growth. It is estimated that up to 80% of microorganisms in soil communities are capable of synthesizing IAA, although the amount of hormone produced may vary depending on the microbial isolate [48]. The studied strains did not demonstrate the ability to produce siderophores. As for biocontrol potential through antibiosis, around 50% of the strains inhibited the growth of the phytopathogenic fungus *Sclerotium rolfsii* by 19 to 50%. *S. rolfsii* was utilized in these tests because it severely impacts many economically important crops, such as legumes, vegetables, and ornamentals, by causing southern blight, which results in significant global yield losses [49].

A multivariate correspondence analysis was performed to reveal patterns and relationships among the PGPR activity in native strains isolated from the *M. verticillata* rhizosphere. This analysis will offer valuable insights into the characteristics and similarities among different strains, leveraging the qualitative data obtained from the PGPR characterization of isolated strains.

**Table 3.** PGPR activity in native strains isolated from the *M. verticillata* rhizosphere. '+' indicates presence, '−' indicates absence.

| Strain | Phosphate Solubilization | Siderophores Production | IAA Production | CHN Production | Antibiosis % |
|---|---|---|---|---|---|
| SM2 | − | − | + | − | 19 |
| SM6 | − | − | − | + | 5 |
| SM8 | − | − | + | − | 21 |
| SM12 | − | − | + | − | 5 |
| SM17 | − | − | − | − | 0 |
| SM18 | + | − | + | − | 12 |
| SM19 | + | − | + | − | 9 |
| SM20 | + | − | + | − | 31 |
| SM21 | − | − | + | + | 14 |
| SM24 | + | − | + | − | 14 |
| SM25 | − | − | − | − | 50 |
| SM27 | + | − | − | − | 29 |
| SM29 | + | − | − | − | 19 |
| SM33 | + | − | + | − | 34 |
| SM35 | − | − | − | − | 38 |
| SM40 | + | − | − | − | 33 |
| SM41 | − | − | + | − | 14 |
| SM42 | + | − | + | − | 12 |
| SM46 | + | − | − | − | 14 |
| SM47 | + | − | + | − | 19 |
| SM49 | + | − | + | − | 19 |
| SM51 | − | − | − | − | 19 |

Correspondence Analysis (CA) enables the visualization of variate observations on planes, facilitating the identification of the strongest associations between the categories of qualitative variables. As shown in Figure 2, three distinct clusters were identified. The first cluster (pink) included strains SM 40, 27, 25, 35, 20, and 33, which were grouped based on their biocontrol effect through antibiosis against pathogenic fungi. The second cluster (green) comprised strains SM 18, 19, 20, 24, 29, 33, 42, 46, 47, and 49, known for their ability to solubilize phosphates. The third cluster (violet) consisted of strains SM 47, 49, 42, 18, 19, 24, 8, 2, 12, and 41, which are characterized by their production of IAA.

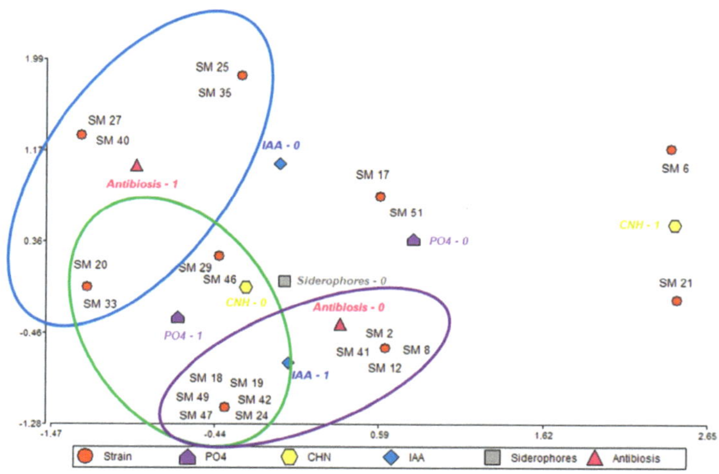


**Figure 2.** Correspondence analysis was constructed with variables resulting from the PGPR characterization of strains isolated from the M. verticillata rhizosphere. $PO_4$, CHN, IAA. The value 0 in the description of each variable indicates negative activity (-), while the value 1 indicates positive activity (+).

### 2.5. Quantitative Values of IAA Production

The strains were selected for their PGPR abilities demonstrated in previous assays. Among these strains, SM 33 exhibited the highest production, with a concentration of 936.3 ng/mL (Table 4). In concordance, Pedraza et al. [50] found that *P. stutzeri* strains produced IAA in a concentration ranging from 1000 to 2900 ng/mL protein in a culture medium supplemented with tryptophan. Due to its superior production capability, strain SM 33 was selected for subsequent plant inoculation assays.

**Table 4.** Quantification of IAA by LC-MS/MS Method produced by strains isolated from the *M. verticillata* rhizosphere.

| Sample | IAA (ng/mL) |
|---|---|
| Control LB | 30.9 |
| SM 8 | 556.7 |
| SM 20 | 578.1 |
| SM 21 | 717.8 |
| SM 33 | 936.3 |

### 2.6. Genotypic Identification

To determine the identity and phylogenetic position of the native strains isolated from the *M. verticillata* rhizosphere, nucleotide sequences of the 16S rRNA gene were analyzed. For this analysis, strains were selected based on their plant growth-promoting activities demonstrated in the previous assays. The chosen strains were SM 8, SM 20, SM 21, and SM 33, selected for their PGPR abilities to solubilize phosphates, produce IAA, and exhibit higher percentages of biocontrol through antibiosis. The 16S rRNA sequences of the strains under study were blasted with available sequences into the NCBI database The homology analysis of the native strains from the *M. verticillata* rhizosphere (Table 5) shows that strains SM 21 and SM 33 share identity levels of around 0.99 for the 16S rRNA gene region with reference sequences such as *Pseudomonas putida* IR7-6 (MW686888.1) and *Stutzerimonas stutzeri* KGS-2 (CP018050.1), respectively. On the other hand, strains SM 8 and SM 20 had around 99% identity with *Pseudomonas* sp. However, the coverage value is insufficient to provide an accession number in NCBI GenBank (low similarity), as it does not meet the necessary coverage standards for reliable and complete taxonomic identification. The low percentage of coverage prevents the verification and formal registration of the sequence.

**Table 5.** Data from BLASTn analysis of 16S rRNA gene sequences from native strains isolated from *M. verticillata* rhizosphere.

| Native Strain | Aligned Species (Access Number) | Homology | Coverage | Annotation (Access Number) |
|---|---|---|---|---|
| SM 8 | *Pseudomonas* sp. | 99% | ~61–63% | *Pseudomonas* sp. |
| SM 21 | *Pseudomona putida* IR7-6 (MW686888.1) | 99% | 99% | *Pseudomonas putida* SM21 (OR343518) |
| SM 33 | *Stutzerimona stutzeri* KGS-2 (CP018050.1) | 99% | 95% | *Stutzerimona stutzeri* SM33 (OR343508) |
| SM 20 | *Pseudomonas* sp. | 99% | ~62–63% | *Pseudomonas* sp. |

Additionally, a phylogenetic tree was obtained (Figure 3), constructed with the 16S rRNA gene nucleotide sequences from the strains isolated from the *M. verticillata* rhizosphere (Figure 3a), and another considering the previous strains and the strains isolated from the *Mentha piperita* rhizosphere [41] (Figure 3b), with the aim of studying and comparing the evolutionary and phylogenetic relationships between the different microbial strains present in the rhizosphere of these two medicinal and aromatic plant species. The phylogenetic tree constructed (Figure 3a) involving four nucleotide sequences of the 16S gene showed that *Pseudomonas* sp. SM 8 is phylogenetically associated with *Pseudomonas* sp. 20, supported by a bootstrap value of 87. Likewise, in Figure 3b, it can be observed that SM 8 and SM 20 strains group into a cluster distinct from 21 and SM 21 strains, indicating that these strains may be phylogenetically closer to *Pseudomonas* sp. SJ 48 isolated from the mint rhizosphere.

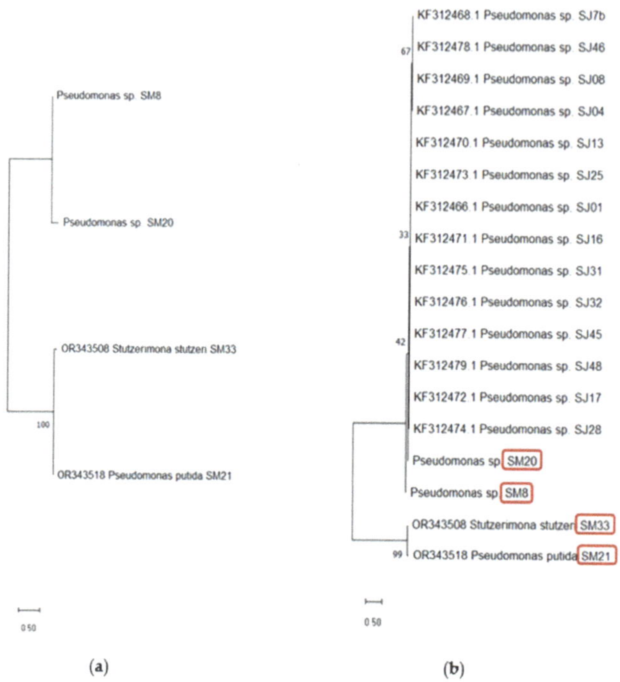


**Figure 3.** Phylogenetic tree of partial 16S rDNA sequences of plant-growth promoting strain: (**a**) four bacterial strains isolated from *M. verticillata* and (**b**) 14 bacterial strains isolated from the *M. piperita* rhizosphere and the strains isolated from *M. verticillata* (red frame). The numbers at the nodes indicate the bootstrap values out of 1000 replicates. The bar represents 0.50 substitutions per site.

### 2.7. Growth-Promoting Effects

PGPR strain can stimulate plant growth through various mechanisms, including the production of substances that act as plant growth regulators or the stimulation of their biosynthesis by plants, improved nutrient assimilation, and the inhibition of fungal pathogens [51–54]. For this analysis, the isolated strain SM 33 was selected based on its plant growth-promoting activities, including phosphate solubilization and IAA production. In the present study, inoculation with the native strain SM 33 positively and significantly modified the evaluated growth parameters such as the number of leaves, number of branches, number of nodes, stem length, fresh shoot weight (Table 6), and dry root weight ($p < 0.05$) (Figure 4). This is consistent with data previously reported by our laboratory in *M. piperita* plants inoculated with PGPR strains of interest [55] and by numerous authors on different plant species belonging to the Lamiaceae family, such as basil, winter savory (*Satureja montana*), rosemary, among others [27,56–61].

The observed increase in all parameters, as a result of inoculation with PGPR strains, may be linked to the growth-promoting mechanisms present in these microorganisms. Specifically, in the case of strain SM 33, this growth enhancement could be related to its ability to produce IAA (Table 4). Although the strain WCS417r is considered a PGPR and various authors have found significant effects with its inoculation [62–65], our study did not show significant differences in the evaluated parameters compared to the control. These contradictory results indicate that the impact of PGPR application on cultivated plants is species and strain-specific [41,66].

**Table 6.** Effect of direct inoculation of native strain SM33 on plant growth parameters of *M. verticillata*. Data shown are means ± SE. Values within a column followed by the same letter are not significantly different according to Fisher's LSD test ($p$ < 0.05).

| | Leaf Number | Ramification Number | Node Number | Shoot Lenght (cm) |
|---|---|---|---|---|
| control | 17.50 ± 1.20 a | 0.5 0± 0.13 a | 7.17 ± 0.22 a | 15.67 ± 0.99 a |
| SM 33 | 24.07 ± 1.12 b | 1.70 ± 0.23 b | 8.30 ± 0.22 b | 20.07 ± 0.83 b |
| WCS417r | 17.27 ± 1.33 a | 0.85 ± 0.21 a | 7.00 ± 0.18 a | 15.75 ± 1.15 a |

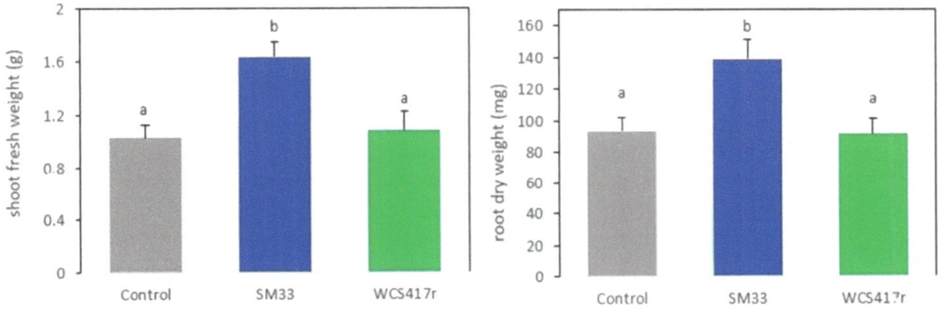


(**a**) (**b**)

**Figure 4.** Effect of direct inoculation of *M. verticillata* with PGPR strains (SM 33 and WCS417r) on growth parameters: (**a**) Shoot fresh weight and (**b**) root dry weight. Different letters indicate significant differences. Letters above bars indicate significant differences according to Fisher's LSD test ($p$ < 0.05).

In response to the direct inoculation of *M. verticillata* plants with PGPR strains, statistically significant differences ($p$ < 0.05) were only evident in the shoot fresh weight when the plant was inoculated with the SM 33 strain (Figure 4a). The aerial fresh weight of inoculated plants was 160% and 129% higher in plants treated with the SM 33 strain compared to plants inoculated with WCS417r or uninoculated, respectively. Additionally, the root dry weight was 48% higher than the control. This increase in root weight is a common response to bacterial IAA production, as inoculation with IAA-producing bacteria induces the proliferation of lateral roots and root hairs [67]. Therefore, the observed increase in root weight may be attributed to the high levels of IAA produced by SM 33. Similarly, Chandra et al., [68] observed an increase in root and shoot biomass in crop plants as a result of inoculation with IAA-producing bacteria isolated from *Stevia rebaudiana*. Additionally, Yousef [69] demonstrated that IAA and PGPR significantly enhanced germination rates, root growth, and shoot growth in plants.

IAA enhances root length by increasing the number of root branches, root hairs, and root laterals that aid in the uptake of nutrients from the surroundings [68]. Synthesizing IAA is considered an effective tool for screening beneficial microorganisms, suggesting that IAA-producing bacteria have a profound effect on plant growth [70].

SM 33 was identified as *Stutzerimonas stutzeri* (formerly *Pseudomonas stutzeri*) [35] according to the study conducted on the 16S rRNA gene region, as previously mentioned (Table 4). Similar to our studies, it has been observed that the *P. stutzeri* strain possesses plant growth-promoting activity. For instance, *P. stutzeri* MJL19 stimulates the growth of soybean plants under saline stress conditions and shows excellent colonization efficiency, likely due to its strong chemotactic response to root exudates and enhanced biofilm formation capacity in the presence of high salt concentrations [36]. Additionally, another study found that inoculation with a certain concentration of *P. stutzeri* significantly promoted tomato growth and induced significant changes in tomato root exudates, which in turn significantly induced growth, swarming motility, and biofilm formation [71]. Furthermore, a nitrogen-fixing *P. stutzeri* strain, A15, isolated from the rice rhizosphere and endosphere, has been shown to promote plant growth as well [72].

## 3. Materials and Methods

### 3.1. Soil Sample Collection and Strain Isolation

*M. verticillata* plants grown in the Santa Maria de Punilla, Punilla department, Córdoba province, Argentina (31°44′43.9″ S 64°22′01.0″ W) were collected. The sample was collected with prior authorization for scientific collection and the use of said biological material by the Secretary of Environment of the Province of Córdoba (GOBDIGI-0158493111-522) in accordance with the Nagoya Protocol. Samples were stored at 4 °C until they were used for isolation of rhizobacteria. Roots were washed in sterile distilled water to remove loosely adherent soil and then rotary shaken for 20 min in 90 mL sterile phosphate buffer, pH 7, to obtain a rhizospheric soil suspension. Serial dilutions were prepared from the suspension. Then, 0.1 mL of each dilution was spread on Nutrient Agar medium (Britania, Buenos Aires, Argentina) and on King B medium [73] and incubated at 28 °C for 48 h. After incubation, total bacteria and fluorescent bacteria under UV light were counted as colony-forming units (CFU) per g rhizospheric soil [74]. Bacterial strains were purified on King B medium, identified, and characterized. After 48 h, bacterial counts were conducted on the dilutions that presented between 30 and 300 CFU. A collection of 44 native strains was generated from King B plates based on fluorescence, color, size, and colony morphology. The selected colonies were re-cultured on KB Agar until a pure culture was obtained. Subsequently, a Gram test using the KOH method was performed on each strain, and only those with Gram-negative characteristics were selected for further tests [63].

### 3.2. Determination of Physical and Chemical Properties of the Rhizosphere Soil

Physico-chemical analyses such as pH, total organic carbon, and total nitrogen were performed at the Laboratorio Integral Agropecuario in Río Cuarto, Córdoba, Argentina, using standard protocols. For this experiment, soil samples were analyzed for organic matter (OM) using the [75] method and for electrical conductivity (EC) with a 1:5 soil-to-water ratio following APHA (1998) guidelines. Nitrogen, in the form of nitrate ions (NO3-N), was extracted from the soil with water and quantified colorimetrically after reacting with phenol sulfonic acid. Potassium content was determined by extracting the soil with Bray's extracting reagent and measuring available potassium using an absorptiometer [76]. Soil pH was measured using a digital pH meter (Mettler-Toledo, Cologne, Germany), and soil texture was determined according to [77].

### 3.3. Phenotypic Characterization

Native strains were phenotypically characterized based on morphological, physiological, and biochemical criteria [46,78,79]. The tests and parameters included colony morphology, Gram determination using KOH test [80], catalase determination [81], oxidase determination (Britania), fluorescence color under UV illumination, pigment production [73], temperature tolerance, growth at 5% and 6.5% salt concentration, Tween 80 hydrolysis [82], starch hydrolysis [83], casein hydrolysis [84] and lecithinase production [85]. Each test was performed twice.

### 3.4. Determination of the Main Direct and Indirect Mechanisms of Plant Promotion

#### 3.4.1. Indole-3-Acetic Acid Production

Qualitative Determination

Indole-3-acetic acid (IAA) level was qualitatively assayed as described in [86]. The test was performed both in the presence and absence of L-tryptophan as a precursor of IAA. One hundred mL LB broth in 250 mL Erlenmeyer flasks were inoculated with 100 μL of overnight bacterial culture, adjusted to optical density 0.5 ($10^7$–$10^8$ CFU mL$^{-1}$) measured at 600 nm. The bacteria were grown at 28 ± 2 °C for 96 h with continuous shaking at 150 rpm. The supernatant was collected by centrifuging at 10,000 rpm for 10 min. IAA was detected by mixing 100 μL of Salkowski reagent (1 mL 0.5 M $FeCl_3$, 30 mL concentrated $H_2SO_4$, and 50 mL distilled $H_2O$) with 100 μL supernatant and allowed to react at room temperature for 30 min. IAA production was confirmed by pink color development. Blanks (LB broth medium and Salkowski reagent) showing yellow color were below the detection limit. *Azospirillum brasilense* Sp7 was used as a positive control. The assay was replicated twice.

Quantitative Determination

As described in [87], the identification and quantification of the analyte under study was carried out using liquid chromatography coupled with tandem mass spectrometry (LC-MS/MS) for four selected strains isolated from the *M. verticillata* rhizosphere. The liquid chromatograph used was Alliance 2695 (Waters Inc., San Jose, CA, USA), and chromatographic conditions were gradient with methanol (100%): water/acetic acid (0.2%) (20:80) at a flow rate of 0.2 mL/min. Subsequently, they were analyzed by a double mass spectrometer coupled with Quatro UltimaTM Pt (Micromass, Manchester, UK) through the MassLynk software v4.0.

#### 3.4.2. Siderophore Production

Siderophore production was qualitatively assayed using the chrome azurol S (CAS) method with some modification [88]. Petri dishes were coated with 15 mL M9 medium. After the medium solidified, one-half was cut off and replaced by 7.5 mL CAS-blue agar [89]. Bacteria were inoculated in the boundary between the two media. A CAS-agar plate was inoculated with *Sinorhizobium meliloti* strain Rm1021 as a negative control [89]. Plates were incubated for 48 h at 28 °C. CAS reaction was determined by a color change from blue to orange. The lack of siderophore production by the negative control was evidenced by the absence of an orange color around the colonies. The technique was validated using color development by colonies of the native fluorescent strains. The assay was carried out twice.

#### 3.4.3. Phosphate Solubilization Ability

Phosphate solubilization ability was qualitatively assayed by inoculating bacteria as a single point on Petri dishes coated with Pikovskaya medium [90]. Plates were incubated for 48 h at 28 °C. The formation of a clear zone around bacterial colonies was considered to be a positive reaction. *S. meliloti* RM1021 was used as a negative control. The lack of phosphate solubilization in the negative control was evidenced by the absence of a clear zone around the colonies. The technique was validated as the formation of clear zones around colonies of the native fluorescent strains. The assay was replicated twice.

#### 3.4.4. Hydrogen Cyanide (HCN) Production

The bacterial strains to be tested were streaked onto Petri dishes containing 10% diluted TSA supplemented with the amino acid glycine (4.4 g/L). Filter paper impregnated with 0.5% picric acid and 2% sodium carbonate was placed on the inner side of the Petri dish lid. The plates were then incubated at 28 °C for 3 days. Plates where the filter paper changed color from yellow to orange-brown are considered to be HCN producers [91].

3.4.5. Biocontrol by In Vitro Antibiosis

The ability of bacterial strains to inhibit the growth of the phytopathogenic fungus *Sclerotium rolfsii* was assessed using a method described by [92]. The bacterial strains were streaked onto 25% Trypticase Soy Agar (TSA) plates, covering one-quarter of the plate After 24 h of incubation, a fungal plug of *S. rolfsii* was placed on the opposite side of the plate. The plates were then incubated for 4 days. The results were expressed as the percentage of fungal growth inhibition compared to a control. The percentage of mycelial growth inhibition was calculated by subtracting the percentage of mycelial growth in the interaction zone from 100%.

### *3.5. Amplification and Bioinformatic Analysis of 16S rRNA Region Sequences*

3.5.1. DNA Extraction

A single colony was inoculated on 3 mL LB medium and incubated at 28 °C for 24 h. A 2-mL culture sample taken at a late exponential growth phase was subjected to DNA extraction using a Genomic DNA Purification Kit (Fermentas, Thermo Fisher Scientific; Pittsburgh, PA, USA) according to the manufacturer's instructions.

3.5.2. 16S rRNA Gene Nucleotide Sequence Analysis

Four strains isolated from the rhizosphere of *M. verticillata* cultivation were selected for their characteristics of promoting plant growth, and the corresponding analysis of the 16S rRNA gene sequences was performed. Direct PCR was performed with 1 µL DNA template in a 20-µL PCR reaction mixture containing universal primers 27F(5′AGAGTTTGATCCT-GGCTCAG3′), 1492R(5′TACGGTTACCTTGTTACGACTT-3′), 518F (5′-CCAGCAGCCGC-GGTAATACG-3′) y 800R (5′-TACCAGGGTATCTAATCC-3′) [93]. Purified PCR products (~1400 bp) were sequenced by Macrogen, Inc. (Seoul, Korea) using an automated DNA sequencing system (model 3730XL, Applied Biosystems; Foster City, CA, USA). The obtained sequences were compared with deposited sequences in to NCBI public database using the BLASTn search program [94]. GeneBank accession numbers for the 16S rRNA sequences for each analyzed strain were obtained when high homology and coverage were found.

3.5.3. Construction of the Phylogenetic Tree

Two phylogenetic trees were constructed: one with the sequences obtained from the four selected strains isolated from *M. verticillata* and another tree including these sequences and those obtained from GenBank of 14 strains previously isolated by Santoro et al. [41]. from the rhizosphere of *Mentha piperita* (*Pseudomonas* sp. SJ01 accession number (No.): KF312466.1, *Pseudomonas* sp. SJ04 accession number (No.): KF312467.1, *Pseudomonas* sp. SJ7b accession number (No.): KF312468.1, *Pseudomonas* sp. SJ08 accession number (No.): KF312469.1, *Pseudomonas* sp. SJ13 accession number (No.): KF312470.1, *Pseudomonas* sp. SJ16 accession number (No.): KF312471.1, *Pseudomonas* sp. SJ17 accession number (No.): KF312472.1, *Pseudomonas* sp. SJ25 accession number (No.): KF312473.1, *Pseudomonas* sp. SJ28 accession number (No.): KF312474.1, *Pseudomonas* sp. SJ31 accession number (No.): KF312475.1, *Pseudomonas* sp. SJ32 accession number (No.): KF312476.1, *Pseudomonas* sp. SJ45 accession number (No.): KF312477.1, *Pseudomonas* sp. SJ46 accession number (No.): KF312478.1, *Pseudomonas* sp. SJ48 accession number (No.): KF312479.1).

Sequence alignment was performed using Clustal W with BioEdit version 7.2.5 [95], and the evolutionary analyses were conducted in MEGA11 [95]. The phylogenetic trees were inferred using the Neighbor-Joining method [96]. The percentage of replicate trees in which the associated taxa clustered together in the bootstrap test (1000 replicates) are shown next to the branches [97]. Only values greater than 50% are indicated. The tree is drawn to scale, with branch lengths in the same units as those of the evolutionary distances used to infer the phylogenetic tree (scale bar). The evolutionary distances are the difference in nucleotide base and were computed using the using the Kimura 2-parameter method [95] and are in the units of the number of base substitutions per site. The rate variation among sites was modeled with a gamma distribution (shape parameter = 1).

### 3.6. Evaluation of Growth-Promoting Effects

To evaluate the effects of *M. verticillata* seedling inoculation, the methodology of [5] was used with some modifications. *M. verticillata* seedlings with three growth nodes were transplanted into pots with soil mixed with perlite (80–20, $v/v$) and inoculated at the root crown with the isolated strain SM 33, selected for its PGPR characteristics, or *Pseudomonas simiae* WCS417r, used as a reference strain due to its PGPR characteristics demonstrated in previous studies [41,55]. Prior to inoculation, bacterial cultures were grown overnight at 28 °C with rotation at 120 rpm until reaching the exponential phase. Each culture was then washed twice in 0.9% NaCl by centrifugation (10,000 rpm, 10 min) in an Eppendorf centrifuge, resuspended in physiological solution, and adjusted to a final concentration of $\sim 10^9$ colony-forming units (CFU)/mL for use as an inoculum. Saturated cultures were diluted in sterile distilled water to a final concentration of $10^9$ CFU $mL^{-1}$. Seedlings were inoculated with 100 µL of the corresponding bacterial suspension, while control plants were inoculated with 100 µL of sterile physiological solution.

After inoculation, plants were grown in a growth chamber with controlled conditions of light (16/8 h light/dark cycle), temperature (24 $\pm$ 2 °C), and relative humidity ($\sim$70%). All plants received distilled water (20 mL/pot) once per week [55]. The experiments were performed under non-sterile conditions. The experiments were replicated (10 pots per treatment; 1 plant per pot), and the pots were arranged randomly in the growth chamber. Forty-five days after inoculation, the plants were removed from the pots, the roots were washed to remove soil, and standard growth parameters (shoot length, leaf number, node number, shoot fresh weight, root dry weight) were measured [55].

### 3.7. Statistical Analysis

Plant data were subjected to analysis of variance (ANOVA) followed by a comparison of multiple treatment levels with the control using the post hoc Fisher Least Significant Difference (LSD) test. Additionally, a multivariate cluster analysis was conducted to observe relationships between the isolated native strains and the determined biochemical variables, and a Correspondence Analysis (CA) was performed to observe associations with growth promotion variables. Infostat software program 2020 (Group Infostat, Universidad Nacional de Córdoba, Argentina) was used for all statistical analyses.

## 4. Conclusions

This study successfully isolated and characterized native *Pseudomonas* strains from the rhizospheric soil of *Minthostachys verticillata* plants, demonstrating their potential as plant growth-promoting rhizobacteria (PGPR). Through a series of biochemical tests, the strains exhibited capabilities such as phosphate solubilization and hydrogen cyanide production alongside biocontrol properties. The strain SM 33, which has a high similarity to *Stutzerimonas stutzeri* based on 16S rRNA gene sequencing, demonstrated significant growth-promoting effects on *M. verticillata* seedlings, increasing both shoot fresh weight and root dry weight of the plants. Future studies will require complete genome sequencing to ensure accurate identification and to fully characterize the strain. These findings underscore the potential application of native *Pseudomonas* strains in enhancing plant growth and health, offering promising avenues for sustainable agricultural practices.

**Author Contributions:** Conceptualization, E.B.; methodology, R.d.V.M., L.d.R.C. and E.B.; validation, R.d.V.M. and E.B.; formal analysis, E.B.; investigation, R.d.V.M., L.d.R.C. and M.F.; resources, W.G., L.d.R.C. and E.B.; data curation, R.d.V.M.; writing—original draft preparation, R.d.V.M. and M.F.; writing—review and editing, E.B. All authors have read and agreed to the published version of the manuscript.

**Funding:** This research was funded by Secretaría de Ciencia y Tecnica de la Universidad Nacional De Río Cuarto Grant Number PICT SSPR 2021, PPI 2020-2022: C529-1, Consejo Nacional de Investigaciones Científicas y Técnicas (CONICET), Agencia Nacional De Promoción Científica Y Tecnológica (ANPCyT).

**Data Availability Statement:** The raw data supporting the conclusions of this article will be made available by the authors upon request.

**Conflicts of Interest:** The authors declare no conflict of interest. The funders had no role in the design of the study, in the collection, analyses, or interpretation of data, in the writing of the manuscript, or in the decision to publish the results.

## References

1. Schmidt-Lebuhn, A.N. Ethnobotany, Biochemistry and Pharmacology of *Minthostachys* (Lamiaceae). *J. Ethnopharmacol.* **2008**, *118*, 343–353. [CrossRef] [PubMed]
2. Juliani, H.R.; Brunetti, P.; Koroch, A.R. *Minthostachys verticillata* (Griseb.) Epling. In *Medicinal and Aromatic Plants of South America Volume 2*; Máthé, Á., Bandoni, A., Eds.; Medicinal and Aromatic Plants of the World, Volume 7; Springer: Cham, Switzerland, 2021; pp. 367–378. [CrossRef]
3. Ojeda, M.; Arroyo, A.; Borgogno, P.; Biderbost, E.; Balzarini, M. Yield of Peperina (*Minthostachys mollis* (Kunth.) Griseb.) Populations in the Year Following Planting: Response to Cropping Regimen. *Span. J. Agric. Res.* **2004**, *2*, 393–399. [CrossRef]
4. Martínez, G.J. Recolección y Comercialización de Plantas Medicinales en el Departamento Santa María, Provincia de Córdoba. *Acta Farm. Bonaer.* **2005**, *24*, 575–584.
5. Ojeda, M.S.; Martínez, G.J.; Massuh, Y.; Torres, L.E.; Chaves, A.G.; Arizio, O.; Curioni, A.O. *Plantas Aromáticas y Medicinales: Modelos para su Domesticación, Producción y usos Sustentables: Carqueja-Orégano-Peperina-Suico-Manzanilla*; Editorial Universidad Nacional de Córdoba: Córdoba, Argentina, 2015.
6. Glinos, E.; Condat, E.; Mulieri, P.; Ashworth, L. Essential Dependence on Wild Pollination Service: A Medicinal Plant under Threat *Minthostachys verticillata* (Lamiaceae). *Arthropod-Plant Interact.* **2019**, *13*, 865–874. [CrossRef]
7. Roig, F.A. *Flora Medicinal Mendocina: Las Plantas Medicinales y Aromáticas, de la Provincia de Mendoza (Argentina): Aborígenes, Exóticas Espontáneas o Naturalizadas y Cultivadas*; EDIUNC: Mendoza, Argentina, 2001; Serie Manuales.
8. Martínez, G.J.; Planchuelo, A.M.; Fuentes, E.; Ojeda, M. A Numeric Index to Establish Conservation Priorities for Medicinal Plants in the Paravachasca Valley, Córdoba, Argentina. *Biodivers. Conserv.* **2006**, *15*, 2457–2475. [CrossRef]
9. Primo, V.; Rovera, M.; Zanón, S.; Oliva, M.; Demo, M.; Daghero, J.; Sabini, L. Determinación de la Actividad Antibacteriana y Antiviral del Aceite Esencial de *Minthostachys verticillata* (Griseb.) Epling. *Rev. Arg. Microbiol.* **2001**, *33*, 113–117.
10. Cariddi, L.N.; Panero, A.; Demo, M.S.; Sabini, L.I.; Maldonado, A.M.; Grosso, M.; Zygadlo, J. Inhibition of Immediate-Type Allergic Reaction by *Minthostachys verticillata* (Griseb.) Epling Essential Oil. *J. Essent. Oil Res.* **2007**, *19*, 190–196. [CrossRef]
11. Cariddi, L.; Moser, M.; Andrada, M.C.; Demo, M.S.; Zygadlo, J.A.; Sabini, L.I.; Maldonado, A.M. The Effect of *Minthostachys verticillata* Essential Oil on the Immune Response of Patients Allergic to Dust Mites. *Boletín Latinoam. Y Del Caribe De Plantas Med. Y Aromáticas* **2009**, *8*, 224–233.
12. Cariddi, L.; Escobar, F.; Moser, M.; Panero, A.; Alaniz, F.; Zygadlo, J.; Sabini, L.; Maldonado, A. Monoterpenes Isolated from *Minthostachys verticillata* (Griseb.) Epling Essential Oil Modulates Immediate-Type Hypersensitivity Responses In Vitro and In Vivo. *Planta Med.* **2011**, *77*, 1687–1694. [CrossRef]
13. Escobar, F.M.; Magnoli, A.; Sabini, M.C.; Cariddi, L.N.; Bagnis, G.; Soltermann, A.; Cavaglieri, L. *Minthostachys verticillata* Essential Oils as Potential Phytogenic Additives and Chemoprotective Strategy on Aflatoxin $B_1$ Toxicity. *J. Appl. Anim. Res.* **2019**, *47*, 217–222. [CrossRef]
14. Escobar, F.M.; Sabini, M.C.; Cariddi, L.N.; Sabini, L.I.; Mañas, F.; Cristofolini, A.; Bagnis, G.; Gallucci, M.N.; Cavaglieri, L.R. Safety Assessment of Essential Oil from *Minthostachys verticillata* (Griseb.) Epling (peperina) 90-days Oral Subchronic Toxicity Study in Rats. *Regul. Toxicol. Pharmacol.* **2015**, *71*, 1–7. [CrossRef] [PubMed]
15. Montironi, I.D.; Cariddi, L.N.; Reinoso, E.B. Evaluation of the Antimicrobial Efficacy of *Minthostachys verticillata* Essential Oil and Limonene Against *Streptococcus uberis* Strains Isolated from Bovine Mastitis. *Rev. Argent. Microbiol.* **2016**, *48*, 210–216. [CrossRef] [PubMed]
16. Cerioli, M.F.; Moliva, M.V.; Cariddi, L.N.; Reinoso, E.B. Effect of the Essential Oil of *Minthostachys verticillata* (Griseb.) Epling and Limonene on Biofilm Production in Pathogens Causing Bovine Mastitis. *Front. Vet. Sci.* **2018**, *5*, 146. [CrossRef] [PubMed]
17. Hiltner, L. Über neuere Erfahrungen und Probleme auf dem Gebiet der Bodenbakteriologie und unterbesonderer Berücksichtigung der Gründüngung und Brachte. *Arb. Der Dtsch. Landwirtsch. Ges.* **1904**, *98*, 59–78.
18. Foster, R.C. Microenvironments of Soil Microorganisms. *Biol. Fert. Soils* **1998**, *6*, 189–203. [CrossRef]
19. McNear, D.H., Jr. The Rhizosphere–Roots, Soil and Everything In Between. *Nat. Sci. Educ.* **2013**, *4*, 1.
20. Tan, S.; Yanga, C.; Meia, X.; Shenb, S.; Razaam, W.; Shena, Q.; Xu, Y. The Effect of Organic Acids from Tomato Root Exudates on Rhizosphere Colonization of *Bacillus amyloliquefaciens* T-5. *Appl. Soil Ecol.* **2013**, *64*, 15–22. [CrossRef]
21. Walker, S.T.; Bais, H.P.; Grotewold, E.; Vivanco, J.M. Root Exudation and Rhizosphere Biology. *Plant Physiol.* **2003**, *132*, 44–51. [CrossRef] [PubMed]
22. Bais, H.P.; Park, S.W.; Weir, T.L.; Callaway, R.M.; Vivanco, J.M. How Plants Communicate Using the Underground Information Superhighway. *Trends Plant Sci.* **2004**, *9*, 26–32. [CrossRef]
23. Ortíz-Castro, R.; Contreras-Cornejo, H.A.; Macías-Rodríguez, L.; López-Bucio, J. The Role of Microbial Signals in Plant Growth and Development. *Plant Signal. Behav.* **2009**, *4*, 701–712. [CrossRef]

24. Babalola, O.O. Beneficial Bacteria of Agricultural Importance. *Biotechnol. Lett.* **2010**, *32*, 1559–1570. [CrossRef]
25. Kloepper, J.W. Plant-Growth-Promoting Rhizobacteria as Biological Control Agents. In *Soil Microbial Ecology: Applications in Agricultural and Environmental Management*; Metting, F.B., Jr., Ed.; Marcel Dekker Inc.: New York, NY, USA, 1993; pp. 255–274.
26. Glick, B.R. Bacteria with ACC Deaminase can Promote Plant Growth and Help to Feed the World. *Microbiol. Res.* **2014**, *169*, 30–39. [CrossRef]
27. Vejan, P.; Abdullah, R.; Khadiran, T.; Ismail, S.; Boyce, A.N. Role of Plant Growth Promoting Rhizobacteria in Agricultural Sustainability–A review. *Molecules* **2016**, *21*, 573. [CrossRef] [PubMed]
28. Backer, R.; Rokem, J.S.; Ilangumaran, G.; Lamont, J.; Praslickova, D.; Ricci, E.; Subramanian, S.; Smith, D.L. Plant Growth-Promoting Rhizobacteria: Context, Mechanisms of Action, and Roadmap to Commercialization of Biostimulants for Sustainable Agriculture. *Front. Plant Sci.* **2018**, *9*, 1473. [CrossRef]
29. Babalola, O.O.; Fadiji, A.E.; Enagbonma, B.J.; Alori, E.T.; Ayilara, M.S.; Ayangbenro, A.S. The Nexus Between Plant and Plant Microbiome: Revelation of the Networking Strategies. *Front. Microbiol.* **2020**, *11*, 548037. [CrossRef] [PubMed]
30. Etesami, H.; Adl, S.M. Plant Growth-Promoting Rhizobacteria (PGPR) and Their Action Mechanisms in Availability of Nutrients to Plants. In *Phyto-Microbiome in Stress Regulation. Environmental and Microbial Biotechnology*; Kumar, M., Kumar, V., Prasad, R., Eds.; Springer: Singapore, 2020; pp. 147–203. [CrossRef]
31. Gupta, G.; Parihar, S.S.; Ahirwar, N.K.; Snehi, S.K.; Singh, V. Plant Growth Promoting Rhizobacteria (PGPR): Current and Future Prospects for Development of Sustainable Agriculture. *J. Microb. Biochem. Technol.* **2015**, *7*, 96–102. [CrossRef]
32. Sivasakthi, S.; Usharani, G.A.; Saranraj, P. Biocontrol Potentiality of Plant Growth Promoting Bacteria (PGPR)–*Pseudomonas fluorescens* and *Bacillus subtilis*: A review. *Afr. J. Agric. Res.* **2014**, *9*, 1265–1277.
33. David, B.V.; Chandrasehar, G.; Selvam, P.N. Chapter 10: *Pseudomonas fluorescens:* A Plant-Growth-Promoting Rhizobacterium (PGPR) with Potential Role in Biocontrol of Pests of Crops. In *New and Future Developments in Microbial Biotechnology and Bioengineering*; Prasad, R., Tuteja, N., Gill, S.S., Eds.; Elsevier: Amsterdam, The Netherlands, 2018; pp. 221–243.
34. Sah, S.; Krishnani, S.; Singh, R. *Pseudomonas* Mediated Nutritional and Growth Promotional Activities for Sustainable Food Security. *Curr. Res. Microb. Sci.* **2021**, *2*, 100084. [CrossRef] [PubMed]
35. Lalucat, J.; Gomila, M.; Mulet, M.; Zaruma, A.; García-Valdés, E. Past, Present and Future of the Boundaries of the *Pseudomonas* Genus: Proposal of *Stutzerimonas* gen. *Nov. Syst. Appl. Microbiol.* **2022**, *45*, 126289. [CrossRef]
36. Lami, M.J.; Adler, C.; Caram-Di Santo, M.C.; Zenoff, A.M.; de Cristobal, R.E.; Espinosa-Urgel, M.; Vincent, P.A. *Pseudomonas stutzeri* MJL19, a Rhizosphere-Colonizing Bacterium That Promotes Plant Growth Under Saline Stress. *J. Appl. Microbiol.* **2020**, *129*, 1321–1336. [CrossRef]
37. Patten, C.L.; Glick, B.R. Role of *Pseudomonas putida* Indole Acetic Acid in Development of the Host Plant Root System. *Appl. Environ. Microbiol.* **2002**, *68*, 3795–3801. [CrossRef]
38. Sergeeva, E.; Hirkala, D.L.; Nelson, L.M. Production of Indole-3-Acetic Acid, Aromatic Amino Acid Aminotransferase Activities and Plant Growth Promotion by *Pantoea agglomerans* Rhizosphere Isolates. *Plant Soil* **2007**, *297*, 1–13. [CrossRef]
39. El-Ramady, H.R.; Alshaal, T.A.; Amer, M.; Domokos-Szabolcsy, É.; Elhawat, N.; Prokisch, J.; Fári, M. *Sustainable Agriculture Reviews*; Ozier-Lafontaine, H., Lesueur-Jannoyer, M., Eds.; Springer: Cham, Switzerland, 2014; pp. 345–447.
40. Calvo Velez, P.; Reymundo Meneses, L.; Zuniga Davila, D. A Study of Potato (*Solanum tuberosum*) Crop Rhizosphere Microbial Population in Highland Zones. *Ecol. Apl.* **2008**, *7*, 141–148.
41. Santoro, V.M.; Bogino, P.C.; Nocelli, N.; Cappellari, L.; Giordano, W.F.; Banchio, E. Analysis of Plant Growth-Promoting Effects of Fluorescent Pseudomonas Strains Isolated from *Mentha piperita* Rhizosphere and Effects of Their Volatile Organic Compounds on Essential Oil Composition. *Front. Microbiol.* **2016**, *7*, 1085. [CrossRef]
42. Ahmed, E.A.; Hassan, E.A.; El Tobgy, K.M.K.; Ramadan, E.M. Evaluation of Rhizobacteria of Some Medicinal Plants for Plant Growth Promotion and Biological Control. *Ann. Agric. Sci.* **2014**, *59*, 273–280. [CrossRef]
43. Memenza-Zegarra, M.; Zúñiga-Dávila, D. Bioprospection of Native Antagonistic Rhizobacteria from the Peruvian Coastal Ecosystems Associated with *Phaseolus vulgaris*. *Curr. Microbiol.* **2021**, *78*, 1418–1431. [CrossRef]
44. Houlden, A.; Timms-Wilson, T.M.; Day, M.J.; Bailey, M.J. Influence of Plant Developmental Stage on Microbial Community Structure and Activity in the Rhizosphere of Three Field Crops. *FEMS Microbiol. Ecol.* **2008**, *65*, 193–201. [CrossRef]
45. Agaras, B.; Wall, L.G.; Valverde, C. Specific Enumeration and Analysis of the Community Structure of Culturable *Pseudomonads* in Agricultural Soils Under No-Till Management in Argentina. *Appl. Soil Ecol.* **2012**, *61*, 305–319. [CrossRef]
46. Palleroni, N.J. Genus *Pseudomonas*. In *Bergey's Manual® of Systematic Bacteriology*, 2nd ed.; Brenner, D.J., Krieg, N.R., Staley, J.T., Eds.; Springer: New York, NY, USA, 2005; pp. 323–379.
47. Garrido-Sanz, D.; Meier-Kolthoff, J.P.; Göker, M.; Martín, M.; Rivilla, R.; Redondo-Nieto, M. Genomic and genetic diversity within the *Pseudomonas fluorescens* complex. *PLoS ONE* **2016**, *11*, e0150183. [CrossRef]
48. Fiodor, A.; Ajijah, N.; Dziewit, L.; Pranaw, K. Biopriming of Seed with Plant Growth-Promoting Bacteria for Improved Germination and Seedling Growth. *Front. Microbiol.* **2023**, *14*, 1142966. [CrossRef] [PubMed]
49. Dwivedi, S.K.; Prasad, G. Integrated management of Sclerotium rolfsii: An overview. *Eur. J. Biomed. Pharm. Sci.* **2016**, *3*, 137–146.
50. Pedraza, R.; Ramirez-Mata, A.; Xiqui, M.; Baca, B. Aromatic Aminoacid Aminotransferase Activity and Indole-3 Acetic Acid Production by Associative Nitrogen-Fixing Bacteria. *FEMS Microbiol. Lett.* **2004**, *233*, 15–21. [CrossRef] [PubMed]
51. Raaijmakers, J.M.; Paulitz, T.C.; Steinberg, C.; Alabouvette, C.; Moënne-Loccoz, Y. The Rhizosphere: A Playground and Battlefield for Soilborne Pathogens and Beneficial Microorganisms. *Plant Soil* **2009**, *321*, 341–361. [CrossRef]

52. Choudhary, D.K. Plant Growth-Promotion (PGP) Activities and Molecular Characterization of Rhizobacterial Strains Isolated from Soybean (*Glycine max* L. Merril) Plants Against Charcoal Rot Pathogen, *Macrophomina phaseolina*. *Biotechnol. Lett.* **2011**, *33* 2287–2295. [CrossRef] [PubMed]
53. Lyu, D.; Backer, R.; Smith, D.L. Three Plant Growth-Promoting Rhizobacteria Alter Morphological Development, Physiology, and Flower Yield of *Cannabis sativa* L. *Ind. Crops Prod.* **2022**, *178*, 114583. [CrossRef]
54. Chepsergon, J.; Moleleki, L.N. Rhizosphere Bacterial Interactions and Impact on Plant Health. *Curr. Opin. Microbiol.* **2023** *73*, 102297. [CrossRef] [PubMed]
55. Cappellari, L.R.; Santoro, M.V.; Reinoso, H.; Travaglia, C.; Giordano, W.; Banchio, E. Anatomical, Morphological, and Phytochemical Effects of Inoculation with Plant Growth-Promoting Rhizobacteria on Peppermint (*Mentha piperita*). *J. Chem. Ecol.* **2015**, *41* 149–158. [CrossRef] [PubMed]
56. Ordookhani, K.; Sharafzadeh, S.; Zare, M. Influence of PGPR on Growth, Essential Oil and Nutrients Uptake of Sweet Basil. *Adv Environ. Biol.* **2011**, *5*, 672–677.
57. Mohammadi, H.; Dashi, R.; Farzaneh, M. Effects of Beneficial Root *Pseudomonas* on Morphological, Physiological, and Phytochemical Characteristics of *Satureja hortensis* (Lamiaceae) Under Water Stress. *Braz. J. Bot.* **2017**, *40*, 41–48. [CrossRef]
58. Dehghani Bidgoli, R.; Azarnezhad, N.; Akhbari, M.; Ghorbani, M. Salinity Stress and PGPR Effects on Essential Oil Changes in *Rosmarinus officinalis* L. *Agric. Food Secur.* **2019**, *8*, 2. [CrossRef]
59. Vessey, J.K. Plant Growth Promoting Rhizobacteria as Biofertilizers. *Plant Soil* **2003**, *255*, 571–586. [CrossRef]
60. van Loon, L.C. Plant Responses to Plant Growth-Promoting Rhizobacteria. In *New Perspectives and Approaches in Plant Growth-Promoting Rhizobacteria Research*; Bakker, P.A.H.M., Raaijmakers, J.M., Bloemberg, G., Höfte, M., Lemanceau, P., Cooke, B.M., Eds.; Springer: Dordrecht, The Netherlands, 2007; pp. 243–254. [CrossRef]
61. Pérez-Montaño, F.; Alías-Villegas, C.; Bellogín, R.A.; del Cerro, P.; Espuny, M.R.; Jiménez-Guerrero, I.; López Baena, F.J.; Ollero, F.J.; Cubo, T. Plant Growth Promotion in Cereal and Leguminous Agricultural Important Plants: From Microorganism Capacities to Crop Production. *Microbiol. Res.* **2014**, *169*, 325–336. [CrossRef] [PubMed]
62. Santoro, M.V.; Zygadlo, J.; Giordano, W.; Banchio, E. Volatile Organic Compounds from Rhizobacteria Increase Biosynthesis of Essential Oils and Growth Parameters in Peppermint (*Mentha piperita*). *Plant Physiol. Biochem.* **2011**, *49*, 1177–1182. [CrossRef]
63. Santoro, M.; Cappellari, L.; Giordano, W.; Banchio, E. Plant Growth Promoting Effects of Native *Pseudomonas* Strains on *Mentha piperita* (peppermint): An in Vitro Study. *Plant Biol.* **2015**, *17*, 1218–1226. [CrossRef] [PubMed]
64. Cappellari, L.; Santoro, M.V.; Schmidt, A.; Gershenzon, J.; Banchio, E. Induction of Essential Oil Production in *Mentha x piperita* by Plant Growth Promoting Bacteria Was Correlated with an Increase in Jasmonate and Salicylate Levels and a Higher Density of Glandular Trichomes. *Plant Physiol. Biochem.* **2019**, *141*, 142–153. [CrossRef] [PubMed]
65. Mekureyaw, M.F.; Pandey, C.; Hennessy, R.C.; Nicolaisen, M.H.; Liu, F.; Nybroe, O.; Roitsch, T. The Cytokinin-Producing Plant Beneficial Bacterium *Pseudomonas fluorescens* G20-18 Primes Tomato (*Solanum lycopersicum*) for Enhanced Drought Stress Responses. *J. Plant Physiol.* **2022**, *270*, 153629. [CrossRef] [PubMed]
66. Kalozoumis, P.; Savvas, D.; Aliferis, K.; Ntatsi, G.; Marakis, G.; Simou, E.; Tampakaki, A.; Karapanos, I. Impact of Plant Growth-Promoting Rhizobacteria Inoculation and Grafting on Tolerance of Tomato to Combined Water and Nutrient Stress Assessed via Metabolomics Analysis. *Front. Plant Sci.* **2021**, *12*, 670236. [CrossRef] [PubMed]
67. Gilbert, S.; Poulev, A.; Chrisler, W.; Acosta, K.; Orr, G.; Lebeis, S.; Lam, E. Auxin-Producing Bacteria from Duckweeds Have Different Colonization Patterns and Effects on Plant Morphology. *Plants* **2022**, *11*, 721. [CrossRef] [PubMed]
68. Chandra, S.; Askari, K.; Kumari, M. Optimization of Indole Acetic Acid Production by Isolated Bacteria from *Stevia rebaudiana* Rhizosphere and its Effects on Plant Growth. *J. Genet. Eng. Biotechnol.* **2018**, *16*, 581–586. [CrossRef]
69. Yousef, N.M. Capability of Plant Growth-Promoting Rhizobacteria (PGPR) for Producing Indole Acetic Acid (IAA) Under Extreme Conditions. *Eur. J. Biol. Res.* **2018**, *8*, 174–182. [CrossRef]
70. Wahyudi, A.T.; Astuti, R.P.; Widyawati, A.; Meryandini, A.; Nawangsih, A.A. Characterization of Bacillus sp. Strains Isolated from Rhizosphere of Soybean Plants for Their Use as Potential Plant Growth for Promoting Rhizobacteria. *J. Microbiol. Antimicrob.* **2011**, *3*, 34–40.
71. Zhang, H.; Zheng, D.; Hu, C.; Cheng, W.; Lei, P.; Xu, H.; Gao, N. Certain Tomato Root Exudates Induced by Pseudomonas stutzeri NRCB010 Enhance Its Rhizosphere Colonization Capability. *Metabolites* **2023**, *13*, 664. [CrossRef] [PubMed]
72. Pham, V.T.; Rediers, H.; Ghequire, M.G.; Nguyen, H.H.; De Mot, R.; Vanderleyden, J.; Spaepen, S. The Plant Growth-Promoting Effect of the Nitrogen-Fixing Endophyte *Pseudomonas stutzeri*. *Arch. Microbiol.* **2017**, *199*, 513–517. [CrossRef] [PubMed]
73. King, E.O.; Ward, M.K.; Randey, D.E. Two Simple Media for the Demonstration of Pyocyanin and Fluorescein. *J. Lab. Clin. Med.* **1954**, *44*, 301–307. [PubMed]
74. Mehnaz, S.; Weselowski, B.; Aftab, F.; Zahid, S.; Lazarovits, G.; Iqbal, J. Isolation, Characterization and Effect of Fluorescent Pseudomonads on Micropropagated Sugarcane. *Can. J. Microbiol.* **2009**, *55*, 1007–1011. [CrossRef] [PubMed]
75. Walkley, A.; Black, I.A. An Examination of the Degtjareff Method for Determining Soil Organic Matter, and a Proposed Modification of the Chromic Acid Titration Method. *Soil Sci.* **1934**, *37*, 29–38. [CrossRef]
76. Whittles, C.L.; Little, R.C. A Colorimetric Method for the Determination of Potassium and its Application to the Analysis of Soil Extracts. *J. Sci. Food Agric.* **1950**, *1*, 323–326. [CrossRef]
77. Bouyoucos, G.J. Hydrometer Method Improved for Making Particle Size Analysis of Soil. *Agron. J.* **1962**, *54*, 464–465. [CrossRef]
78. Lysenko, O. *Pseudomonas*–an attempt at a general classification. *J. Gen. Microbiol.* **1961**, *25*, 379–408. [CrossRef]

79. Stanier, R.Y.; Palleroni, N.J.; Doudoroff, M. The Aerobic Pseudomonads: A Taxonomic Study. *J. Gen. Microbiol.* **1966**, *43*, 159–271. [CrossRef]
80. Moaledj, K. Comparison of Gram-Staining and Alternate Methods, KOH Test and Aminopeptidase Activity in Aquatic Bacteria: Their Application to Numerical Taxonomy. *J. Microbiol. Methods* **1986**, *5*, 303–310. [CrossRef]
81. Urzí, C.; Brusetti, L.; Salamone, P.; Sorlini, C.; Stackebrandt, E.; Daffonchio, D. Biodiversity of Geodermatophilaceae Isolated from Altered Stones and Monuments in the Mediterranean Basin. *Environ. Microbiol.* **2001**, *3*, 471–479. [CrossRef] [PubMed]
82. Sierra, G. A Simple Method for the Detection of Lipolytic Activity of Microorganisms and Some Observations on the Influence of the Contact Between Cells and Fatty Substrates. *Antonie Van Leeuwenhoek* **1957**, *23*, 15–22. [CrossRef] [PubMed]
83. Gordon, R.E.; Mihm, J.M. A comparative study of some strains received as Nocardia. *J. Bacteriol.* **1956**, *73*, 15–27. [CrossRef] [PubMed]
84. Salisbury, W.A.; Lykos, J.J. Hydrolysis of Casein: A Differential Aid for the Identification of *Serratia marcescens*. *J. Clin. Path.* **1972**, *25*, 1083–1085. [CrossRef] [PubMed]
85. Sneath, P.H.A. Cultural and Biochemical Characteristics of the Genus Chromobacterium. *J. Gen. Microbiol.* **1956**, *15*, 70–98. [CrossRef]
86. Sachdev, D.P.; Chaudhari, H.G.; Kasture, V.M.; Dhavale, D.D.; Chopade, B.A. Isolation and Characterization of Indole Acetic Acid (IAA) Producing *Klebsiella pneumoniae* Strains from Rhizosphere of Wheat (*Triticum aestivum*) and Their Effect on Plant Growth. *Indian J. Exp. Biol.* **2009**, *47*, 993–1000. [PubMed]
87. Masciarelli, O.; Llanes, A.; Luna, V. A New PGPR Co-Inoculated with *Bradyrhizobium japonicum* Enhances Soybean Nodulation. *Microbiol. Res.* **2014**, *169*, 609–615. [CrossRef] [PubMed]
88. Milagres, A.M.; Machuca, A.; Napoleao, D. Detection of siderophore production from several fungi and bacteria by a modification of chrome azurol S (CAS) agar plate assay. *J. Microbiol. Methods* **1999**, *37*, 1–6. [CrossRef]
89. Schwyn, B.; Neilands, J.B. Universal Chemical Assay for the Detection and Determination of Siderophores. *Anal. Biochem.* **1987**, *160*, 47–56. [CrossRef]
90. Pikovskaya, R. Mobilization of Phosphorus in Soil in Connection with Vital Activity of Some Microbial Species. *Mikrobiologiya* **1948**, *17*, 362–370.
91. Kang, B.R.; Anderson, A.J.; Kim, Y.C. Hydrogen Cyanide Produced by *Pseudomonas* chlororaphis O6 is a Key Aphicidal Metabolite. *Can. J. Microbiol.* **2019**, *65*, 185–190. [CrossRef] [PubMed]
92. Hamid, S.; Lone, R.; Mohamed, H.I. Production of Antibiotics from PGPR and Their Role in Biocontrol of Plant Diseases. In *Plant Growth-Promoting Microbes for Sustainable Biotic and Abiotic Stress Management*; Mohamed, H.I., El-Beltagi, H.E.-D.S., Abd-Elsalam, K.A., Eds.; Springer: Cham, Switzerland, 2021; pp. 441–461. [CrossRef]
93. Weisburg, W.G.; Barns, S.M.; Pelletier, D.A.; Lane, D.J. 16S Ribosomal DNA amplification for phylogenetic study. *J. Bacteriol.* **1991**, *173*, 697–703. [CrossRef] [PubMed]
94. Altschul, S.; Madden, T.; Schäffer, A.; Zhang, J.; Zhang, Z.; Miller, W.; Lipman, D.J. Gapped BLAST and PSI-BLAST: A New Generation of Protein Database Search. *Nucleic Acids Res.* **1997**, *25*, 3389–3402. [CrossRef] [PubMed]
95. Tamura, K.; Stecher, G.; Kumar, S. MEGA 11: Molecular Evolutionary Genetics Analysis Version 11. *Mol. Biol. Evol.* **2021**, *38*, 3022–3027. [CrossRef] [PubMed]
96. Saitou, N.; Nei, M. The Neighbor-Joining Method: A New Method for Reconstructing Phylogenetic Trees. *Mol. Biol. Evol.* **1987**, *4*, 406–425. [CrossRef]
97. Felsenstein, J. Confidence limits on phylogenies: An approach using the bootstrap. *Evolution* **1985**, *39*, 783–791. [CrossRef]


**Disclaimer/Publisher's Note:** The statements, opinions and data contained in all publications are solely those of the individual author(s) and contributor(s) and not of MDPI and/or the editor(s). MDPI and/or the editor(s) disclaim responsibility for any injury to people or property resulting from any ideas, methods, instructions or products referred to in the content.

plants

MDPI

*Article*

# Histological and Physiological Study of the Effects of Biostimulants and Plant Growth Stimulants in *Viburnum opulus* 'Roseum'

**Dezső Kovács [1], Katalin Horotán [2,*], László Orlóci [1], Marianna Makádi [3], István Mosonyi [1], Magdolna Sütöri-Diószegi [1] and Szilvia Kisvarga [1]**

1 Institute of Landscape Architecture, Urban Planning and Garden Art, Hungarian University of Agriculture and Life Sciences (MATE), 1223 Budapest, Hungary; kovacsdezso.zsztgy@gmail.com (D.K.); orloci.laszlo@uni-mate.hu (L.O.); mosonyi.istvan.daniel@uni-mate.hu (I.M.); sutorine.dioszegi.magdolna@uni-mate.hu (M.S.-D.); kisvarga.szilvia@uni-mate.hu (S.K.)

2 Institute of Biology, Eszterházy Károly Catholic University, 3300 Eger, Hungary

3 Research Institute of Nyíregyháza, IAREF, University of Debrecen, 4400 Nyíregyháza, Hungary; makadim@agr.unideb.hu

* Correspondence: horotan.katalin@uni-eszterhazy.hu

**Abstract:** Biostimulants and other plant growth promoters can provide an effective solution to the challenge of urbanisation and climate change. *Viburnum opulus* 'Roseum' is a globally popular deciduous shrub species that can be made more resistant to urban influences by using natural growth-promoting substances. In our study, we investigated the effects of growth promoters Kelpak®, Bistep and Yeald Plus on the species, both histologically and physiologically (proline stress hormone measurement). Our measurements were complemented using the analysis of rhizosphere alkaline phosphatase, β-glucosidase and β-glucosaminidase enzymes, to obtain a more complete picture of the combined effect of biostimulants and species. We found that the Bistep biostimulant had an outstanding effect on the leaf tissue culture results of the variety. The transpiration and evapotranspiration findings also confirmed the efficacy of biostimulants. In the case of POD activity and rhizosphere enzyme measurements, Bistep and Yeald Plus obtained statistically higher values than the control group. Kelpak produced better results than the control group in several measurements (alkaline phosphatase levels; evapotranspiration results), but in other cases it resulted in lower values than the control treatment. The use of Bistep and Yeald Plus can greatly assist growers in the cultivation of *V. opulus* 'Roseum' in an urban environment.

**Keywords:** biostimulants; ornamental; woody plants; shrubs; green surface; sustainable; Viburnum; physiology; growth; rhizosphere

**Citation:** Kovács, D.; Horotán, K.; Orlóci, L.; Makádi, M.; Mosonyi, I.; Sütöri-Diószegi, M.; Kisvarga, S. Histological and Physiological Study of the Effects of Biostimulants and Plant Growth Stimulants in *Viburnum opulus* 'Roseum'. *Plants* **2024**, *13*, 1446. https://doi.org/10.3390/plants13111446

Academic Editors: Spyridon A. Petropoulos and Mohamad Hesam Shahrajabian

Received: 17 April 2024
Revised: 17 May 2024
Accepted: 21 May 2024
Published: 23 May 2024

**Copyright:** © 2024 by the authors. Licensee MDPI, Basel, Switzerland. This article is an open access article distributed under the terms and conditions of the Creative Commons Attribution (CC BY) license (https://creativecommons.org/licenses/by/4.0/).

## 1. Introduction

Continuing to secure and maximise crop yields under the threat of climate change is a major challenge for the agricultural sector, which is under further pressure from demographic projections (9.5 billion global population by 2050) [1]. According to current projections, global agricultural production will decrease by 50% and 30% respectively due to abiotic and biotic stresses [2]. To reduce and counteract this, several technological innovations have been proposed over the past three decades to enhance the sustainability of agricultural production systems, with the aim of reducing the use of synthetic agricultural chemicals (pesticides, fertilisers). Biostimulants have emerged as an alternative to replace these, and are often used in several horticultural sectors, including ornamental horticulture. In the latter sector, the application is aimed at increasing the vitality and resistance of plants while achieving commercial standards, which is a prerequisite for producing quality commodities [3,4]

Based on what has been described so far, the use of natural plant biostimulants (PBs) is a promising and environmentally friendly trend that is gaining increasing importance [5]. PBs also promote and enhance flowering, growth, crop setting, yield and nutrient use efficiency (NUE), in addition to their ability to improve tolerance to environmental factors [6–9]. The benefits provided also have implications for the use of biostimulants, with their use projected to increase by 10.4% per year globally until 2030 [10].

Plant biostimulants from natural products are attracting increasing interest among researchers, farmers and industrial companies as an effective tool to increase crop yields [11]. They can be derived from various organic materials and may contain humic substances, complex organic substances, useful chemical elements, peptides and amino acids, inorganic salts, seaweed extracts, chitin and chitosan derivatives, transpirants, and other nitrogen-containing substances. The use of biostimulants in plants leads to higher nutrient content in their tissues and positive metabolic changes [12]. Biostimulants are substances that incorporate one or more substances and/or microorganisms that promote plant nutrient uptake and use efficiency, increasing the tolerance of plants to abiotic/biotic stresses, and, when applied in small quantities, improve crop quality or morphological and physiological characteristics [13]. In addition to these benefits, biostimulants can enhance the activity of rhizosphere microbes and soil enzymes, the production of hormones and/or growth regulators in soil and plants, and photosynthetic processes [14]. The mechanisms of the action of biostimulants are often unknown and difficult to identify, as they are usually derived from more complex sources containing multiple bioactive compounds that may collectively combine to contribute to specific effects on plants.

Biostimulants can also positively influence other plant traits beyond the stress response. The application of *Ascophyllum nodosum* algae as biostimulants has been shown to increase photosynthetic rate, stomatal conductance, and transpiration in treated plants, as well as to increase stress tolerance [15]. In addition, the beneficial effects of biostimulants on leaf tissue and the intestinal tissue system are also observed [16]; several types of biostimulants, including humic acids, have been shown to possess cell wall strengthening effects [17]. Pylak et al. [18] have shown that the low molecular weight of fulvic acids allows them to penetrate tissues through membrane pores and form complexes with cations that help transport nutrients into the cell.

Sharma et al. [19] have shown that marine algal extracts stimulate chlorophyll synthesis due to their high hormone levels. Higher chlorophyll levels have been observed when algal extracts were applied to potato plants [20]. Apart from algae extracts, other types of biostimulants can also induce an increase in chlorophyll storage, which has been shown to induce significant chlorophyll growth [21]. Nardi and his colleagues [22] stated that humic acids can affect photosynthesis, resulting in an increase in shoot and root mass in the cultivar studied. On the other hand, a direct relationship between chlorophyll and carotenoid content has been demonstrated [23].

Plant responses to abiotic stresses are the result of the interaction between different signalling molecules and hormonal balance, which exert their effects by altering metabolic and gene expression levels. These effects could explain that the studied biostimulants significantly reduced the levels of stress enzymes, including peroxidase (POD) and superoxide dismutase [24,25]. Biostimulants can increase biomass, photosynthetic pigments and modified chlorophyll fluorescence-related parameters [26]. A correlation between stress enzymes and proline levels has also been observed [27], which may also be of interest for specific measurements, as increased proline levels are associated with increased transpiration and evapotranspiration results [28]. The transpiration rate showed a decreasing trend during the plant growth period, with increasing leaf temperature and stomatal diffusion resistance [29]. Proline accumulation has been shown in plants exposed to various abiotic stresses and, according to Tripathi and Gaur [30], proline protects cells by scavenging free radicals during oxidative stress caused by heavy metals. The literature suggests that proline protects plants by acting as a cellular osmotic regulator between the cytoplasm and the vacuole and detoxifies reactive oxygen species (ROS), thus ensuring membrane integrity and stabilising antioxidant enzymes [31].

In the case of ornamental plants, several studies have investigated how biostimulants affect them under different conditions, for example in urban plantations. Akensous and colleagues [32] have studied plant growth-promoting rhizobacteria (PGPR) and biostimulants extracted from compost materials that can increase drought tolerance or morphological traits [33]. The latter effects have been described in several herbaceous ornamental plant species grown in containers [34]. Zhang in his study [35] found that *Festuca arundinacea*, also used as an ornamental plant, enhances the defensive response to heat stress through lipid conversion. Annual ornamental plants were treated with a humic acid-based biostimulant, which improved the plants' tolerance to thermal stress and their nutrient uptake from the soil with a concomitant reduction in fertiliser application requirements, consequently reducing environmental stress and improving overall plant quality. The acidic pH peat used as a medium for the plants was not suitable for β-glucosidase and β-glucosaminase enzymes, which are very sensitive to soil pH [36]. Another possible reason could be the involvement of the enzyme β-glucosaminase in maintaining plant health, which plays an important role in the protection against pathogenic fungi and animal pests, as described by Parham and Deng [37].

Their application increases leaf and flower number as well as chlorophyll content and net photosynthesis, parameters that can be of particular importance in ornamental plant production and urban applications [38]. Seaweed extracts and humic and fulvic acids, used as biostimulants, can also be a solution to increase the flower number in woody-stemmed species with ornamental flowers and to increase vase life in cut flowers [39–41]. The use of a combination of different types of biostimulants and fertilisers may also be a good option in ornamental plant applications [42,43]. In ornamental shrub applications, biostimulant improved root morphology, increased leaf number and thus leaf area and dry biomass accumulation [44].

*Viburnum opulus* L. belongs to the plant family *Caprifoliaceae* [45] and is a valuable ornamental, medicinal and food plant. The basic species *V. opulus* shrub is common in natural habitats in Western and Central Europe, Asia, Caucasus and Asia Minor [46–48]. It is a fast-growing deciduous shrub up to 4–5 m tall. The leaves are transverse, three-lobed, rounded at the base and have roughly serrated margins. The upper part of the leaves is bare and dark green, and the leaf filaments are lighter and slightly hairy with star-shaped hairs. Leaves develop with the flowers, then turn scarlet-reddish-purple [49]. White flowers develop in cups 4–11 cm in diameter at the top of the stem. Each flower is composed of an outer ring of large sterile flowers and an inner ring of small fertile flowers. The ornamental cultivar 'Roseum' has only sterile flower spikes, which give it a snowball-like appearance [50].

In their study, Kajszczak et al. [48] and Levent Altun et al. [51] explain that the main publications on *V. opulus* so far are on its morphology, botany and medicinal uses, with the main focus on the fruits of the species [51] and its cultivation processes as a fruit [52]. However, the species is an important and frequently used ornamental shrub, especially because of its varieties, and has been little studied from this point of view. 'Roseum' (syn. 'Sterilis', 'Snowball'), also a little studied ornamental shrub, although a long established and popular species, is a traditional ornamental plant in Hungary and Europe. This choice was supported by the fact that the species was known as 'Sambucus Rose' as early as the 16th century, so its conservation and maintenance as a heritage species may be important for the future. It is still a commonly used cultivar, which prefers a temperate climate, making it often difficult to maintain and keep healthy in urban areas. The use of growth regulators and biostimulants during cultivation and rearing can be an important and environmentally friendly method of successful plant production.

Research into the use of growth regulators that can be used in the horticultural production of future varieties may be important, including the testing and incorporation of biostimulants into horticultural cultivation. Therefore, the aim of treating *V. opulus* 'Roseum' as a popular urban shrub with plant growth regulators is to demonstrate their effects on the variety from several aspects, including morphological, histological and physiological aspects, and to find a biostimulant for the variety with the potential for effective and favourable utilisation.

Due to the lack of literature in this field, the aim of our study was to contribute (i) to the knowledge about this species from an ornamental horticultural point of view, and (ii) the rising importance and expansion of sustainable horticulture and agriculture, which will also contribute to the knowledge of the potential urban uses of this species.

## 2. Results

### 2.1. *Histological Results*

The microscopic images show that the effects of the treatments on the transport tissues are clearly visible (Figure 1). In the control group (Figure 1a), the epidermis cells are homogeneous, the cell walls are intact, the transport cells are formed and the intestinal tissue cells are healthy. In the case of the Kelpak® biostimulant (Figure 1b), there was no detectable effect on leaf tissue, thus correlating with other results of Kelpak® biostimulant-treated plants. The leaf cross-section shows the same shape as the control group. In Yeald Plus (Figure 1c) and Bistep (Figure 1d), more significant differences in the transport tissues were observed compared to the control; in addition, the epidermis cells were thickened, the cell lines were uniform and strong, and the collenchyma cells were homogeneous and delimited. The transport tissue is regularly outlined by beams and secondary thickening, with individual areas being more distinct (woody and collenchyma) than in the Kelpak® and control groups.

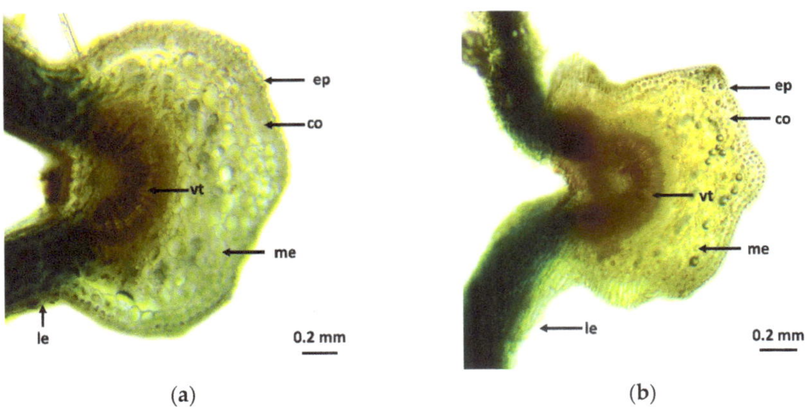


**Figure 1.** *Cont.*

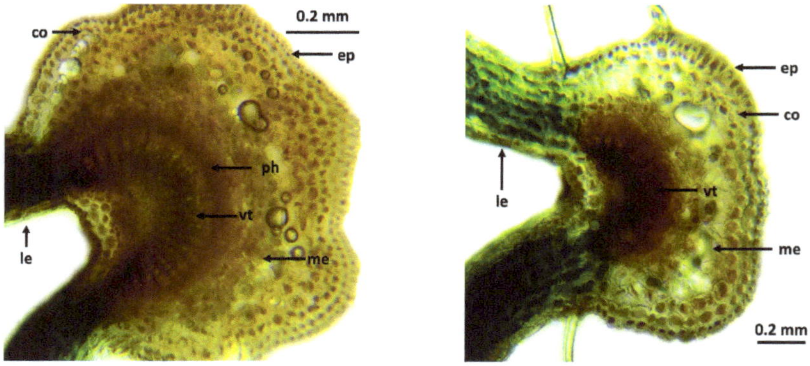


(c) (d)

**Figure 1.** Microscopic images of leaf tissue and transport rays in *V. opulus* 'Roseum' as a result of treatments. (**a**) Control; (**b**) Kelpak; (**c**) Yeald Plus; (**d**) Bistep. The abbreviations shown in the pictures mean the following: ep—epidermis; co—collenchyma; vt—vascular tissue; le—leaflet; me—mesophyll, ph—phloem.

## 2.2. Physiological Results

### 2.2.1. Transpiration and Evapotranspiration

Measuring and accurately determining the transpiration and evapotranspiration values will provide a more nuanced picture when investigating the effects of growth regulators (Figure 2). Based on the measured results, the evapotranspiration of the control group, Kelpak® (3%) and Yeald Plus (4.8%) were statically indistinguishable, followed by the transpiration data. However, for the Bistep biostimulant, both the evapotranspiration and transpiration data show a significant decrease. These values indicate a decrease in evapotranspiration for the Bistep biostimulant, which can be described as an indication of improved water balance in the soil and the plant. A strong correlation can be observed when these results are combined with the transpiration results per leaf.

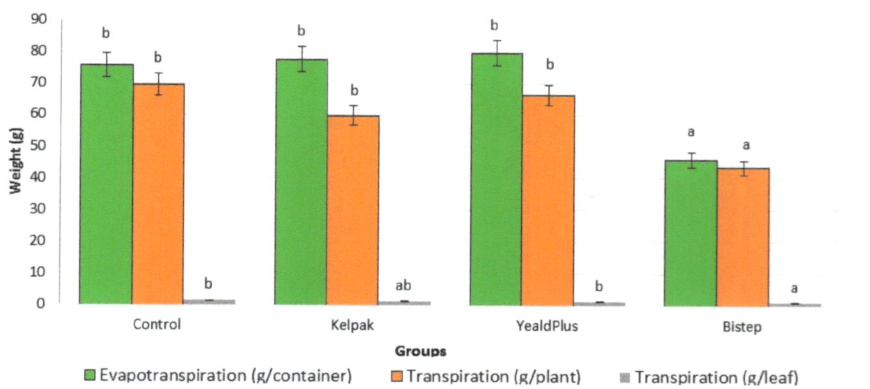


**Figure 2.** Evapotranspiration and transpiration in *V. opulus* 'Roseum' cultivars under the influence of bistimulants and growth regulators. To compare differences, ANOVA and one-way ANOVA with Duncan test were performed ($p < 0.05$). Different letters indicate different statistical groups.

### 2.2.2. Chlorophyll and Carotenoid Content

Significant differences were observed for both the chlorophyll and carotenoid content of the biostimulants tested (Figure 3). The chlorophyll content of the control group was significantly higher than that of the Kelpak®-treated group (18%) which did not exceed the other two biostimulants, and Kelpak® produced the lowest chlorophyll results. The growth regulators Yeald Plus (36%) and Bistep (20.5%) had statistically higher chlorophyll contents than the control group.

Such a strong difference was not observed for carotenoid content; however, the values measured in the control group show a statistically significant difference only with the growth regulator Yeald Plus (11.2% higher). The lowest carotenoid content was found for the Kelpak® biostimulant (13.5% lower), which is significantly lower than the values measured in the other groups.

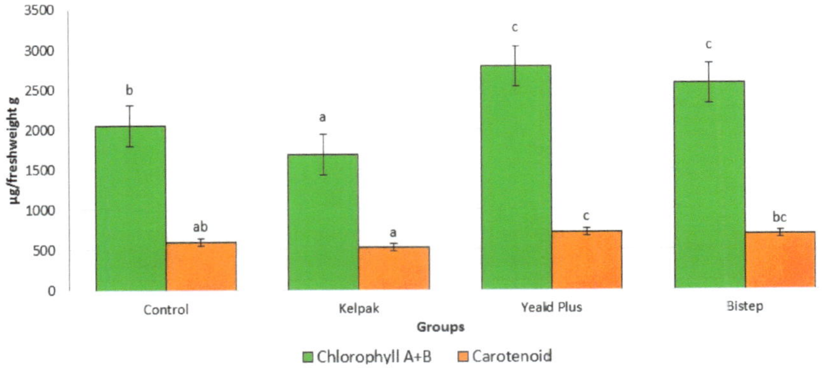


**Figure 3.** Chlorophyll (A + B) and carotenoid content in *V. opulus* 'Roseum' cultivars under the influence of growth regulators and bistimulants. To compare differences, ANOVA and one-way ANOVA with Duncan test were performed ($p < 0.05$). Different letters indicate different statistical groups.

### 2.2.3. Proline Level

Figure 4 summarises the results of the proline level measurements, showing significant differences between the control group and the treated groups. All treated groups showed significantly lower levels compared to the control group. There were no statistical differences between treatments. The lowest proline level was measured with the Kelpak® biostimulant (58.2% lower proline level).

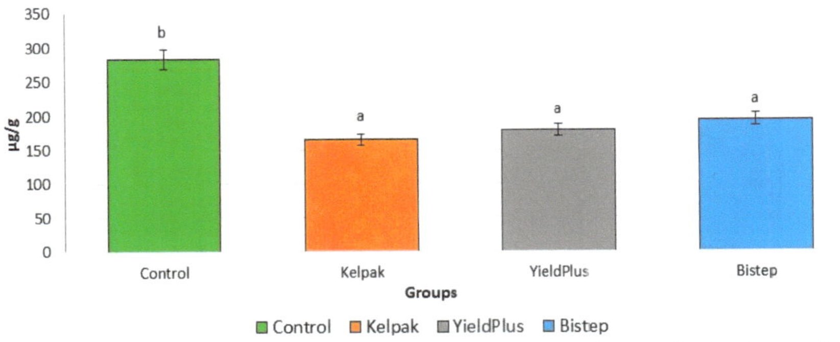


**Figure 4.** Proline level in *V. opulus* 'Roseum' cultivars under the influence of bistimulants and growth regulators. To compare differences, ANOVA and one-way ANOVA with Duncan test were performed ($p < 0.05$). Different letters indicate different statistical groups.

### 2.2.4. Peroxidase Enzyme

Plants with higher antioxidant levels are more resistant to oxidative stress, and these enzymes are considered biomarkers of biotic and abiotic stress effects [53]. In the *V. opulus* 'Roseum' cultivar, the POD enzymes present in the control group did not show a significant difference compared to the Kelpak® biostimulant (12.9%) (Figure 5). In the Yeald Plus biostimulant, the POD value was significantly increased (73.5%), while the highest value was detected in the Bistep-treated group (78.6%).

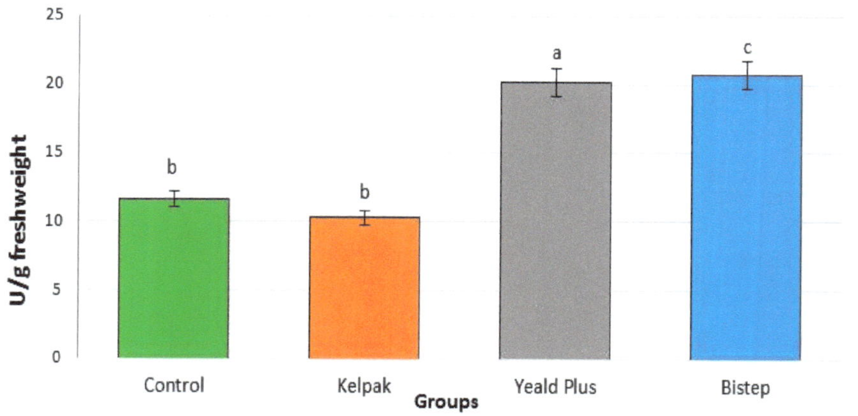


**Figure 5.** Peroxidase enzyme activity level in *V. opulus* 'Roseum' cultivars under the influence of growth regulators and biostimulants. To compare differences, one-way ANOVA with Duncan test was performed ($p < 0.05$). Different letters indicate different statistical groups.

2.2.5. Rhizosphere Enzyme Activities

The soil life associated with microorganisms in the rhizosphere, the environment of the root system of *V. opulus* 'Roseum', can be an important parameter for the study of the effects of biostimulants and mineral elements on plant and soil life.

The alkaline phosphatase enzyme level was lowest in the control group, while it was significantly higher in all treated groups (Figure 6). However, the β-glucosidase enzyme levels were significantly lower in the Kelpak®- (29.2%) and Yeald Plus-treated groups (20.1%) compared to the control group. The highest enzyme level was measured for Bistep (77.3% higher), which represents a statistically different group compared to the other treatments. The levels and significance data for the enzyme β-glucosidase and β-glucosaminidase showed a similar pattern in the treatment with Kelpak®. Kelpak® resulted in the lowest levels of these enzymes. The highest β-glucosidase and β-glucosaminidase activities were measured in the Bistep treatment.

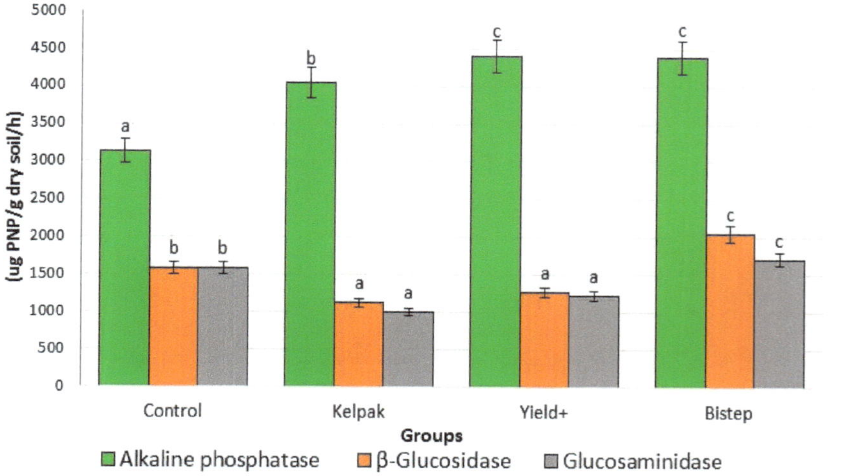


**Figure 6.** Activity of some enzymes of the rhizosphere during the treatment of *V. opulus* 'Roseum' with bistimulants. To compare differences, ANOVA and one-way ANOVA with Duncan test were performed ($p < 0.05$). Different letters indicate different statistical groups.

## 3. Discussion

There are several ornamental plant species that have been very popular and widely used since their first commercialisation, but their tolerance needs to be supported at several levels as environmental conditions change. The traditional species *V. opulus* 'Roseum' is a conspicuous ornamental shrub, both in its habit and inflorescence, as described by Kajszczak et al. [50]. The species can currently be grown safely in urban conditions, but in a few years its cultivation may become endangered and as it is less resistant to the changing urban environment. To prevent this, treatment with biostimulants and other growth regulators and plant conditioners that do not cause environmental stress could be a solution.

We have been running experiments with three biostimulant products named Bistep, Yeald Plus and Kelpak®. Bistep, also known as Ferbanat L, is a biostimulant containing humic and fulvic acids [54,55]. Yeald Plus is a rooting stimulant containing a growth regulator in the form of zinc ammonium acetate [56]. Kelpak® is a biostimulant containing marine algae extracts with the crushed parts of *Ecklonia maxima* [57].

The plant growth regulators had a beneficial effect on leaf tissue condition, both epidermal and intestinal, as discussed by Soppelsa and et al. [16]. The effect of Bistep and Yeald Plus was particularly effective, whereas there was no outstanding difference for Kelpak® compared to the control group [58]. These results also relate to weight gain outcomes, which follow a similar trend. It can be said that the application of Bistep and Yeald Plus strengthened the transport tissues. In parallel, a decrease in transpiration and evapotranspiration values can be explained. This is in agreement with the measurements of Conselvan et al. [17] that humic acid-based biostimulants have cell wall strengthening effects. Consistent with the findings of Pylak et al. [18], the results on histological measurements can be explained for the humic acid-containing Bistep biostimulant, as Kelpak® resulted in the lowest chlorophyll and carotenoid content. Kelpak was also found to have a negative effect in terms of histological results, which cannot be correlated with the results of Kularathne et al. [59], as the plants were in a strong cell division and elongation phase during the growing period under treatment.

The findings of Wadas et al. [20] on *V. opulus* 'Roseum' also provide evidence that humic acids can affect both respiration and photosynthesis, as the application of humic acid-containing Bistep significantly increased both chlorophyll and carotenoid content. Nardi et al. [22] found a similar result and also found higher rates of photosynthesis increase compared to the control group. On the other hand, a direct relationship between chlorophyll and carotenoid content was found, as their levels increase or decrease in the same proportion under the influence of growth regulators, as shown by the results of Ngoroyemoto et al. [23].

Assuming that the use of biostimulants and other plant growth regulators reduces the stress state of the plant, we completed our series of measurements by determining the proline levels and peroxidase enzyme activity levels associated with stress response. The proline level produced by the control group was significantly higher than in the treated groups, where *V. opulus* 'Roseum' was found to be beneficially affected by biostimulants, with the biostimulant Bistep being the most prominent. However, peroxidase enzyme activity indicated that stress levels were significantly increased using Yeald Plus and Bistep and reached statistically higher values than in the control group. As peroxidase enzyme activity increased, lower proline levels were detected, confirming the claim of Ozden et al. [27] that this was observed in *V. opulus* 'Roseum' in both control and treated plants. The results of Rajametov et al. [28] are in agreement with our results on measured respiration parameters; evapotranspiration, transpiration and evaporation per leaf were the highest in the control group plants, while the lowest values were found in the Bistep biostimulant-treated groups. The decrease in transpiration rate, in comparison with the results of Surendar et al. [29], suggests that at lower transpiration values, the stress cell resistance is higher. Proline regulates cellular changes by preventing water loss through osmotic regulation, which, in addition to higher proline levels, may explain the higher transpiration values in the culture strain studied. It was hypothesised that the increase in proline levels indicates higher stress levels following the results of Bohnert and Jensen [31], which may be explained by the beneficial effects of biostimulants on stress responses.

Rhizosphere is a special environment where strong communication takes place between plant and microbes [60]. Since biostimulants and plant growth regulators can affect the whole plant, including its root system and rhizosphere, it was considered essential to conduct soil microbiological studies. We proposed increasing enzyme activities in the rhizosphere of the studied plant variety because alkaline phosphatase, β-glucosidase and β-glucosaminidase enzymes help plants in the uptake of macroelements, but significant differences were found among the effects of biostimulants and enzyme activities. In the case of alkaline phosphatase activity, which is produced only by microbes, significant increases were found in the treatments compared to the control one. All the applied biostimulants contained microelements and/or organic matter, and nitrogen but not phosphorous (P); therefore, the increased P demand of plants resulted in an increased alkaline phosphatase activity in the rhizosphere [61]. The addition of nutrients (e.g., nitrogen) to the soil with the biostimulants could decrease the β-glucosidase and β-glucosaminidase activities [62], as it was found in the case of Kelpak and Yield Plus treatments. However, the main effects of humic substances, which have also been described for Bistep, are increased root mass and improved morphological properties, increased nutrient uptake and utilisation efficiency, and consequently improved yields [54]. These effects of Bistep could be strengthened by the increased enzyme activities in the rhizosphere.

Based on the results measured and evaluated during the studies, it can be concluded that the *V. opulus* 'Roseum' cultivar can be promoted and its wide use can be protected in the future through the application of biostimulants. Of the three biostimulants used, Bistep has shown outstanding results and its use is recommended for the cultivation of the variety.

## 4. Materials and Methods

### *4.1. Biostimulants Used for Treatment*

#### 4.1.1. Bistep

Bistep, internationally known as Ferbanat L plant conditioner, is a naturally derived liquid nano-fertiliser specialised in stress management. It contains micro-humusates coupled with microelements and hydrologically active natural substances. It contains a combination of humic acid and fulvic acid, which stimulate root development. Its use increases the amount of soil humus, since it has a 2-year after-effect after use thanks to the microorganisms. The manufacturer guarantees that the use of the product reduces the use of fertilisers by about 50% and also the need for pesticides. It is most effective when applied at appropriate intervals from the very first phenophase of the plant.

To better understand the mechanism of action of Bistep, a detailed knowledge of its constituents is necessary. Humic acids, described as dark-coloured, heterogeneous aggregates of organic matter, are the result of microbiotic metabolism and can be characterised as one of the most abundant organic substances, highly resistant to various environmental stresses [63]. The traditional, albeit nowadays criticised, classification of this material distinguishes three groups: humic acids, the water-insoluble part; humic acids (or HAs), which are soluble in pH > 2; and fulvic acids (also FAs), which are soluble in water [64]. This classical definition is still present in the older scientific literature, but chemically HS is nothing more than the product of saponification reactions from soils and sediments through alkaline extraction [14]. However outdated, this division is the most commonly used in all the literature reviewed, as it allows a meaningful distinction between the products, i.e., a clear distinction between humic acid-based and fulvic acid-based products, and is therefore appropriate [65]. Fulvic acids (FAs) improve the structure and fertility of heterogeneous textured soils and play a crucial role in increasing biomass and yields in crop production [66]. Humic acid and fulvic acids together convert minerals into organic compounds that are readily taken up by plants [64]. The main effects of humic substances, which have also been described for Bistep, are increased root mass and improved morphological properties, increased nutrient uptake and utilisation efficiency, and consequently improved yields [54].

#### 4.1.2. Kelpak®

Kelpak® is made from *Ecklonia maxima* dark brown algae, which grow in the cold waters off the west coast of South Africa. This fast-growing algae species is characterised by elevated levels of auxin and cytokinin-like substances. Kelpak® contains plant cells extracted from freshly cut and crushed plants without heating, freezing or the use of chemicals [57]. Marine algae extracts are generally the sum of the products of aqueous or solvent extraction and hydrolysis of the biomass of algal species from the genera *Ascophyllum, Ecklonia, Macrocystis* and *Durvillea* [67,68]. Among these, preparations containing extracts of algal species *Ecklonia maxima* and *Ascophyllum nodosum* are the most widely used [69], e.g., Kelpak® and Goëmar BM 86® biostimulants [70].

The production methods are not standardised but are patented, and eight extraction methods are currently in use [71]. Seaweed extracts are widely used in agricultural and organic farming, and their use in the ornamental horticulture sector is also significant, as they have a positive effect on plant quality by promoting plant development. Their widespread use started in the mid-2000s [72]. The degree of effectiveness of algal extracts depends on two factors: one is the hormone level of the plants and the other is the number of micronutrients (mainly cytokines) present in the crude extract [59].

#### 4.1.3. Yeald Plus

Yield Plus is marketed specifically as a rooting promoter [56] and is produced by De Sangose Ltd. in the UK. Its application can achieve up to 30–35% root mass increase in two weeks [73]. Its application is facilitated by the plant's ability to take it up and utilise it adequately at root and leaf temperatures even at lower temperatures (+5–7 °C), thus providing flexibility of use, further supported by its good root and leaf surface utilisation. For transplanted vegetable crops, dipping the roots in a 0.05–0.1% solution before transplanting and foliar fertilisation at 1–1.5 litres/ha after transplanting may be recommended. As a generally accepted rule of application, the younger the plant, the lower the dose [74].

Yield Plus owes its effectiveness to the ammonium acetate form of zinc, a proprietary technology. Composition: 6% nitrogen, 5% zinc, 1% potassium peroxide, 0.03% boron, 0.25% copper, 0.25% iron, 0.25% manganese and 0.001% molybdenum. In terms of its effects, it stimulates auxin synthesis, supports root development, improves water and nutrient uptake, increases chlorophyll levels and accelerates carbohydrate metabolism [73]. Based on the literature, its use in nursery production of woody stem plants is not yet widespread (Table 1).

**Table 1.** Main characteristics of used biostimulants.

| Key Characteristics | Bistep (syn. Ferbanat L) | Kelpak® | Yeald Plus |
|---|---|---|---|
| Manufacturer | UAB ALJARA<br>LT-11219 Vilnius, Geniu str. 16-38.<br>Lithuania | KELP PRODUCTS (PTY) LTD<br>7975 Simon's Town, Blue Water Close<br>South Africa | De Sangosse Ltd.<br>Hillside Mill Quarry Lane, Swaffham Bulbeck, Cambridge CB5 0LU,<br>UK |
| Base elements | Microhumates and coupled microelements | Aqueous extracts of Ecklonia maxima | 6% nitrogen, 5% zinc, 1% potassium peroxide, 0.03% boron, 0.25% copper, 0.25% iron, 0.25% manganese, 0.001% molybdenum |
| Effects | Increase soil humus volume, support root development, increase yield and green weight | Support shoot and bud development, support root development | Leaf fertilisation, irrigation, watering |
| Economic benefit | 50% reduction in fertiliser use | It also has a beneficial effect on seed development | Increases root mass by 30–35% in two weeks |

### 4.2. Plant Material Used in the Experiment

Plant selection was based on a complex set of criteria, which included choosing plants that were widely available in woody nurseries. In order to schedule the trials and monitor development, the plant had to be selected from a range of shrubby plants that have long been used and favoured in private gardens in municipalities and in urban green spaces. The cultivar *V. opulus* 'Roseum' has these characteristics. The shrubs were propagated in our nursery in Hungary (Figure 7).

(**a**) (**b**) (**c**)

**Figure 7.** *V. opulus* 'Roseum' experimental specimens at (**a**) basal stand before treatment and potting (**b**) potted and first treated plants before transport to the nursery (**c**) condition after last treatment at the nursery (Kovács, 2023).

### 4.3. Planting Parameters

The plants in the experiment were planted in 1.5 litre containers. The composition of the planting medium consisted of 30% milled black peat (pH: 6.24; humus content: 17.89%) and 70% milled white peat (pH: 5.81; humus content: 25.58%) with a particle size of 10–30 mm and EC levels ranging from 2.0 to 4.0.

The measurements were made with 25 individuals in 5 replicates. Plants were uniformly top watered with sprinkler heads. No pruning was performed during their development. Plant protection treatments were applied only and exclusively against aphid damage, using Mospilan 20SG (manufacturer and distributor Sumi agro Hungary Kft.).

The first treatment was applied at the same time as planting; the planting was carried out during the whole week of 1 May each year, and thereafter repeated spraying was carried out once every 5 weeks until the shoots stopped growing. The end of July is the end of shoot growth in the Hungarian climate. Accordingly, the plants received a total of 3 treatments per season. The concentrate rates applied were as follows:

- Kelpak® 0.4% solution;
- Yeald Plus 0.3% solution;
- Bistep 0,5% solution;
- Control group received only water at the same time as the treatments.

### 4.4. Histological Measurements

Based on the literature background, the agents used and their active ingredients can modify a number of morphological parameters, so it was considered important to assess the treated plants at tissue level. For this purpose, leaf cross-sectional images were taken and the leaf material was collected in the 5th week after the last treatment. Leaf tissue samples were cleaned with distilled water and mechanically sectioned with a blade, and fresh sections were used to eliminate any distortion that might occur during fixation and to obtain as accurate a picture of changes as possible.

A Euromex bScope BS.1153-PLi biological microscope with a compatible camera (Levenhuk m1400 plus) was used for the study. Due to the nature of the sections, PLi 4/0.1 lenses were used, which provided a magnification of 40×. The eyepiece was of type and size WF120×/20. Samples were unstained. The images were post-corrected using GIMP 2.10 (property of Spencer Kimball, Peter Mattis).

### 4.5. Physiological Measurements

Measurement of leaf transpiration and calculation of leaf conductance to water vapour are important in almost all studies of plant water relations [75]. To assess the physio-logical status, essential for a complex study of the effects of growth stimulants, transpiration and evapotranspiration parameters were recorded. To carry out the survey, the tanks were filled evenly with water the day before. In order to measure only evaporation through the canopy, the tanks were covered with waterproof material to prevent other water loss from the canopy. Measurements were taken the following morning and evening using this lysimeter. The fresh weight change, the combination of plant growth and water status changes can be calculated from evapotranspiration and water uptake. The system consists of two contacting containers filled with nutrient solution: each is placed on an electronic balance. One of these vessels carries the plant and is connected to the other by a flexible tube. Water uptake in each vessel reduces the solution level equally. The mass loss of the vessel without the plant gives the rate of water uptake, the total mass loss on both scales gives transpiration. The difference between the two data gave the evaporation rate, i.e., the amount of water evaporated through the canopy and the surface of the medium, as described by Van Leperen et al. [76], which also showed the rate of transpiration and evapotranspiration.

### 4.6. Chlorophyll and Carotenoid Content Measurement

Chlorophyll and carotenoid content were determined according to the method of Arnon [77]. Accordingly, a 150 mg fresh weight plant sample, weighed to 3 decimal places (grams), was homogenised on an analytical balance (Explorer Pro 64, OHAUS Europe, Nänikon, Switzerland) in a rubbing flask cooled to below 4 °C with quartz particles, in the presence of a knife-edge of $Na_2CO_3$ in 80 *v/v*% acetone. The homogenised samples were filled to 10 mL final volume with 80 *v/v*% acetone and centrifuged at 4 °C at 1000 rpm (5418 R, Eppendorf AG., Hamburg, Germany) for 10 min.

The absorbance of the supernatant was determined using a spectrophotometer (GeneSys VIS-10, Thermo Fisher Scientific Inc., Waltham, MA, USA) at wavelengths λ1 = 663, λ2 = 644, λ3 = 480 nm, and the chlorophyll content was calculated from the values obtained using the following formula:

$$\mu\text{g chlorophyll/g feed} = (20.2 \times A644 + 8.02 \times A663) \times V/w$$

where:

- Ax—absorbance at a given wavelength;
- V—final volume of the sample;
- w—the sampled and homogenised plant fresh weight.

For the determination of dry matter content, the plant samples were dried at 80 °C for 24 h before drying, and afterwards their mass was measured on an analytical balance (Explorer Pro 64, OHAUS Europe, Switzerland) to 3 decimal places.

### 4.7. Peroxidase Enzyme Activity Measurement

Peroxidase (POD) activity was measured spectrophotometrically (λ = 460 nm) in the presence of $H_2O_2$ substrate in the method of Shannon et al. [78] using orthodianisidine chromogenic reagent (molar absorption coefficient for oxidised orthodianisidine: $\varepsilon = 11.3\ \text{mM}^{-1}\ \text{cm}^{-1}$). For the measurement, samples were prepared from 300 mg of fresh plant tissue in a rubbing flask cooled to below 4 °C using quartz sand and phosphate buffer at 0 °C (pH = 6.5) and centrifuged at 4 °C at 13,500 rpm for 20 min. Quantities of 10 µL of the supernatant were weighed out to give a reaction mixture with a final volume of 1760 µL, containing 20 µL of orthodianisidine solution (10 mg/mL dissolved in methanol), 30 µL of 0.3% $H_2O_2$ and sodium acetate–acetic acid buffer (pH = 4.5).

Enzyme activity was measured at room temperature for 2 min and absorbance values were recorded every 10 s. To calculate the enzyme activity, the value of the absorbance change during 1 min was multiplied by the dilution of the test sample in the reaction mixture and divided by the molar absorption coefficient of orthodianisidine. The resulting enzyme activity unit U/mL was converted to U/mg using the sample fresh weight.

### *4.8. Proline Level Measurement*

The proline content was determined spectrophotometrically. The original method is that of Chinard [79], modified by Ábrahám [80], which shows that at acidic pH, proline forms a red product with ninhydrin. In a first step, 3% sulphosalicylic acid is prepared as follows: 3 g of 5-sulphosalicylic acid is prepared in 100 mL of solution with DV (can be stored freely for weeks). In a second step, acidic ninhydrin is prepared: 0.625 g of ninhydrin, 15 mL of cc. acetic acid, 20 mL of 6 M orthophosphoric acid. After vortexing and heating, it dissolves and the solution can be stored for a few weeks. The method requires 100 mg of plant sample. The plant sample is homogenised with 3% sulphosalicylic acid (5 microl/mg fresh weight), stored frozen, and centrifuged for 5 min at room temperature. The reaction mixture consists of 100 microl of 3% sulphosalicylic acid, 200 microl of cc. acetic acid, 200 microl of acid ninhydrin and 100 microl of sample. The solutions are weighed in test tubes, sealed and kept in a drying oven at 96 °C for 1 h. The reaction is stopped in cold water, then dissolved with 1.5 mL toluene, vortexed for 20 s, and left to stand for 5 min. The absorbance of the supernatant is measured in a glass cuvette at 520 nm. The value obtained is compared with the L-proline standard.

### *4.9. Rhizosphere Measurements*

Rhizosphere soil samples were collected from the pots according to Tong et al. [81]. Remaining plant particles were removed with forceps and samples were stored at −20 °C until the analysis. Enzyme activities were measured photometrically at 405 nm, based on the released *p*-nitrophenol. β-glucosidase enzyme activity was calculated from the concentration of *p*-nitrophenol released from *p*-nitrophenyl-β-D-glucoside as a substrate [82]. Alkaline phosphatase activity was measured using the method of Tabatabai and Bremner [83] using *p*-nitrophenyl phosphate for substrate. The substrate used for the measurement of β-glucosaminidase activity was *p*-nitrophenyl-N-acetyl-β -D-glucosaminidase [37].

### *4.10. Statistical Evaulation*

The measured data were sorted into a format suitable for statistical evaluation using the Excel data management software package (Office 2021 Professional Plus), and then the IBM SPSS Statistics 29.0.1.0.software package was used for one-way analysis of variance (ANOVA), followed by Duncan test ($p < 0.05$).

## 5. Conclusions

Vegetation is affected by a wide range of stresses individually and in combination with others, which have a major impact on plant vitality. These effects are more pronounced in the urban environment, where air pollution, thermal insulation and higher temperatures, as well as a high proportion of paved surfaces, are all identified as stress factors. Organic biostimulants could decrease the environmental stress effects on horticultural plants. Our aim was to test three biostimulants on *Viburnum opulus* 'Roseum' as a popular urban shrub in plant nursery conditions. Our results indicated that the biostimulant Bistep had positive effects on the studied histological and physiological parameters of the test plant; moreover, the measured enzyme activities in the rhizosphere were also higher compared to the other treatments. Biostimulants and plant growth regulators, which are increasingly dominating the sustainable agriculture and horticulture sector, can provide a solution if they are properly selected and applied. These products could help to adapt widely used varieties to stress conditions without compromising their functions and benefits (ornamental value and pollutant sequestration).

**Author Contributions:** Conceptualisation M.S.-D.; methodology, L.O.; software, I.M.; validation, S.K., K.H. and L.O.; formal analysis, K.H.; investigation, D.K.; resources, M.M.; data curation, S.K.; writing—original draft preparation, D.K.; writing—review and editing, S.K. and K.H.; visualisation, D.K.; supervision, S.K.; project administration, K.H. All authors have read and agreed to the published version of the manuscript.

**Funding:** The research was supported by the project 'The feasibility of the circular economy during national defense activities' of the 2021 Thematic Excellence Programme of the National Research, Development and Innovation Office under grant no.: TKP2021-NVA-22, led by the Centre for Circular Economy Analysis.

**Data Availability Statement:** Data is not public due to the measurement is in progress.

**Acknowledgments:** We are grateful to László Papp and Károly Bóka for their help in accurately marking the microscopic photographs.

**Conflicts of Interest:** The authors declare no conflicts of interest. The funders had no role in the design of the study; in the collection, analyses, or interpretation of data; in the writing of the manuscript; or in the decision to publish the results.

## References

1. Lucini, L.; Colla, G.; Moreno, M.B.M.; Bernardo, L.; Cardarelli, M.; Terzi, V.; Bonini, P.; Rouphael, Y. Inoculation of Rhizoglomus irregulare or Trichoderma atroviride differentially modulates metabolite profiling of wheat root exudates. *Phytochemistry* **2019**, *157*, 158–167. [CrossRef]
2. Kumar, A.; Verma, J.P. Does plant—Microbe interaction confer stress tolerance in plants: A review. *Microbiol. Res.* **2018**, *207*, 41–52. [CrossRef] [PubMed]
3. Rajabi Hamedani, S.; Rouphael, Y.; Colla, G.; Colantoni, A.; Cardarelli, M. Biostimulants as a tool for improving environmental sustainability of greenhouse vegetable crops. *Sustainability* **2020**, *12*, 5101. [CrossRef]
4. Kisvarga, S.; Farkas, D.; Boronkay, G.; Neményi, A.; Orlóci, L. Effects of biostimulants in horticulture, with emphasis on ornamental plant production. *Agronomy* **2022**, *12*, 1043. [CrossRef]
5. Mukherjee, A.; Patel, J.S. Seaweed extract: Bistimulant of plant defense and plant productivity. *Int. J. Environ. Sci. Technol.* **2020**, *17*, 553–558. [CrossRef]
6. Colla, G.; Rouphael, Y. Biostimulants in horticulture. *Sci. Hortic.* **2015**, *196*, 1–134. [CrossRef]
7. Padmaja, V.V.; Pavani, K.; Srilatha, P.; Lalitha, K.; Sarada, G.; Naik, M.R.; Kiran, Y.D.; Gopal, K. Role of Biostimulants in Horticulture: A Review. *Int. J. Environ. Clim. Change* **2023**, *13*, 1146–1157. [CrossRef]
8. Bell, J.C.; Bound, S.A.; Buntain, M. Biostimulants in agricultural and horticultural production. *Hortic. Rev.* **2022**, *49*, 35–95.
9. Joly, P.; Calteau, A.; Wauquier, A.; Dumas, R.; Beuvin, M.; Vallenet, D.; Crovadore, J.; Cochard, B.; Lefort, F.; Berthon, J.Y. From strain characterization to field authorization: Highlights on Bacillus velezensis strain B25 beneficial properties for plants and its activities on phytopathogenic fungi. *Microorganisms* **2021**, *9*, 1924. [CrossRef]
10. Research and Markets. Global Biostimulants Market Report 2022–2030—Compound Annual Growth of 10.4% with Market Set to Reach $6.79 Billion by 2030. 2022. Available online: https://www.globenewswire.com/en/news-release/2022/06/17/2464668/28124/en/Global-Biostimulants-Market-Report-2022-2030-Compound-Annual-Growth-of-10-4-with-Market-Set-to-Reach-6-79-Billion-by-2030.html (accessed on 6 March 2024).
11. Brown, P.; Saa, S. Biostimulants in agriculture. *Front. Plant Sci.* **2015**, *6*, 155882. [CrossRef]
12. Nardi, S.; Pizzeghello, D.; Schiavon, M.; Ertani, A. Plant biostimulants: Physiological responses induced by protein hydrolyzed-based products and humic substances in plant metabolism. *Sci. Agric.* **2016**, *73*, 18–23. [CrossRef]
13. Calvo, P.; Nelson, L.; Kloepper, J.W. Agricultural uses of plant biostimulants. *Plant Soil* **2014**, *383*, 3–41. [CrossRef]
14. Giannattasio, M.; Vendramin, E.; Fornasier, F.; Alberghini, S.; Zanardo, M.; Stellin, F.; Squartini, A. Microbiological features and bioactivity of a fermented manure product (preparation 500) used in biodynamic agriculture. *J. Microbiol. Biotechnol.* **2013**, *23*, 644–651. [CrossRef] [PubMed]
15. Kałużewicz, A.; Krzesiński, W.; Spiżewski, T.; Zaworska, A. Effect of biostimulants on several physiological characteristics and chlorophyll content in broccoli under drought stress and re-watering. *Not. Bot. Horti Agrobot. Cluj-Napoca* **2017**, *45*, 197–202. [CrossRef]
16. Soppelsa, S.; Kelderer, M.; Testolin, R.; Zanotelli, D.; Andreotti, C. Effect of biostimulants on apple quality at harvest and after storage. *Agronomy* **2020**, *10*, 1214. [CrossRef]

17. Conselvan, G.B.; Fuentes, D.; Merchant, A.; Peggion, C.; Francioso, O.; Carletti, P. Effects of humic substances and indole-3-acetic acid on Arabidopsis sugar and amino acid metabolic profile. *Plant Soil* **2018**, *426*, 17–32. [CrossRef]
18. Pylak, M.; Oszust, K.; Frąc, M. Review report on the role of bioproducts, biopreparations, biostimulants and microbial inoculants in organic production of fruit. *Rev. Environ. Sci. Bio/Technol.* **2019**, *18*, 597–616. [CrossRef]
19. Sharma, U.; Pal, D.; Prasad, R. Alkaline phosphatase: An overview. *Indian J. Clin. Biochem.* **2014**, *29*, 269–278. [CrossRef] [PubMed]
20. Wadas, W.; Dziugieł, T. Changes in assimilation area and chlorophyll content of very early potato (*Solanum tuberosum* L.) cultivars as influenced by biostimulants. *Agronomy* **2020**, *10*, 387. [CrossRef]
21. Abubakar, A.R.; Ashraf, N.; Ashraf, M. Effect of plant biostimulants on growth, chlorophyll content, flower drop and fruit set of pomegranate cv. Kandhari Kabuli. *Int. J. Agric. Environ. Biotechnology* **2013**, *6*, 305–309.
22. Nardi, S.; Pizzeghello, D.; Muscolo, A.; Vianello, A. Physiological effects of humic substances on higher plants. *Soil Biol. Biochem.* **2002**, *34*, 1527–1536. [CrossRef]
23. Ngoroyemoto, N.; Gupta, S.; Kulkarni, M.G.; Finnie, J.F.; Van Staden, J. Effect of organic biostimulants on the growth and biochemical composition of *Amaranthus hybridus* L. *South Afr. J. Bot.* **2019**, *124*, 87–93. [CrossRef]
24. Cramer, G.R.; Urano, K.; Delrot, S.; Pezzotti, M.; Shinozaki, K. Effects of abiotic stress on plants: A systems biology perspective. *BMC Plant Biol.* **2011**, *11*, 163. [CrossRef] [PubMed]
25. Peripolli, M.; Dornelles, S.H.; Lopes, S.J.; Tabaldi, L.A.; Trivisiol, V.S.; Rubert, J. Application of biostimulants in tomato subjected to water deficit: Physiological, enzymatic and production responses. *Rev. Bras. De Eng. Agrícola E Ambient.* **2021**, *25*, 90–95. [CrossRef]
26. Johnson, R.; Puthur, J.T. Biostimulant priming in Oryza sativa: A novel approach to reprogram the functional biology under nutrient-deficient soil. *Cereal Reasearch Commun.* **2022**, *50*, 45–52. [CrossRef]
27. Ozden, M.; Demirel, U.; Kahraman, A. Effects of proline on antioxidant system in leaves of grapevine (*Vitis vinifera* L.) exposed to oxidative stress by $H_2O_2$. *Sci. Hortic.* **2009**, *119*, 163–168. [CrossRef]
28. Rajametov, S.N.; Yang, E.Y.; Cho, M.C.; Chae, S.Y.; Jeong, H.B.; Chae, W.B. Heat-tolerant hot pepper exhibits constant photosynthesis via increased transpiration rate, high proline content and fast recovery in heat stress condition. *Sci. Rep.* **2021**, *11*, 14328. [CrossRef]
29. Surendar, K.K.; Devi, D.D.; Ravi, I.; Jeyakumar, P.; Velayudham, K. Effect of water stress on leaf temperature, transpiration rate, stomatal diffusive resistance and yield of banana. *Plant Gene Trait.* **2013**, *4*, 43–47. [CrossRef]
30. Tripathi, B.N.; Gaur, J.P. Relationship between copper-and zinc-induced oxidative stress and proline accumulation in Scenedesmus sp. *Planta* **2004**, *219*, 397–404. [CrossRef]
31. Bohnert, H.J.; Jensen, R.G. Strategies for engineering water-stress tolerance in plants. *Trends Biotechnol.* **1996**, *14*, 89–97. [CrossRef]
32. Akensous, F.Z.; Anli, M.; Boutasknit, A.; Ben-Laouane, R.; Ait-Rahou, Y.; Ahmed, H.B.; Nasri, N.; Hafidi, M.; Meddich, A. Boosting date palm (*Phoenix dactylifera* L.) growth under drought stress: Effects of innovative biostimulants. *Gesunde Pflanz.* **2022**, *74*, 961–982. [CrossRef]
33. Santacruz-García, A.C.; Senilliani, M.G.; Gómez, A.T.; Ewens, M.; Yonny, M.E.; Villalba, G.F.; Nazareno, M.A. Biostimulants as forest protection agents: Do these products have an effect against abiotic stress on a forest native species? Aspects to elucidate their action mechanisms. *For. Ecol. Manag.* **2022**, *522*, 120446. [CrossRef]
34. Miler, N.; Tymoszuk, A.; Woźny, A.; Michalik, T.; Wiśniewska, J.; Kulus, D. Microbiological Biostimulants in the Improvement of Extended Storage Quality of In Vitro-Derived Plants of Popular Ornamental Perennials. *Agronomy* **2024**, *14*, 289. [CrossRef]
35. Zhang, X.; Xu, Y.; Huang, B. Lipidomic reprogramming associated with drought stress priming-enhanced heat tolerance in tall fescue (*Festuca arundinacea*). *Plant Cell Environ.* **2019**, *42*, 947–958. [CrossRef] [PubMed]
36. Wade, J.; Li, C.; Vollbracht, K.; Hooper, D.G.; Wills, S.A.; Margenot, A.J. Prescribed pH for soil β-glucosidase and phosphomonoesterase do not reflect pH optima. *Geoderma* **2021**, *401*, 115161. [CrossRef]
37. Parham, J.A.; Deng, S.P. Detection, quantification and characterization of β-glucosaminidase activity in soil. *Soil Biol. Biochem.* **2000**, *32*, 1183–1190. [CrossRef]
38. Cristiano, G.; De Lucia, B. Petunia performance under application of animal-based protein hydrolysates: Effects on visual quality, biomass, nutrient content, root morphology, and gas exchange. *Front. Plant Sci.* **2021**, *12*, 640608. [CrossRef]
39. Kumar, S.B.; Prasanth, P.; Sreenivas, M.; Gouthami, P.; Sathish, G. Effect of NPK, biofertilizers and biostimulants on flowering, flower quality and yield of China aster cv.'Arka Kamini'. *Pharma Innov. J.* **2022**, *11*, 1239–1243.
40. El-Baset, A.; Kasem, M.M. Improving growth characteristics and vase life of Dendranthema grandiflorum 'Flyer'using humic and fulvic acids as biostimulants substances. *Sci. J. Flowers Ornam. Plants* **2022**, *9*, 87–102. [CrossRef]
41. Suchitha, N.; Babu, K.K.; Lakshminarayana, D.; Kumar, S.P. Studies on the Effect of Biostimulants on Quality of Cut Flower of Chrysanthemum (Dendranthema grandiflora) cv. *Denjigar Whitec. Int. J. Environ. Clim. Change* **2023**, *13*, 687–692. [CrossRef]
42. Smolin, N.V.; Ivoilov, A.V.; Volgin, V.V.; Potapova, N.V.; Nikolskiy, A.N.; Nedayborshch, J.N.; Potapov, I.V. Growth regulators and mineral fertilizers effect on morphometric indicators and decorative qualities of Zinnia elegans varieties. *Ornam. Hortic.* **2023**, *29*, 163–170. [CrossRef]
43. Lin, Y.; Jones, M.L. Evaluating the growth-promoting effects of microbial biostimulants on greenhouse floriculture crops. *Hortic. Sci.* **2022**, *57*, 97–109. [CrossRef]
44. Loconsole, D.; Sdao, A.E.; Cristiano, G.; De Lucia, B. Different responses to adventitious rhizogenesis under indole-3-butyric acid and seaweed extracts in ornamental's cuttings: First Results in Photinia x fraseri 'Red Robin'. *Agriculture* **2023**, *13*, 513. [CrossRef]

45. Maciejewska, I. Pollen morphology of the Polish species of the family Caprifoliaceae. Part 1. *Acta Soc. Bot. Pol.* **1997**, *66*, 133–142. [CrossRef]
46. Mazur, M.; Szperlik, J.; Salejda, A.M.; Krasnowska, G.; Kolniak-Ostek, J.; Bąbelewski, P. Description of the guelder rose fruit in terms of chemical composition, antioxidant capacity and phenolic compounds. *Appl. Sci.* **2021**, *11*, 9221. [CrossRef]
47. Rop, O.; Reznicek, V.; Valsikova, M.; Jurikova, T.; Mlcek, J.; Kramarova, D. Antioxidant properties of European cranberrybush fruit (*Viburnum opulus* var. edule). *Molecules* **2010**, *15*, 4467–4477. [CrossRef]
48. Kajszczak, D.; Zakłos-Szyda, M.; Podsędek, A. *Viburnum opulus* L.—A review of phytochemistry and biological effects. *Nutrients* **2020**, *12*, 3398. [CrossRef] [PubMed]
49. Bubulica, V.M.; Anghel, I.; Grumezescu, A.M.; Saviuc, C.; Anghel, G.A.; Chifriuc, M.C.; Gheorghe, I.; Lazar, V.; Popescu, A. In vitro evaluation of bactericidal and antibiofilm activity of Lonicera tatarica and *Viburnum opulus* plant extracts on Staphylococcus strains. *Farmacia* **2012**, *60*, 80–91.
50. Kajszczak, D.; Kowalska-Baron, A.; Sosnowska, D.; Podsędek, A. In vitro inhibitory effects of *Viburnum opulus* bark and flower extracts on digestion of potato starch and carbohydrate hydrolases activity. *Molecules* **2022**, *27*, 3118. [CrossRef]
51. Levent Altun, M.; Saltan Çitoğlu, G.; Sever Yilmaz, B.; Çoban, T. Antioxidant properties of *Viburnum opulus* and *Viburnum lantana* growing in Turkey. *Int. J. Food Sci. Nutr.* **2008**, *59*, 175–180. [CrossRef]
52. Akbulut, M.; Calisir, S.; Marakoglu, T.; Coklar, H. Chemical and technological properties of European cranberrybush (*Viburnum opulus* L.) fruits. *Asian J. Chem.* **2008**, *20*, 1875.
53. Jouili, H.; Bouazizi, H.; El Ferjani, E. Plant peroxidases: Biomarkers of metallic stress. *Acta Physiol. Plant.* **2011**, *33*, 2075–2082. [CrossRef]
54. Vujinović, T.; Zanin, L.; Venuti, S.; Tomasi, N.; Pinton, R.; Cesco, S. Biostimulant action of dissolved humic substances from a conventionally and an organically managed soil on nitrate acquisition in maize plants. *Front. Plant Sci.* **2020**, *10*, 502598. [CrossRef] [PubMed]
55. Lehmann, J.; Kleber, M. A talaj szerves anyagainak vitás természete. *Nature* **2015**, *528*, 60–68. [CrossRef] [PubMed]
56. Kwizda. 2008. Available online: https://kwizda.hu/yeald-plus~p36277 (accessed on 16 March 2024).
57. Masny, A.; Basak, A.; Żurawicz, E. Effects of foliar applications of Kelpak SL and Goëmar BM 86®preparations on yield and fruit quality in two strawberry cultivars. *J. Fruit Ornam. Plant Res.* **2004**, *12*, 23–27.
58. Kumar, R.; Trivedi, K.; Anand, K.V.; Ghosh, A. Science behind biostimulant action of seaweed extract on growth and crop yield: Insights into transcriptional changes in roots of maize treated with Kappaphycus alvarezii seaweed extract under soil moisture stressed conditions. *J. Appl. Phycol.* **2020**, *32*, 599–613. [CrossRef]
59. Kularathne, M.N.; Srikrishnah, S.; Sutharsan, S. Effect of Seaweed Extracts on Ornamental Plants: Article Review. *Curr. Agric. Res. J.* **2021**, *9*, 149–160. [CrossRef]
60. Davet, P. *Microbial Ecology of Soil and Plant Growth*; CRC Press: Boca Raton, FL, USA, 2004.
61. Li, J.; Xie, T.; Zhu, H.; Zhou, J.; Li, C.; Xiong, W.; Xu, L.; Wu, Y.; He, Z.; Li, X. Alkaline phosphatase activity mediates soil organic phosphorus mineralization in a subalpine forest ecosystem. *Geoderma* **2021**, *404*, 115376. [CrossRef]
62. Zhang, J.; Jin, K.; Luo, Y.; Du, L.; Tian, R.; Wang, S.; Shen, Y.; Zhang, J.; Li, N.; Shao, W.; et al. Responses of soil enzyme activity to long-term nitrogen enrichment and water addition in a typical steppe. *Agronomy* **2023**, *13*, 1920. [CrossRef]
63. Stevenson, F.J. *Humus Chemistry: Genesis, Composition, Reactions*; John Wiley & Sons: Hoboken, NJ, USA, 1994.
64. Canellas, L.P.; Olivares, F.L.; Aguiar, N.O.; Jones, D.L.; Nebbioso, A.; Mazzei, P.; Piccolo, A. Humic and fulvic acids as biostimulants in horticulture. *Sci. Hortic.* **2015**, *196*, 15–27. [CrossRef]
65. Cristofano, F.; El-Nakhel, C.; Rouphael, Y. Biostimulant substances for sustainable agriculture: Origin, operating mechanisms and effects on cucurbits, leafy greens, and nightshade vegetables species. *Biomolecules* **2021**, *11*, 1103. [CrossRef]
66. Kumar Sootahar, M.; Zeng, X.; Wang, Y.; Su, S.; Soothar, P.; Bai, L.; Kumar, M.; Zhang, Y.; Mustafa, A.; Ye, N. The short-term effects of mineral-and plant-derived fulvic acids on some selected soil properties: Improvement in the growth, yield, and mineral nutritional status of wheat (*Triticum aestivum* L.) under soils of contrasting textures. *Plants* **2020**, *9*, 205. [CrossRef]
67. Kumar, H.D.; Aloke, P. Role of biostimulant formulations in crop production: An overview. *Int. J. Appl. Res. Vet. M* **2020**, *8*, 38–46.
68. Khan, W.; Rayirath, U.P.; Subramanian, S.; Jithesh, M.N.; Rayorath, P.; Hodges, D.M.; Critchley, A.T.; Craigie, J.S.; Norrie, J.; Prithiviraj, B. Seaweed extracts as biostimulants of plant growth and development. *J. Plant Growth Regul.* **2009**, *28*, 386–399. [CrossRef]
69. Zodape, S.T. Seaweeds as a biofertilizer. *J. Sci. Ind. Reaserch* **2001**, *60*, 378–382.
70. Basak, A. Effect of preharvest treatment with seaweed products, Kelpak® and Goëmar BM 86®, on fruit quality in apple. *Int. J. Fruit Sci.* **2008**, *8*, 1–14. [CrossRef]
71. Shukla, P.S.; Mantin, E.G.; Adil, M.; Bajpai, S.; Critchley, A.T.; Prithiviraj, B. Ascophyllum nodosum-based biostimulants: Sustainable applications in agriculture for the stimulation of plant growth, stress tolerance, and disease management. *Front. Plant Sci.* **2019**, *10*, 462648. [CrossRef] [PubMed]
72. Hurtado, A.Q.; Critchley, A.T. A review of multiple biostimulant and bioeffector benefits of AMPEP, an extract of the brown alga Ascophyllum nodosum, as applied to the enhanced cultivation and micropropagation of the commercially important red algal carrageenophyte Kappaphycus alvarezii and its selected cultivars. *J. Appl. Phycol.* **2018**, *30*, 2859–2873.
73. Szabó, V.; Magyar, L.; Hrotkó, K. Effect of leaf spray treatments on rooting and quality of *Prunus mahaleb* (L.) cuttings. *Acta Sci. Polonorum. Hortorum Cultus* **2016**, *15*, 77–87.

74. Kovács, D.; Magyar, L.; Diószegi, M.; Hrotkó, K. Biostimulátorok hatása Forsythia x intermedia Zabel." Beatrix Farrand" konténeres díszcserjék növekedésére = Treatments Affecting the Growth of Forsythia x Intermedia Zabel.'Beatrix Farrand'Container Grown Shrubs. *GRADUS* **2017**, *4*, 284–289.
75. Pearcy, R.W.; Schulze, E.D.; Zimmermann, R. Measurement of transpiration and leaf conductance. In *Plant Physiological Ecology: Field Methods and Instrumentation*; Springer Netherlands: Dordrecht, The Netherlands, 2000; pp. 137–160.
76. Van Leperen, W.; Madery, H. A new method to measure plant water uptake and transpiration simultaneously. *J. Exp. Bot.* **1994**, *45*, 51–60. [CrossRef]
77. Arnon, D.I. Copper enzymes in isolated chloroplasts. Polyphenoloxidase in *Beta vulgaris*. *Plant Physiol.* **1949**, *24*, 1–15. [CrossRef]
78. Shannon, L.M.; Kay, E.; Lew, J.Y. Peroxidase isozymes from horseradish roots. I. Isolation and physical properties. *J. Biol. Chem.* **1966**, *241*, 2166–2172. [CrossRef]
79. Chinard, F.P. Photometric estimation of proline and ornithine. *J. Biol. Chem.* **1952**, *199*, 91–95. [CrossRef] [PubMed]
80. Ábrahám, E.; Hourton-Cabassa, C.; Erdei, L.; Szabados, L. Methods for determination of proline in plants. In *Plant Stress Tolerance: Methods and Protocols*; Humana: Totowa, NJ, USA, 2010; pp. 317–331.
81. Tong, A.Z.; Liu, W.; Liu, Q.; Xia, G.Q.; Zhu, J.Y. Diversity and composition of the Panax ginseng rhizosphere microbiome in various cultivation modesand ages. *BMC Microbiol.* **2021**, *21*, 18. [CrossRef] [PubMed]
82. Eivazi, F.; Tabatabai, M.A. Glucosidases and galactosidases in soils. *Soil Biol. Biochem.* **1988**, *20*, 601–606. [CrossRef]
83. Tabatabai, M.A.; Bremner, J.M. Use of p-nitrophenyl phosphate for assay of soil phosphatase activity. *Soil Biol. Biochem.* **1969**, *1*, 301–307. [CrossRef]


**Disclaimer/Publisher's Note:** The statements, opinions and data contained in all publications are solely those of the individual author(s) and contributor(s) and not of MDPI and/or the editor(s). MDPI and/or the editor(s) disclaim responsibility for any injury to people or property resulting from any ideas, methods, instructions or products referred to in the content.

*plants*

MDPI

*Article*

# Unlocking the Potential of Pepper Plants under Salt Stress: Mycorrhizal Effects on Physiological Parameters Related to Plant Growth and Gas Exchange across Tolerant and Sensitive Genotypes

**Ozlem Altuntas [1,*], Hayriye Yildiz Dasgan [2], Yelderem Akhoundnejad [3] and Yahya Nas [3]**

[1] Department of Horticulture, Faculty of Agriculture, University of Malatya Turgut Ozal, 44900 Malatya, Turkey

[2] Department of Horticulture, Faculty of Agriculture, University of Çukurova, 01330 Adana, Turkey; dasgan@cu.edu.tr

[3] Department of Horticulture, Faculty of Agriculture, University of Sirnak, 73000 Sirnak, Turkey; yakhoundnejad@sirnak.edu.tr (Y.A.); yahya.nas@sirnak.edu.tr (Y.N.)

* Correspondence: ozlem.altuntas@ozal.edu.tr

**Abstract:** Agriculture is confronted with the challenge of ensuring global food security, yet the rapid expansion of salinity stress undoubtedly restricts plant productivity in cultivable areas, posing a significant threat to crop yields. Arbuscular mycorrhizal fungi (AMFs) have emerged as a biological tool for enhancing plant salt stress tolerance. To utilize this biological tool, this study evaluated the response in growth and physiological parameters of tolerant (Karaisali) and sensitive (Demre) pepper genotypes. The experiment involved mycorrhizal-treated (*Glomus clarium*) and non-mycorrhizal (control) plants of both the tolerant and sensitive pepper genotypes. The plants were subjected to two salt doses: 75 and 150 mM. The plant growth and physiological parameters were measured at 40 days after transplanting. The mycorrhizal activity was found to be significantly more effective in the sensitive genotype. We found notable differences in mycorrhizal activity between the pepper genotypes under salt stress conditions, with the sensitive genotype "Demre" showing greater responsiveness to mycorrhizal association compared with the "Karaisali" variety. Under both moderate (75 mM NaCl) and higher salt stress levels (150 mM NaCl), both the "Karaisali" and "Demre" varieties exhibited substantial increases in their shoot dry weights. However, these increases were consistently higher in the "Demre" plants. Moreover, the AMFs demonstrated significant enhancements in photosynthesis rates under both moderate and high salinity levels in both genotypes. Overall, our findings suggest that AMFs play a crucial role in improving plant growth, water status, and photosynthesis characteristics, particularly in salt-sensitive pepper genotypes, under moderate-to-high salinity levels. In conclusion, the plant growth, water status, and photosynthesis characteristics of the salt-sensitive pepper genotype were significantly improved by AMFs at medium and high salinity levels, such as 75 mM and 150 mM NaCl, respectively.

**Keywords:** Arbuscular mycorrhizal fungi (AMFs); *Capsicum annuum* L.; NaCl stress; photosynthesis; stomatal conductance; leaf water potential; plant growth

**Citation:** Altuntas, O.; Dasgan, H.Y.; Akhoundnejad, Y.; Nas, Y. Unlocking the Potential of Pepper Plants under Salt Stress: Mycorrhizal Effects on Physiological Parameters Related to Plant Growth and Gas Exchange across Tolerant and Sensitive Genotypes. *Plants* **2024**, *13*, 1380. https://doi.org/10.3390/plants13101380

Academic Editors: Spyridon A. Petropoulos and Mohamad Hesam Shahrajabian

Received: 14 March 2024
Revised: 6 May 2024
Accepted: 13 May 2024
Published: 16 May 2024

**Copyright:** © 2024 by the authors. Licensee MDPI, Basel, Switzerland. This article is an open access article distributed under the terms and conditions of the Creative Commons Attribution (CC BY) license (https://creativecommons.org/licenses/by/4.0/).

## 1. Introduction

Mycorrhizal symbiosis is the most common symbiosis in nature. This symbiosis is an evolutionarily primitive microbiota and, according to the fossil record and molecular data, occurred with the first terrestrial plants [1,2]. Paleobotanical and molecular sequence data indicate that the first land plants established associations with Glomalean fungi in the Glomeromycota approximately 460 million years ago [3]. The symbiotic relationship formed by fungi with plant roots functions like an extended root system. In this symbiotic relationship, fungi help the plant in the intake of water and nutrients with their hyphae, while the fungi receive carbohydrates, the product of photosynthesis, from the host plant [4] The most important type of mycorrhiza for vegetable growers is the arbuscular mycorrhizal symbiosis. Employing mycorrhizal symbiosis is an effective strategy for enhancing plant tolerance to salt stress [5]. The outer hyphae of mycorrhiza can form a massive mycelium network in the soil, which can expand the absorption range of the root system [6], helping host plants absorb low-concentration nutrients and enhancing the exchange of substances between AMFs in the soil and plants [7].

Numerous ongoing research studies focus on AMFs and their symbiotic association with plants, aiming to elucidate their essential roles in plant growth, plant protection, and soil quality improvement. The reports of these studies provided much evidence that association with AMFs facilitates better nutrient uptake, enhancing plant growth under harsh soil (salinity) and climate conditions (drought). It has been reported by many researchers that AMFs promote plant growth and resistance to salt stress. Many authors have reported that mycorrhizal fungi increase tolerance to salt stress through various mechanisms, among which increasing nutrient uptake, producing plant growth hormones, and improving rhizospheric and soil conditions are to be mentioned [8–10].

The abiotic stresses resulting from climate change, driven by global warming, pose significant challenges to crop production worldwide. The costs of dealing with abiotic stresses are potentially enormous, and robust, affordable, environmentally friendly approaches are needed to reduce the negative effects of these on plants. Effective use of mycorrhizae that provides tolerance to stress is of critical importance to ensuring sustainable agriculture in everchanging environmental conditions. The association of AMFs enhances the ability of vegetable plants to withstand various types of stress to some extent. Mycorrhizae assist host plants to adjust the photosynthesis process through leaving $CO_2$ regulation and maintaining the leaf water content during the high-pressure osmotic potential [11]. This phenomenon is provided by performing mycorrhizal spores when it germinates and penetrates the plant root hairs en route to establishing symbioses. An essential harness forms in between the rhizosphere and plant leaves when mycorrhizal hyphae uptake water from the soil, which non-mycorrhizal plants are unable to use through high-salt concentrations. It was found that the percentage variation in the growth rate (i.e., root and shoot lengths and root and shoot dry weights) and nutrient uptake in plants under drought and different levels of salinity stress was directly proportional to the percentage of mycorrhization [12]. AMF symbiosis can affect the stomatal behavior and photosynthesis of host leaves and is confirmed to increase both the transpiration and photosynthetic rates along with the chlorophyll concentration [13]. It has been observed that inoculation with AM fungus alone or in combination with Rhizobium brought a significant increase in the chlorophyll "a", chlorophyll "b", and total chlorophyll content in ground nut, thereby increasing the rate of photosynthesis.

While Turkey boasts fertile land suitable for a variety of crops and horticulture in the Middle East, over time, its soil has experienced an accumulation of many ions and elements. This gradual buildup has caused a significant shift in soil composition toward higher salinity concentrations. Soil beneficial microorganisms such as AMFs are among the most promising elements to encourage growth of crops under this compelling circumstance. For several centuries, pepper (*Capsicum annuum* L.) has been one of the most important field and greenhouse crops in the Mediterranean region due to its economic profit. In particular, greenhouse cultivation demands higher levels of fertilization compared with field cultivation. However, the salt from these fertilizers does not leach deep into the soil with rainfall, necessitating measures to mitigate salt stress during cultivation. Since they establish associations with roots, mycorrhizae-infected plants are able to uptake more nutrients and water. Consequently, mycorrhizae treatments have great potential to maintain ion regulation and thus hinder yield loss under salt stress conditions.

Our hypothesis seeks to investigate the impact of mycorrhizae on the plant growth and physiological parameters in salt-sensitive genotypes, particularly in mitigating salt stress damage. While using salt-resistant genotypes is a feasible strategy for saline environments, harnessing beneficial microorganisms like mycorrhizae offers a more rational approach when cultivating productive and high-quality salt-sensitive varieties. The objective of our study is to evaluate the physiological support provided by mycorrhizae when growing salt-sensitive varieties in saline conditions.

## 2. Materials and Methods

### *2.1. Plant and Mycorrhizal Inoculation and Experimental Conditions*

The experiment was conducted in the growth chamber of Cukurova University's Department of Horticulture in the Faculty of Agriculture. To investigate the alleviation of harmful effects of salt stress on sensitive pepper plants through mycorrhiza inoculation, a pot experiment was designed with both non-mycorrhizal and mycorrhizal treatments. The salt-tolerant (Karaisali) and salt-sensitive (Demre) local pepper genotypes identified in previous studies were used [14]. The plants were treated with three different salt dosages of 0 (control), 75 mM, and 150 mM NaCl. Seedlings of both pepper varieties with the true leaf stage were inoculated with *Glomus clarum*, and a non-inoculated set was used to be left as the control group. The climatic conditions inside the growth chamber were 22 ± 2 °C during the day and 18 ± 2 °C at night, with a 16 h photoperiod and 60% relative humidity. The amount of photosynthetically active radiation received by the upper plant surfaces was 300 mol $m^{-2}$ $s^{-1}$. Four-liter pots filled with vermiculite were utilized for plant cultivation. Three pepper seedlings were transplanted into each pot. The experiment was designed based on a randomized block design with three replicates. The mycorrhizal treatments involved inoculating each plant with 1000 AMF spores during the seedling transplanting process. The nutrient solution used for plant nutrition was as follows (M): $Ca(NO_3)_24H_2O$, 3.0 × $10^{-3}$; $K_2SO_4$, 0.90 × $10^{-3}$; $MgSO_47H_2O$, 1.0 × $10^{-3}$; $KH_2PO_4$, 0.2 × $10^{-3}$; $H_3BO_3$, 1.0 × $10^{-5}$; $10^{-4}$ M Fe EDTA, $MnSO_4H_2O$, 1.0 × $10^{-5}$; $CuSO_45H_2O$, 1.0 × $10^{-7}$; $(NH)_6Mo_7O_{24}4H_2O$, 1.0 × $10^{-8}$; $ZnSO_47H_2O$, 1 × $10^{-6}$. To enhance the effectiveness of the mycorrhizae, the nutrient solution was diluted by 50% to fulfill the nutrient requirements for plant nutrition. The salinity treatments were imposed upon the plants 14 days after transplanting the seedlings. Salt concentrations of 75 mM and 150 mM were gradually reached over 3 days. The plants were exposed to salt stress for 26 days. The morphological and physiological features were measured 40 days after transplanting (Figure 1).

**Figure 1.** *Cont.*

**Figure 1.** Responses of pepper plants to 75 mM and 150 mM salt stresses and AMF inoculation after 40 days of seedling transplanting and 26 days under salinity.

### 2.2. Determinatione of Plant Physiological Parameters

Leaf water potential (MPa): This was measured by the pressure chamber with a Plant Water Status Console mark and 3005-1412 model instrument (Soil Moisture Equipment Corp., Goleta, CA, USA) [15]. Measurements were performed with the third leaf from the tips of the plants.

Osmotic potential (MPa): This was determined according to the process laid out by Kuçukkomurcu [16], where 1 g of fresh leaves from the fourth leaves of the plants was weighed and homogenized with 19 mL of distilled water. The homogenized leaf samples were kept at −20 °C. The homogenized samples were passed through 0.45 µm precision filters. These samples were measured with a freezing point osmometer (Gonotec Osmomat 3000, ELITechGroup Inc., Logan Utah, UT, USA). The osmotic potential was calculated according to the Van't Hoff equation [17].

Photosynthetic rate, stomatal conductance, and transpiration rate: The youngest fully expanded leaves (from the fifth leaves from the tops of the pepper plants) of individual plants were used for gas exchange measurements. The gas exchange characteristics were measured during the 16 h photoperiod using an LICOR LI-6400 portable photosynthesis system (LI-COR, Inc., Lincoln, NE, USA) (block temperature: 25 °C; $CO_2$ reference: 360 µmol $CO_2$ $mol^{-1}$; PAR: 500 µmol $m^{-2}s^{-1}$; flow rate: 300 µmol $s^{-1}$) [18].

Membrane injury index (%) (ion leakage): Tissue ion (electrolyte) leakage was measured as an indicator of the ability of the cellular membranes to maintain integrity or recover from imposed stresses [19]. Electrolyte leakage was measured by taking five leaf discs from the second top leaf and placing them in tubes containing 10 mL of DI water [20]. After incubation for 5 h at 25 °C, the conductivity of the solution was measured using a (WTW and model Cond3110) portable conductivity meter. The tubes were then heated at 100 °C for 10 min, and the conductivity was remeasured (considered the total cellular electrolytes).

The injury index was estimated with the following formula:

$$I = (1 - (1 - T1/T2)/(1 - C1/C2)) \times 100$$

where T1 and T2 are the initial and second measurements of conductivity, respectively, and C1 and C2 are the respective values of the controls [20].

Chlorophyll SPAD measurements: A portable chlorophyll meter (SPAD–502, Konica Minolta Sensing, Inc., Tokyo, Japan) was used to measure the leaf greenness of the pepper plants. For each plant, measurements were taken at four locations on each leaf, with three on each side of the midrib on all fully expanded leaves, and then averaged [20].

Shoot and root parameters: Three plants from each replicate were harvested, and the shoot dry weights and root dry weights were determined. The plant material for the dry weight was dried at 70 °C for two days [15].

Leaf Area: Three plants from each replicate were harvested and measured using a LICOR 3100 (LI-COR, Inc., Lincoln, NE, USA) leaf area meter [15].

### 2.3. Statistical Analysis

The data were subjected to factorial analysis of variance using the Statistical Package for Social Sciences (SPSS version 20), and the means were separated by Duncan's post hoc tests at the 5% level ($p < 0.05$).

## 3. Results

### 3.1. Effects of AMFs under Salt Stress on Pepper's Plant Growth Parameters

The "Karaisali" variety achieved the higher shoot biomass (almost up to double the value) compared with that of the "Demre" variety without mycorrhizal inoculation 40 days after transplanting at a 150 mM salinity level (Table 1). The root dry weight also showed a similar pattern in the "Karaisali" and "Demre" varieties when the plants were exposed to a 150 mM salinity concentration 40 days after transplanting. Our observations indicate that the mycorrhization network provided a beneficial relationship with the "Karaisali" plant rather than the "Demre" plant in order keep growing through salinity situations and a higher osmotic pressure. The symbiotic relationship between plants and mycorrhizal fungi may enhance plant growth, as evidenced by the significant increase in the shoot dry weight observed in the "Karaisali" variety. This feature was also observed for the root dry weight compared with the "Demre" plant. The "Demre" plant did not become more productive as the salinity level increased during its growth time. However, the root dry weight clearly confirmed the mycorrhizal symbioses with an enhanced dry mass 40 days after transplanting the "Karaisali" plant. The dry mass of the "Karaisali" plant, either in terms of shoot weight or root weight, was significantly different compared with those in the "Demre" plant for both the mycorrhizal and non-mycorrhizal "Karaisali" plants under the salinity circumstances.

The mycorrhizal activity was found to be significantly more effective in the sensitive genotype. Specifically, in the "Karaisali" variety, there was a 9.8% increase in the shoot dry weight under the 75 mM salt stress, while in the "Demre" variety, this increase was 45.2% under the same conditions. The effect of mycorrhizae on the increase in the shoot dry weight under the 150 mM salt stress was recorded to be 8.9% in the "Karaisali" variety, while it was recorded to be 124.3% in the "Demre" variety, which is sensitive. Although mycorrhiza activity's effect on the root dry weight under salt stress was lower than the rates of increase in the shoot dry weight, the AMFs were more effective on the susceptible genotype "Demre" than on the tolerant "Karaisali" genotype. At the 75 mM salt stress, the AMFs increased the root dry weight by 20% in the "Karaisali" plant and 25% in the "Demre" plant. At the 150 mM salt stress, the AMFs increased the root dry weight by 13% in the "Karaisali" genotype and 19% in the "Demre" genotype.

**Table 1.** "Karaisali" and "Demre" biomass productions, including shoot and root dry weights, during different salinity concentrations with and without AMF inoculation 40 days after transplanting and after 26 days under salinity (g plant$^{-1}$).

| Variety | Mycorrhyzae | Salt Doses | Shoot Dry Weight | Root Dry Weight |
|---|---|---|---|---|
| "Karaisali" salt-tolerant | M (+) | 0 (control) | 9.34 a | 2.18 a |
| | | 75 mM | 5.59 b | 1.26 a |
| | | 150 mM | 3.30 b | 0.67 c |
| | M (−) | 0 (control) | 8.58 a | 2.16 a |
| | | 75 mM | 5.09 b | 1.05 b |
| | | 150 mM | 3.03 b | 0.59 c |
| "Demre" salt-sensitive | M (+) | 0 (control) | 7.73 a | 1.35 a |
| | | 75 mM | 4.43 bc | 0.75 b |
| | | 150 mM | 3.05 cd | 0.50 b |
| | M (−) | 0 (control) | 6.74 ab | 1.24 b |
| | | 75 mM | 3.05 cd | 0.60 b |
| | | 150 mM | 1.36 d | 0.42 b |
| $p \leq 0.05$ | | | 0.152 | 0.067 |

Different letters within a column indicate significant differences at $p \leq 0.05$.

The implied salinity showed that the leaf chlorophyll content decreased (Table 2). The effect of AMFs on chlorophyll's increase in the "Karaisali" genotype was 1.9% and 14.3% under 75 mM and 250 mM of salt stress, respectively. In the "Demre" genotype, under 75 mM and 250 mM of salt stress, the increasing effect of AMFs on chlorophyll was 3.7% and 7.4%, respectively.

**Table 2.** Comparison of mycorrhizal association's effects on leaf chlorophyll content, leaf area, and membrane damage in both mycorrhizal and non-mycorrhizal plants of the two pepper genotypes under different saline concentrations 40 days after transplanting and after 26 days under salinity.

| Variety | Mycorrhyzae | Salt Doses | Chlorophyll (SPAD) | Leaf Area (cm$^2$ plant$^{-1}$) | Membran Injury Index (%) |
|---|---|---|---|---|---|
| "Karaisali" salt-tolerant | M (+) | 0 (control) | 72.10 a | 837.93 a | |
| | | 75 mM | 68.60 a | 605.48 ab | 2.64 |
| | | 150 mM | 67.33 a | 491.04 b | 5.87 |
| | M (−) | 0 (control) | 71.93 a | 837.93 a | |
| | | 75 mM | 67.33 a | 502.86 b | 3.77 |
| | | 150 mM | 58.93 b | 407.61 b | 6.73 |
| "Demre" salt-sensitive | M (+) | 0 (control) | 69.30 a | 761.68 a | |
| | | 75 mM | 64.30 ab | 579.89 ab | 2.92 |
| | | 150 mM | 57.03 bc | 340.10 ab | 10.62 |
| | M (−) | 0 (control) | 70.23 a | 805.09 a | |
| | | 75 mM | 62.03 ab | 508.56 ab | 5.71 |
| | | 150 mM | 53.10 c | 226.01 b | 11.69 |
| $p \leq 0.05$ | | | 0.199 | 0.069 | 0.094 |

Different letters within a column indicate significant differences at $p \leq 0.05$.

The leaf area was not significantly affected by salt in either genotype at the beginning of the salt treatments. However, in the measurements made after 40 days of transplanting, the leaf area of the plants to which salt was applied significantly decreased. At 75 mM of salt stress, the AMFs increased the leaf area by 20.4% in the "Karaisali" plant and 14.0% in the "Demre" plant. However, at 150 mM of salt stress, the beneficial effects of AMFs on the leaf area were more pronounced, with increases of 20.5% in the "Karaisali" genotype and 50.5% in the "Demre" genotype (Table 2).

The membrane damage was lower in the tolerant genotype and greater in the sensitive genotype, but the damage was lower in the plants treated with mycorrhizae than in the plants without mycorrhizae (Table 2). The salt stress's alleviating effect of AMFs on the cell membranes was particularly more successful at the 75 mM salt stress compared with the 150 mM salt stress. Accordingly, in the "Karaisali" genotype, the reduction in membrane damage by the AMFs was 30% and 12.8% under 75 mM and 150 mM of salt stresses, respectively. In the sensitive "Demre" genotype, the reduction rates of the membrane damage by the AMFs were 48.9% and 9.15% under 75 mM and 150 mM of salt stresses, respectively.

### 3.2. *Effects of AMFs under Salt Stress on Pepper's Physilogical Parameters*

The "Karaisali" plant was able to adjust its leaf water potential by uptaking much more water via AMF contribution rather than the non-mycorrhizal "Karaisali" plant and consequently control the osmotic adjustment. Ultimately, the "Karaisali" genotype showed better performance in terms of holding water in a way to keep plants surviving in higher-salinity conditions (Table 3). Consequently, the "Karaisali" and "Demre" genotypes showed various responses to the salt concentrations in terms of leaf water potential as well as osmotic pressure. Despite neither plant having any struggles during the control treatment, as the salinity increased, significant differences in leaf water potential and osmotic potential became more visible during plant growth.

**Table 3.** Comparison of mycorrhizal association effects on leaf water potential and osmotic potential in both mycorrhizal and non-mycorrhizal plants of two pepper genotypes under different saline concentrations 40 days after transplanting and after 26 days under salinity.

| Variety | Mycorrhyzae | Salt Doses | Leaf Water Potential (MPa) | Osmotic Potential (MPa) |
|---|---|---|---|---|
| "Karaisali" salt-tolerant | M (+) | 0 (control) | −1.60 f | −1.24 e |
| | | 75 mM | −3.47 d | −1.70 b |
| | | 150 mM | −5.13 bc | −1.87 ab |
| | M (−) | 0 (control) | −2.07 e | −1.33 d |
| | | 75 mM | −3.90 cd | −1.77 b |
| | | 150 mM | −5.73 bc | −1.96 ab |
| "Demre" salt-sensitive | M (+) | 0 (control) | −2.27 de | −1.31 d |
| | | 75 mM | −4.40 c | −1.85 ab |
| | | 150 mM | −6.07 b | −2.03 ab |
| | M (−) | 0 (control) | −3.67 d | −1.65 c |
| | | 75 mM | −4.70 c | −1.98 ab |
| | | 150 mM | −7.13 a | −2.34 a |
| $p \leq 0.05$ | | | 0.001 | 0.367 |

Different letters within a column indicate significant differences at $p \leq 0.05$.

When we examined the leaf osmotic potential values, it was determined that the mycorrhiza applications improved the plants under salt stress. The leaf water potential (LWP) and leaf osmotic potential (LOP) values between the mycorrhizal and non-mycorrhizal groups with two different combinations created from "Karaisali" and "Demre" pepper plants were found to be statistically significant at the $p \leq 0.05$ level. The best results for the LWP and LOP values, which were determined to be indicators of water status in the leaf, were obtained in the "Karaisali" control mycorrhizal applications, while the lowest values were obtained in the "Demre" non-AMF 150 mM application. The best results for the LWP and LOP values, which were determined to be indicators of the water status in the leaf, were obtained in the "Karaisali" control with mycorrhizal −1.60 MPa and −1.24 MPa applications, while the lowest values were obtained in the "Demre" non-AMF 150 mM −7.13 MPa and −2.34 MPa applications, respectively.

### 3.3. Photosynthetic Rate, Stomatal Conductance, and Transpiration Rate

The interaction between mycorrhizal fungi and plants occurred at higher photosynthesis activities, resulting in stomatal conductivity (Table 4). With a short glance at the previous results, the leaves' chlorophyll content and size of area data could be roughly predictable in terms of having correlation with the photosynthesis rates achieved. In the meantime, the "Karaisali" plant with mycorrhizal associations performed better in terms of photosynthetic activity than the non-mycorrhizal plants. These phenomena enabled the plants to attain higher amounts of $CO_2$ and $H_2O$ in terms of generating growth enzymes and energy for plant survival. At an initial glance, there were different photosynthetic rates between the mycorrhizal plants and non-mycorrhizal ones for both the "Karaisali" and "Demre", plants where they were exposed to 75 and 150 mM salinity concentrations 40 days after being transplanted. The values of the photosynthetic rate almost doubled in the mycorrhizal plants compared with those in the controlled ones.

**Table 4.** Comparison of mycorrhizal association effects on photosynthetic and gas exchange parameters in both mycorrhizal and non-mycorrhizal plants of two pepper genotypes under different saline concentrations 40 days after transplanting and after 26 days under salinity.

| Variety | Mycorrhyzae | Salt Doses | Photosynthetic Rate (μmol $CO_2$ $m^{-2}s^{-1}$) | Stomatal Conductance (mmol $H_2O$ $m^{-2}$ $s^{-1}$) | Transpiration Rate (mmol $H_2O$ m $m^{-2}$ $s^{-1}$) |
|---|---|---|---|---|---|
| "Karaisali" salt-tolerant | M (+) | 0 (control) | 9.94 a | 0.13 a | 1.71 a |
| | | 75 mM | 8.53 ab | 0.07 ab | 1.21 ab |
| | | 150 mM | 5.20 cd | 0.04 b | 0.66 c |
| | M (−) | 0 (control) | 9.65 a | 0.11 a | 1.57 a |
| | | 75 mM | 7.38 b | 0.06 ab | 0.85 b |
| | | 150 mM | 3.43 d | 0.03 c | 0.50 cd |
| "Demre" salt-sensitive | M (+) | 0 (control) | 8.71 ab | 0.09 a | 1.30 ab |
| | | 75 mM | 6.61 bc | 0.06 ab | 1.04 ab |
| | | 150 mM | 3.97 d | 0.03 c | 0.44 cd |
| | M (−) | 0 (control) | 7.71 ab | 0.07 ab | 1.29 ab |
| | | 75 mM | 5.52 c | 0.03 b | 0.56 c |
| | | 150 mM | 2.68 e | 0.02 c | 1.71 a |
| $p \leq 0.05$ | | | 0.047 | 0.011 | 0.020 |

Different letters within a column indicate significant differences at $p \leq 0.05$.

In the gas exchange parameter measurements conducted 40 days after transplantation, it was observed that the photosynthesis rate, stomatal conductance, and transpiration rate decreased in both genotypes and under both salt doses. Based on the results of the gas exchange parameters, it can be concluded that the mycorrhizae significantly supported the pepper genotypes. The application of AMFs to the pepper plants under salt stress increased their photosynthesis. The effect of the AMFs was higher at 150 mM of salt stress compared with 75 mM. The photosynthesis-enhancing effect of the AMFs was 15.6% and 51.6% under 75 mM and 150 mM of salt stress, respectively, in the "Karaisali" genotype. In the "Demre" genotype, under the same saline conditions, the rates of photosynthesis increase were 19.7% and 48.13%, respectively. The other photosynthetic parameters, namely the stomatal conductance and transpiration rate, were also consistent with the photosynthesis rates in both peppers under both salt stresses.

## 4. Discussion

The study revealed that all treatments involving mycorrhizae effectively mitigated salt stress and facilitated the water status, resulting in improved plant growth and physiological parameters.

Our study determined that mycorrhizae were more effective in terms of plant growth in sensitive genotypes under salt stress. The improvement of water relations in plants through mycorrhizal contributions to host plants is consistent with several studies [11]. Mycorrhizal fungi are confirmed to uptake up to 20% of the unavailable water captured in a plant's rhizosphere [21]. In our study, under salt stress conditions, the pepper plants reduced their osmotic potential and increased their water uptake (Table 3). As a result, the turgor pressure increased, facilitating cell development and stomatal opening (Table 4). Studies have reported that salt stress reduces the leaf water potential in tomatoes [22,23], melons [24], and eggplants [25]. The osmotic pressure caused by high concentrations of salt and its damaging impact on plants can be significantly mitigated by the absorption of additional water through mycorrhiza hyphae during this phenomenon.

In mycorrhizal symbiosis, the fungus can mitigate or alleviate the detrimental effects of stress on the plant by inducing changes in the biochemical and physiological properties of the host plant under salt stress conditions. The mycorrhizae induce biochemical alterations in host plants under salt stress conditions, potentially resulting in increased accumulation of antioxidant enzymes, proline, betaine, or soluble carbohydrates compared with non-mycorrhizal plants. The sugar accumulation facilitated by mycorrhizae in plants serves as a defense mechanism against salinity. Elevated accumulation of soluble sugars in mycorrhiza-inoculated plants has been evidenced as a consequence of mycorrhizae enhancing photosynthesis [26,27]. In our study, the observed heightened rate of photosynthesis in mycorrhizae-inoculated pepper plants (Table 4), coupled with reduced plant damage, can be attributed to the accumulation of photosynthesis products such as sugar. This accumulation likely contributes to plant resilience against stressors. Previous studies have reported that mycorrhizal inoculation enables host plants to enhance water and nutrient uptake through hyphal networks and improve their gas exchange capacity under saline conditions [28,29]. In plants under saline conditions, mycorrhizal fungal colonization additionally (1) enhances the hydraulic conductivity of the root at low water potentials, (2) stimulates and alters the root system morphology, and (3) increases stomatal conductance. Mycorrhiza-inoculated plants exhibit high chlorophyll contents and increased N and Mg uptake, while mycorrhizae prevent Na transport in saline conditions [30–37].

Physiological changes in pepper plants inoculated with mycorrhizae compared with non-inoculated plants under salt stress include improved photosynthetic activity, stomatal conductance, and water relations (Table 4). Mycorrhizal inoculation enhances the water uptake capacity of the host plant (Table 3) by increasing hydraulic conductivity in the roots and positively affecting the osmotic balance of photosynthesis product carbohydrates, thereby contributing positively to physiological processes in the host plant. This enhancement in the water uptake capacity not only leads to increased plant growth in mycorrhizal plants but may also mitigate the toxic ion effect through dilution as the water uptake increases. Exemplary studies demonstrating this phenomenon include those by Al Karaki [38], who observed increased root dry weights and yields in salt-stressed tomatoes inoculated with mycorrhizae, resulting in higher fruit quantities and weights. Similarly, El-Sarkassy, Ibrahim, and Desoky [39] found that mycorrhizae application improved all growth parameters in pepper plants across various salt stress levels. It is known that mycorrhizal symbiosis enhances the uptake of nutrient elements and influences the phytohormone balance in plants, leading to increased plant growth and alleviation of environmental stresses. Consequently, it promotes biomass and yields in plants and has a positive effect on some quality parameters [40–42]. Under salt stress, photosynthesis decreases in plants. This may be attributed to disruption of the water balance, alteration of the enzyme structure, and inhibition of protein synthesis by specific ions (e.g., Na+) [43] In our study, we observed higher plant growth parameters (Tables 1 and 2) and higher photosynthetic properties in plants treated with mycorrhizae (Table 4).

In salt stress conditions, reduced stomatal conductance [44] and enzyme inactivation [45] result in $CO_2$ accumulation among cells. Elevation in the intercellular $CO_2$ concentration indirectly suggests damage to photosynthetic organs [27]. Zhang et al. demonstrated in their study on tomatoes that mycorrhizal inoculation increased photosynthesis by enhancing $CO_2$ assimilation in tomato plants through non-stomatal factors [46]. Mycorrhizae significantly elevated the concentration of photosynthetic pigments in tomato leaves grown in saline-alkaline soil, thereby facilitating the capture and utilization of light energy. As is known, the efficiency of plant photosynthesis is closely linked to growth and development. According to the findings of our study, the photosynthesis rate was higher in pepper plants treated with mycorrhizae, leading to increased leaf areas and shoot dry weights. A similar study demonstrated that mycorrhizal symbiosis enhances photosynthesis in plants, leading to increased assimilate accumulation, as well as improved mineral nutrition and fruit quality [47].

In this study, mycorrhizal symbiosis altered the plants' physiological aspects, such as water relations in leaves, consequently enhancing photosynthetic activities in both the "Karaisali" and "Demre" plants subjected to salinity stress. These results suggest that the association (symbiosis) of both plants with mycorrhizal fungi may improve growth capability by increasing water uptake through external hyphae and enhancing the osmotic potential through the provision of salt and high concentrations of minerals.

## 5. Conclusions

Climate change, which is increasingly evident worldwide, necessitates cultivation in saline soils due to improper soil management practices. While the use of salt-tolerant varieties seems to be an immediate solution, a more sustainable approach involves harnessing nature-friendly beneficial microorganisms like mycorrhizal fungi to cultivate productive and high-quality, salt-sensitive varieties and mitigate salt-induced plant damage. Moreover, the dwindling availability of fertilizer resources and rising production costs forecast serious efficiency challenges in crop production over the next 50 years. Therefore, leveraging mycorrhizal fungi, which reduces reliance on chemical fertilizers, proves ecologically sound and vital for agricultural sustainability. Recognizing the potential of mycorrhizae as a natural fertilizer in saline soils for promoting mycorrhizal inoculation in mycorrhizae-dependent plants has emerged as an increasingly crucial agricultural strategy. In this study, it was shown that the growth, water status, and photosynthesis characteristics of the "Demre" salt-sensitive pepper genotype were significantly improved by AMFs at medium and high salinity levels such as 75 mM and 150 mM NaCl, respectively.

**Author Contributions:** O.A., H.Y.D. and Y.A. planned and designed the research and wrote the manuscript; Y.A. performed the sampling and data analysis; O.A., H.Y.D. and Y.N. edited the manuscript. All authors have read and agreed to the published version of the manuscript.

**Funding:** This research received no external funding.

**Data Availability Statement:** Data are contained within the article.

**Conflicts of Interest:** The authors declare no conflicts of interest.

## References


1. Lee, B.R.; Muneer, S.; Avice, J.C.; Jin Jung, W.; Kim, T.H. Mycorrhizal Colonisation and P-Supplement Effects on N Uptake and N Assimilation in Perennial Ryegrass under Well-Watered and Drought-Stressed Conditions. *Mycorrhiza* **2012**, *22*, 525–534. [CrossRef] [PubMed]
2. Fusconi, A.; Berta, G. Environmental Stress and Role of Arbuscular Mycorrhizal Symbiosis. In *Abiotic Stress Responses in Plants*; Ahmad, P., Prasad, M.N.V., Eds.; Springer: Berlin/Heidelberg, Germany, 2012; pp. 197–214.
3. Finlay, R.D. Ecological Aspects of Mycorrhizal Symbiosis: With Special Emphasis on the Functional Diversity of Interactions Involving the Extraradical Mycelium. *J. Exp. Bot.* **2008**, *59*, 1115–1126. [CrossRef] [PubMed]
4. Ikiz, O. The Effects of Mycorrhiza on Soilless Pepper Cultivation. Ph.D. Thesis, Code number: 730. Department of Horticulture, Institute of Natural and Applied Sciences, University of Cukurova, Adana, Turkey, 2003.
5. Birhane, E.; Sterck, F.J.; Fetene, M.; Bongers, F.; Kuyper, T.W. Arbuscular Mycorrhizal Fungi Enhance Photosynthesis, Water use Efficiency, and Growth of Frankincense Seedlings under Pulsed Water Availability Conditions. *Oecologia* **2012**, *169*, 895–904. [CrossRef] [PubMed]
6. Smith, S.E.; Facelli, E.; Pope, S.; Andrew Smith, F. Plant Performance in Stressful Environments: Interpreting New and Established Knowledge of the Roles of Arbuscular Mycorrhizas. *Plant Soil* **2010**, *326*, 3–20. [CrossRef]
7. Igiehon, N.O.; Babalola, O.O. Biofertilizers and Sustainable Agriculture: Exploring Arbuscular Mycorrhizal Fungi. *Appl. Microbiol. Biotechnol.* **2017**, *101*, 4871–4881. [CrossRef] [PubMed]
8. Baum, C.; El-Tohamy, W.; Gruda, N. Increasing the Productivity and Product Quality of Vegetable Crops Using Arbuscular Mycorrhizal Fungi: A review. *Sci. Hortic.* **2015**, *187*, 131–141. [CrossRef]
9. Heikham, E.; Rupam, K.; Bhoopander, G. Arbuscular mycorrhizal Fungi in Alleviation of Salt Stress: A Review. *Ann. Bot.* **2009**, *104*, 1263–1280.
10. Sun, W.; Shahrajabian, M.H. The application of Arbuscular Mycorrhizal Fungi as Microbial Biostimulant, Sustainable Approaches in Modern Agriculture. *Plant* **2023**, *12*, 3101. [CrossRef]
11. Khalvati, M.A.; Hu, Y.; Mozafar, A.; Schmidhalter, U. Quantification of Water Uptake by Arbuscular Mycorrhizal Hyphae and its Significance for Leaf Growth, Water Relations, and Gas Exchange of Barley Subjected to Drought Stress. *Plant Biol.* **2005**, *7*, 706–712. [CrossRef] [PubMed]
12. Anilkumar, K.K.; Muraleedhara, K.G. Effect of Vesicular Arbuscular Mycorrhizal Fungi Inoculation on Growth and Nutrient Uptake of Leguminous Plants Under Stress Conditions. *J. Mycol. Pl. Pathol.* **2003**, *33*, 33–36.
13. Devi, M.C.; Reddy, M.N. Effect of Arbuscular Mycrrhizal Fungi and Rhizobium Association on Chlorophyll Content of Ground Nut (*Arachis hypogaea* L.). *Mycorrhiza News* **2004**, *6*, 15–17.
14. Altuntas, O.; Dasgan, H.Y.; Kutsal, I.K. Effects of mycorrhiza on pepper plant growth and nutrients under salinity stress. In Proceedings of the XVIth EUCARPIA Capsicum and Eggplant Working Group Meeting in memoriam Dr. Alain Palloix, Kecskemét, Hungary, 12–14 September 2016; Diamond Congress Ltd.: Budapest, Hungary, 2016; pp. 194–202.
15. Altuntas, O.; Dasgan, H.Y.; Akhoundnejad, Y. Silicon-Induced Salinity Tolerance Improves Photosynthesis, Leaf Water Status, Membrane Stability, and Growth in Pepper (*Capsicum annuum* L.). *HortScience* **2018**, *53*, 1820–1826. [CrossRef]
16. Kucukkomurcu, S. Screening of the Okra Genotypes for Their Tolerance to Salinity and Drought Stress. Master's Thesis, Department of Horticulture, Institute of Natural and Applied Sciences, University of Çukurova, Adana, Turkey, 2017; p. 177.
17. Akhoundnejad, Y.; Dasgan, H.Y. Effect of Different Irrigation Levels on Physiological Performance of Some Drought Tolerant Melon (*Cucumis melo* L.) Genotypes. *Appl. Ecol. Environ. Res.* **2019**, *17*, 9997–10012. [CrossRef]
18. Akhoundnejad, Y.; Dasgan, H.Y. Photosynthesis, Transpiration, Stomatal Conductance of Some Melon (*Cucumic melo* L.) Genotypes under Different Drought Stress. *Fresenius Environ. Bull.* **2020**, *29*, 10974–10979.
19. Akhoundnejad, Y.; Dasgan, H.Y. Physiological Performance of Some High Temperature Tolerant Tomato Genotypes. *Int. J. Sci. Technol. Res.* **2018**, *4*, 57–74.
20. Dasgan, H.Y.; Bayram, M.; Kusvuran, S.; Coban, G.A.; Akhoundnejad, Y. Screening of Tomatoes for Their Resistance to Salinity and Drought Stress. *J. Biol. Agric. Healthc.* **2018**, *8*, 31–37.
21. Ruth, B.; Khalvati, M.; Schmidhalter, U. Quantification of Mycorrhizal Water Uptake Via High-Resolution on-Line Water Content Sensors. *Plant Soil* **2011**, *342*, 459–468. [CrossRef]
22. Romero-Aranda, M.R.; Soria, T.; Cuartero, J. Tomato Plant-Water Uptake and Plant-Water Relationships under Saline Growth Conditions. *Plant Sci.* **2001**, *160*, 265–272. [CrossRef] [PubMed]
23. Hossain, M.M.; Nonami, H. Effect of Salt Stress on Physiological Response of Tomato Fruit Grown in Hydroponic Culture System. *Hortic. Sci.* **2012**, *39*, 26–32. [CrossRef]
24. Kuşvuran, Ş.; Daşgan, H.Y.; Abak, K. Responses of Different Melon Genotypes to Drought Stress. *Yuz. Yıl Univ. J. Agric. Sci.* **2011**, *21*, 209–219.
25. Talhouni, M.; Kuşvuran, Ş.; Kiran, S.; Ellialtioğlu, Ş. Tuz Stresi Altında Yetiştirilen Patlıcan Bitkilerinde Klorofil, Yaprak Su Potansiyeli ve Bazı Meyve Özellikleri Üzerine Aşılı Bitki Kullanımının Etkisi. *Toprak. Su Dergisi.* **2019**, *8*, 29–38.
26. Yang, G.; Guo, L.P.; Huang, L.Q.; Chen, M. Inoculation Methods of AM Fungi in Medicinal Plant. *Resour. Sci.* **2008**, *30*, 778–785.
27. Sheng, M.; Tang, M.; Chen, H.; Yang, B.; Zhang, F.; Huang, Y. Influence of Arbuscular Mycorrhizae on Photosynthesis and Water Status of Maize Plants under Salt Stress. *Mycorrhiza* **2008**, *18*, 287–296. [CrossRef] [PubMed]

28. Porcel, R.; Aroca, R.; Ruíz-Lozano, J.M. Salinity Stress Alleviation Using Arbuscular Mycorrhizal Fungi. A review. *Agron. Sustain. Dev.* **2012**, *32*, 181–200. [CrossRef]
29. Hameed, A.; Dilfuza, E.; Abd-Allah, E.F.; Hashem, A.; Kumar, A.; Ahmad, P. Salinity Stress and Arbuscular Mycorrhizal Symbiosis in Plants. In *Use of Microbes for the Alleviation of Soil Stresses*; Springer: New York, NY, USA, 2014; Volume 1, pp. 139–159.
30. Abdel Latef, A.A.; Chaoxing, H.J. Does the İnoculation with *Glomus mosseae* İmprove Salt Tolerance in Pepper Plants? *Plant Grow Regul.* **2014**, *33*, 644–653. [CrossRef]
31. Aroca, R.; Ruiz-Lozano, J.M.; Zamarreño, A.M.; Paz, J.A.; García-Mina, J.M.; Pozo, J.A.; López-Ráez, J.A. Arbuscular Mycorrhizal Symbiosis Influences Strigolactone Production Under Salinity and Alleviates Salt Stress in Lettuce Plants. *J. Plant Physiol.* **2013**, *170*, 47–55. [CrossRef] [PubMed]
32. Ahmad, P.; Abdul Latef, A.A.; Hashem, A.; Abd_Allah, E.F.; Gucel, S.; Tran, L.S.P. Nitric Oxide Mitigates Salt Stress by Regulating Levels of Osmolytes and Antioxidant Enzymes in Chickpea. *Front. Plant Sci.* **2016**, *7*, 347. [CrossRef]
33. Hashem, A.; Abd_Allah, E.F.; Alqarawi, A.A.; El-Didamony, G.; Alwhibi, M.; Egamberdieva, D.; Ahmad, P. Alleviation of Adverse İmpact of Salinity on Faba Bean (*Vicia faba* L.) by Arbuscular Mycorrhizal Fungi. *Pak. J. Bot.* **2014**, *46*, 2003–2013.
34. Hasanuzzaman, M.; Gill, S.S.; Fujita, M. Physiological Role of Nitric Oxide in Plants Grown Under Adverse Environmental Conditions. In *Plant Acclimation to Environmental Stress*; Tuteja, N., Gill, S.S., Eds.; Springer Science + Business Media: New York, NY, USA, 2013; pp. 269–322.
35. Kapoor, R.; Sharma, D.; Bhatnagar, A.K. Arbuscular Mycorrhizae in Micropropagation Systems and Their Potential Applications. *Sci. Hortic.* **2008**, *116*, 227–239. [CrossRef]
36. Kapoor, R.; Evelin, H.; Mathur, P.; Giri, B. Arbuscular mycorrhiza: Approaches for Abiotic Stress Tolerance in Crop Plants For Sustainable Agriculture. In *Plant Acclimation to Environmental Stress*; Springer: New York, NY, USA, 2013; pp. 359–401. [CrossRef]
37. Talaat, N.B.; Shawky, B.T. Protective Effects of Arbuscular Mycorrhizal Fungi on Wheat (*Triticum aestivum* L.) Plants Exposed to Salinity. *Environ. Exp. Bot.* **2014**, *98*, 20–31. [CrossRef]
38. Al-Karaki, G.N. The role of Mycorrhizae in the Reclamation of Degraded Lands in Arid Environments. In *Developments in Soil Classification, Land Use Planning and Policy Implications*; Shahid, S., Taha, F., Abdelfattah, M., Eds.; Springer: Dordrecht, The Netherlands, 2013.
39. El-Sarkassy, N.M.; Ibrahim, S.A.; Desoky, E.M. Salinity Stress Amelioration Using Humic Acid and Mycorrhizae on Pepper Plants. *Zagazig J. Agric. Res.* **2017**, *44*, 2515–2527.
40. Mantelin, S.; Touraine, B. Plant Growth-Promoting Rhizo-Bacteria and Nitrate Availability: Impacts On Root Develop-Ment and Nitrate Uptake. *J. Exp. Bot.* **2004**, *55*, 27–34. [CrossRef] [PubMed]
41. Stewart, L.I. Phosphorus Effects on Arbuscular Mycorrhizal Fungi. PhD Dissertation, McGill University, Montreal, QC, Canada, 2006.
42. Badda, N.; Aggarwal, A.; Kadian, N.; Sharma, N. Influence of Arbuscular Mycorrhizal Fungi and Different Salinity Levels on Growth Enhancement and Nutrient Uptake of *Gossypium arboretum* L. *Kavaka* **2014**, *43*, 14–21.
43. Ruiz-Lozano, J.M.; Porcel, R.; Azcon, C.; Aroca, R. Regulation by Arbuscular Mycorrhizae of the Integrated Physiological Response to Salinity in Plants: New Challenges in Physiological and Molecular Studies. *J. Exp. Bot.* **2012**, *63*, 4033–4044. [CrossRef] [PubMed]
44. Powles, S.B. Photoinhibition of Photosynthesis Induced by Visible Light. *Annu. Rev. Plant Physiol.* **1984**, *35*, 15–44. [CrossRef]
45. Munns, R. Comparative Physiology of Salt and Water Stress. *Plant Cell Environ.* **2002**, *25*, 239–250. [CrossRef] [PubMed]
46. Zhang, F.; Jia-Dong, H.E.; Qiu-Dan, N.I.; Qiang-Sheng, W.U.; Zou, Y.N. Enhancement of Drought Tolerance in Trifoliate Orange by Mycorrhiza: Changes in Root Sucrose and Proline Metabolisms. *Not. Bot. Horti. Agrobot. Cluj-Napoca* **2018**, *46*, 270. [CrossRef]
47. Nzanza, B.; Marais, D.; Soundy, P. Effect of arbuscular Mycorrhizal Fungal Inoculation and Biochar Amendment on Growth and Yield of Tomato. *Int. J. Agric. Biol.* **2012**, *14*, 965–969.

**Disclaimer/Publisher's Note:** The statements, opinions and data contained in all publications are solely those of the individual author(s) and contributor(s) and not of MDPI and/or the editor(s). MDPI and/or the editor(s) disclaim responsibility for any injury to people or property resulting from any ideas, methods, instructions or products referred to in the content.

plants MDPI

Article

# *Codium fragile* (Suringar) Hariot as Biostimulant Agent to Alleviate Salt Stress in Durum Wheat: Preliminary Results from Germination Trials

**Angelo Rossini [1], Roberto Ruggeri [1], Nada Mzid [1], Francesco Rossini [1,*] and Giuseppe Di Miceli [2]**

[1] Department of Agriculture and Forest Sciences, University of Tuscia, Via S. Camillo de Lellis, 01100 Viterbo, Italy; angelo.rossini@unitus.it (A.R.); r.ruggeri@unitus.it (R.R.); nada.mzid@unitus.it (N.M.)
[2] Department of Agricultural, Food and Forestry Sciences (SAAF), University of Palermo, Viale delle Scienze, Ed. 5, 90128 Palermo, Italy; giuseppe.dimiceli@unipa.it
* Correspondence: rossini@unitus.it; Tel.: +39-0761-357541

**Abstract:** Soil salinization is a critical environmental problem in arid and semiarid regions of the world. The aim of the present study was to evaluate the effect of an algae-based biostimulant on germination and seedling vigour of durum wheat (*Triticum turgidum* L. subsp. *durum* (Desf.) Husn.), under different saline conditions (0, 100, and 200 mM NaCl). The experiment was carried out under controlled-environment conditions. Seeds were sprayed with a solution containing a combination of fungicide and different concentrations of *Codium fragile* (Suringar) Hariot algae ($0\%_{w/v}$, $10\%_{w/v}$, $20\%_{w/v}$, and $30\%_{w/v}$). All experimental units were placed in a germination cabinet. The effect of the seaweed extract (SWE) on seed germination and seedling performance under salinity stress was evaluated over a period of 8 days. Coleoptile length and biomass were found to be significantly and positively affected by the application of different SWE doses as compared to the control treatment (0% algae). As for germination traits, seeds treated with SWE showed a final germination (from 82% to 88%), under severe saline conditions, significantly higher than that observed in the control treatment (61%). Our findings indicate that the appropriate dose of biostimulant can markedly improve the germination and the seedlings vigour of durum wheat seeds under saline conditions. Additional studies will be needed to understand the mechanism of action of this biostimulant and its effectiveness in longer studies under field conditions.

**Keywords:** *Codium fragile* (Suringar) Hariot; durum wheat; salt stress; seaweed extract; *Triticum turgidum* L. subsp. *durum* (Desf.) Husn

**Citation:** Rossini, A.; Ruggeri, R.; Mzid, N.; Rossini, F.; Di Miceli, G. *Codium fragile* (Suringar) Hariot as Biostimulant Agent to Alleviate Salt Stress in Durum Wheat: Preliminary Results from Germination Trials. *Plants* **2024**, *13*, 283. https://doi.org/10.3390/plants13020283

Academic Editors: Spyridon A. Petropoulos and Mohamad Hesam Shahrajabian

Received: 15 December 2023
Revised: 15 January 2024
Accepted: 16 January 2024
Published: 18 January 2024

**Copyright:** © 2024 by the authors. Licensee MDPI, Basel, Switzerland. This article is an open access article distributed under the terms and conditions of the Creative Commons Attribution (CC BY) license (https://creativecommons.org/licenses/by/4.0/).

## 1. Introduction

Soil salinization is one of the major factors contributing to the loss of agricultural productivity [1]. Every year, around 10 million hectares of world agricultural land are degraded by salinity [2]. In Europe, salt-affected soils cover an area of about 31 million ha, particularly concentrated in the Mediterranean Basin [3–5]. Salinization due to seawater intrusion, or due to irrigation with poor quality water, is cause of serious concern in rural areas of Cyprus, Greece, Italy, Portugal, and Spain [4,6]. In the European Union alone, 1 million hectares are affected by soil salinity [7]. It was estimated that 25% of irrigated Mediterranean cropland is prone to moderate or high salinization [8], and by 2050, 50% of the world's arable land will be subjected to salinity [9]. High concentrations of salts have detrimental effects on crop development, especially in the phases of seed germination and seedling growth [10–12].

Among cereals, wheat represents the staple crop for 40% of the world's population, with a current annual production of 790.6 million tons [13], particularly for people living in Europe, North America, the Western and Northern parts of Asia, and the Americas [14]. Although durum wheat (*Triticum turgidum* L. subsp. *durum* (Desf.) Husn.) accounts for a small proportion of the global wheat production (around 7%), it has ever been a crucial commodity for countries surrounding the Mediterranean Sea [15]. Durum wheat was used as the model plant because of its importance as a crop in arid and semiarid areas of the Mediterranean Basin [16], and because is one of the most salt-sensitive cereal crops, especially during seed germination [17–21]. A high salinity level can result in 50% grain yield reduction [22]. Salt tolerance during germination and seedling establishment is fundamental for the future development and growth of crops, particularly in arid and semi-arid areas [20,23]. For that reason, studies aiming to find a sustainable way to guarantee a satisfying yield in moderately saline soils are not only warranted but also strongly sought after [24,25].

World population growth and climate change are severely threatening the global food security and environmental sustainability [26,27]. Therefore, the ambitious target of the primary sector for the future (and present) is to produce more food while impacting less on natural resources. In agriculture, efforts are being made in different directions including the reduction of chemical fertilizers' use [28–30]. In this context, biostimulants are receiving a lot of attention from both researchers and farmers as natural substances able to enhance crop performances [31–33]. The use of biostimulants to alleviate the effects of abiotic stressors such as salinity is a promising frontier for crop management.

Biostimulants, especially seaweed extracts (SWE), showed good results in promoting seed germination and seedlings vigour of different plant species under both optimal and saline conditions [34–44]. The positive effect is probably due to the high content of phytohormones, which tend to play a major role in seed germination [45]. However, few studies have been conducted until now on the effect of SWE on seed germination of durum wheat under saline conditions. Moreover, indications on the optimal dose of SWE-based biostimulants for seed treatments are scarce in the scientific literature. Hydropriming proves to be the method most used to treat seeds in experimental trials. However, this technique, consisting of soaking seed for a period in a solution containing a certain amount of SWE [46], does not represent a viable method for large-scale applications in seed industry because primed seeds are extremely difficult to manage [47]. Therefore, methods of delivering biostimulants as seed coatings, possibly together with other active ingredients, need to be explored to speed the technology transfer [48].

Among algae species from which SWE are obtained, *Codium fragile* (Suringar) Hariot, a green seaweed that lives in some salt marshes located in Tunisia, is one of the most invasive of the Mediterranean coasts [49,50]. Some studies showed promising results derived from the application of *Codium* spp. extract to enhance seed germination and seedlings performance of different crop species [46,51–53].

The objective of this study was to assess the effect of different concentrations of *Codium fragile* extract ($0\%_{w/v}$, $10\%_{w/v}$, $20\%_{w/v}$, and $30\%_{w/v}$) delivered as a seed coating on germination and early development of durum wheat seeds, under optimal (0 mM NaCl) and rising saline conditions (100 and 200 mM NaCl). In particular, the following hypotheses were tested: (i) seed treatment with SWE improves germination features and seedling vigour as compared to control and, above all, under salt stress conditions; (ii) the effect is proportional to the applied dose.

To verify these hypotheses, an experiment under controlled environment conditions was set up, randomizing the different treatments (salinity levels and SWE application doses) within the germination cabinet and evaluating germination and seedling traits at the completion of the trial.

## 2. Materials and Methods

### 2.1. Algal Collection and Extract Preparation

*Codium fragile* was collected in the salt marshes of South-East Tunisia. Samples were washed several times, first with fresh water and then with distilled water, air-dried, cut into pieces, and finally powdered with a mixed grinder. Ten grams of the obtained powder were suspended with 1 L of distilled water under continuous stirring at 300 rpm with a heating magnetic stirrer for 2 h at 50 °C [54]. Subsequently, the sample was put into an ultrasonic extractor for 45 min, then filtered using Whatman No. 1 filter paper to obtain 100% algal extract and stored at 4 °C for further analysis and treatments [55].

### 2.2. Experimental Design

Three different artificial growth conditions were used during the experiment:

- OPTIMAL: 0 mM of NaCl, control treatment;
- MID-SALINE: 100 mM of NaCl, a salinity level at which wheat begins to experience distress [56];
- HIGH-SALINE: 200 mM of NaCl, a salinity level considered an extreme condition [57].

Durum wheat seeds (cultivar Monastir) were selected for apparent uniformity in size and shape and treated with a solution containing the following ingredients:

- A liquid fungicide, usually used for cereals coating in Europe, at a dose of 200 g for 100 kg of seed, containing 2.34% Fludioxonil, 2.34% Difenoconazole, and 0.93% Tebuconazole;
- *Codium fragile* extract at 10 g dry matter ($g_{dm}$) $L^{-1}$;
- Distilled water.

Monastir was chosen because it is a modern durum wheat cultivar extensively cultivated around the Mediterranean basin [58] and it is reported to be a cultivar with medium susceptibility to salinity stress [24].

Four seed dressing solutions were prepared, changing just the amount of *Codium fragile* extract as follows: control ($0\%_{w/v}$), 100 g of *Codium fragile* extract (at 10 $g_{dm}$ $L^{-1}$ algae concentration) per 1 L of dressing solution ($10\%_{w/v}$), 200 g of *Codium fragile* extract (at 10 $g_{dm}$ $L^{-1}$ algae concentration) per 1 L of dressing solution ($20\%_{w/v}$), and 300 g of *Codium fragile* extract (at 10 $g_{dm}$ $L^{-1}$ algae concentration) per 1 L of dressing solution ($30\%_{w/v}$). These four treatments were applied for each growth condition and respecting the application rate of 1 kg solution per 100 kg seed to be industrially viable. Seed treatment was applied by mixing 1 g of dressing solution with 100 g of seeds and shaking them in a weighing bottle. The choice of these percentages was based on reference values found in the literature [51] and with the aim of discovering a possible dose–response effect and to find the best application dose. Four replicates (each consisting of 100 seeds) were used for each treatment. The treated seeds were placed into a Petri dish (120 mm diameter), with three layers of Whatman Grade 1 filter paper, soaked in the different saline solutions, and left for 4 days in a seed germination cabinet, in dark conditions, at a constant temperature of 20 ± 1 °C [59]. The Petri dishes were randomly allocated within the cabinet. After 4 days, every replicate was transferred into another Petri dish with three new layers of filter paper soaked in the same saline solutions in order to maintain a constant salinity level, and then put back in the germination cabinet at the same temperature for another 4 days, thus obtaining a total germination time of 8 days.

### 2.3. Measured Traits

Germination was determined based on radicle emergence. The germinated seeds of each replicate were counted every day to calculate the following:

- Final germination percentage: (seeds germinated/total seeds) × 100, after 8 days;
- T50: time to achieve 50% of final/maximum germination [60], calculated according to the following formula [61,62]:

$$T50 = t_i + \frac{\left(\frac{N}{2} - n_i\right)\left(t_j - t_i\right)}{\left(n_j - n_i\right)} \quad (1)$$

where N is the final number of germinated seeds and $n_j$ and $n_i$ are the cumulative number of seeds germinated by adjacent counts at times $t_j$ and $t_i$, respectively, when $n_i < N/2 < n_j$;

- Germination time course.

At the end of the experiment (after 8 days), all germinated seeds were analysed, and the following parameters were measured for each replicate:

- Number of roots, manually counted;
- Length of the main root (cm), measured with a ruler;
- Coleoptile length (cm), measured with a ruler;
- Root biomass (dry weight, g), for which all the roots of each replicate were separated from the germinated seeds and dried off for 2 days at 60 °C;
- Coleoptile biomass (dry weight, g), for which all the coleoptiles of each replicate were separated from the germinated seeds and dried off for 2 days at 60 °C.

### 2.4. Statistical Analysis

Germination data were analysed as time-to-event data, thus allowing a better interpretation of germination experiments [63]. Specifically, relevant parameter estimates (e.g., T50 and final germination) and corresponding standard errors were extracted according to the event-time approach by fitting a three-parameter log-logistic model to data from each germination curve separately [64]. The model equation used in this study was as follows:

$$F(T) = \frac{d}{1 + \exp[b\{\log(T) - \log(T50)\}]} \tag{2}$$

where d denotes the proportion of seeds that germinated during the experiment (upper limit or final maximum germination) and T50 is the time (number of days) to reach 50% of final germination. The parameter b is proportional to the slope of F at time T equal to the parameter T50.

Quantitative data obtained from germinated seeds were subjected to analysis of variance (ANOVA) to test the effect of salinity, SWE, and their interaction. Significantly different means were separated at the 95% probability level using Fisher's protected least significant difference.

Statistical analysis was carried out using the open-source environment R [65] with the add-on packages 'drcte', for event-time models [66].

## 3. Results

### 3.1. Germination Percentage and Median Germination Time (T50)

As shown in Table 1 and Figure 1, the priming with SWE strongly influenced the germination performance of durum wheat seeds, especially at high salinity. The application of SWE in the control treatment allowed seeds to significantly enhance their final germination. Specifically, final germination increased by 3% at 10% SWE concentration, by 4% at 20% SWE concentration, and by 5% at 30% SWE concentration. At the medium salinity level (100 mM NaCl), the seeds primed at 20% and 30% SWE concentration reached a final germination (96%) significantly higher than that of the control and the treatment with the lowest SWE concentration (90 and 93%, respectively).

The effect of the seed treatment dramatically increased at the highest salinity level (200 mM NaCl). Indeed, under that growth condition, while all the treated seeds showed a final germination percentage ranging from 82% (10% SWE) to 88% (30% SWE), untreated seeds stopped around 61% germination rate.

As for median germination time, expressed as T50, the SWE coating did not significantly affect the seeds' response in optimal conditions. Conversely, at medium and high salinity levels, the median germination time of the seeds treated with SWE was significantly reduced as compared to that of the control treatment, except for 20% SWE at 200 mM NaCl.

**Table 1.** Final germination percentage and median germination time (T50) for each combination of growth condition and SWE treatment ± standard errors.

| Growth Condition | SWE | Final Germination (%) | T50 (d) |
|---|---|---|---|
| Salinity 0 mM NaCl | 0% | 89.34 ± 1.28 b | 1.65 ± 0.02 a |
| | 10% | 92.70 ± 1.45 a | 1.62 ± 0.05 a |
| | 20% | 93.13 ± 0.96 a | 1.71 ± 0.05 a |
| | 30% | 94.82 ± 0.79 a | 1.73 ± 0.03 a |
| Salinity 100 mM NaCl | 0% | 90.10 ± 0.55 c | 2.15 ± 0.01 a |
| | 10% | 92.79 ± 0.77 b | 2.03 ± 0.04 b |
| | 20% | 96.05 ± 0.60 a | 2.01 ± 0.03 b |
| | 30% | 96.05 ± 0.60 a | 2.04 ± 0.01 b |
| Salinity 200 mM NaCl | 0% | 60.67 ± 1.08 c | 3.43 ± 0.11 a |
| | 10% | 82.21 ± 1.22 b | 2.95 ± 0.07 bc |
| | 20% | 85.65 ± 0.51 ab | 3.13 ± 0.07 ab |
| | 30% | 87.70 ± 1.54 a | 2.77 ± 0.12 c |

Within each growth condition, means sharing letters are not significantly different at $p < 0.05$.

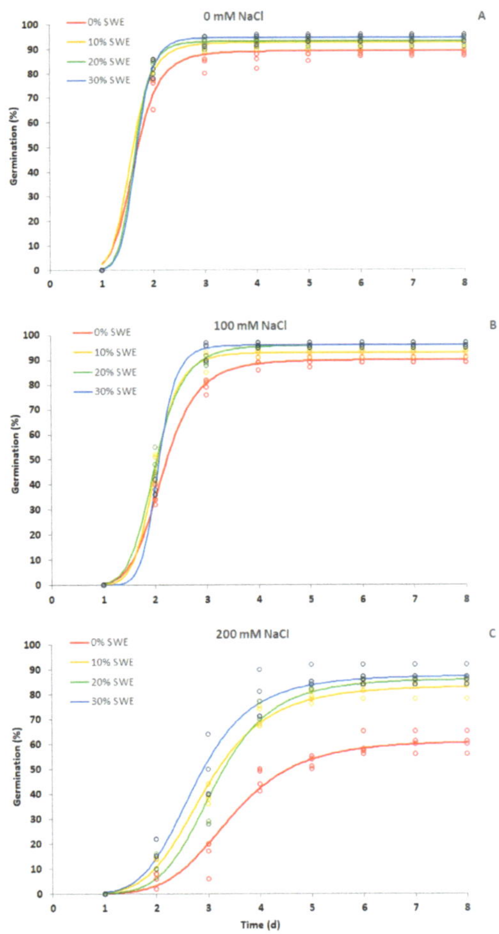


**Figure 1.** Germination curves generated with the three-parameter log-logistic model of seeds treated with different seaweed extracts (SWE) doses under the following conditions: (**A**) optimal conditions (0 mM NaCl); (**B**) mid-saline conditions (100 mM NaCl); (**C**) high-saline conditions (200 mM NaCl).

### 3.2. Number of Roots

As shown in Table 2, the number of seminal roots was significantly influenced by both SWE and salinity. In more detail, salinity markedly reduced the number of roots, which dropped from 5.6 at optimal conditions to 5.0 and 2.67 at medium and high salt concentrations, respectively (Figure 2A); meanwhile, the application of *Codium fragile* extracts significantly increased the number of roots as compared to the control (up to 12%), irrespective of dose (Figure 2B).

**Table 2.** Results of the ANOVA for quantitative traits of germinated seeds.

| | Number of Roots | Length of the Main Root | Root Biomass | Coleoptile Length | Coleoptile Biomass |
|---|---|---|---|---|---|
| SWE | *** | *** | *** | *** | *** |
| Salinity | *** | *** | *** | *** | *** |
| SWE × Salinity | n.s. | n.s. | n.s. | *** | *** |

ANOVA signif. codes: '***' < 0.001; n.s.: not significant.

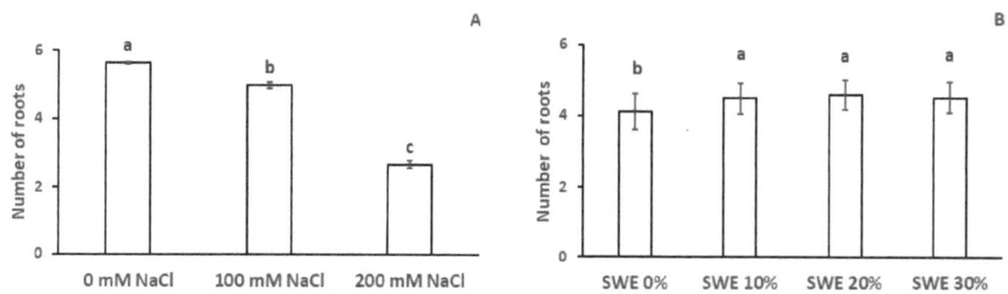


**Figure 2.** Number of roots as affected by (**A**) salinity and (**B**) seaweed extracts (SWE). The vertical bars represent the standard error of the mean. Letters above the histograms correspond to the ranking of the Fisher's protected test at $p < 0.05$.

### 3.3. Length of the Main Root

As with the number of seminal roots, the length of the main root was also significantly affected by salinity and SWE application (Table 2). Salinity reduced the root length by 37% at 100 mM NaCl and by 67% at 200 mM NaCl (Figure 3A). The presence of *Codium fragile* extract in the seed coating markedly increased the root length by 27% at the lowest SWE dose (from 4.8 cm to 6.1 cm) and by 33% at the highest one (from 4.8 cm to 6.4 cm), with a significantly different effect between these two SWE concentrations (Figure 3B).

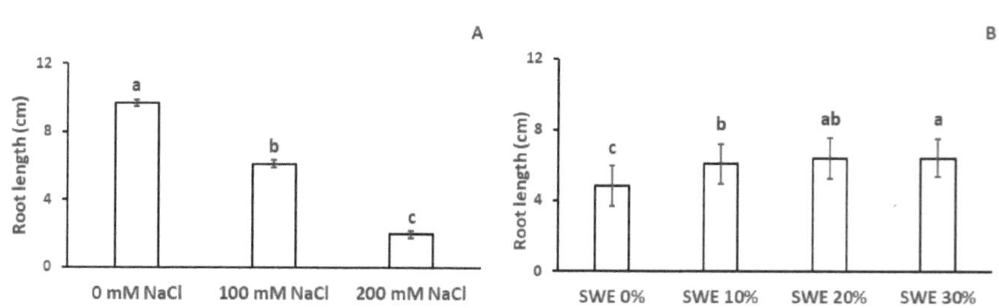


**Figure 3.** Length of the main root as affected by (**A**) salinity and (**B**) seaweed extracts (SWE). The vertical bars represent the standard error of the mean. Letters above the histograms correspond to the ranking of the Fisher's protected test at $p < 0.05$.

### 3.4. Root Biomass

As reported in Table 2, root biomass was significantly influenced by both treatments, like the previously reported traits. Specifically, the increasing salt concentration produced a 26% (from 0.61 g to 0.45 g) and 66% (from 0.61 g to 0.21 g) reduction of root dry weight (Figure 4A). The application of SWE increased root biomass by 35% and 42% at the lowest and highest concentrations, respectively (Figure 4B). No significant difference was detected applying SWE at 20% or 30% concentration.

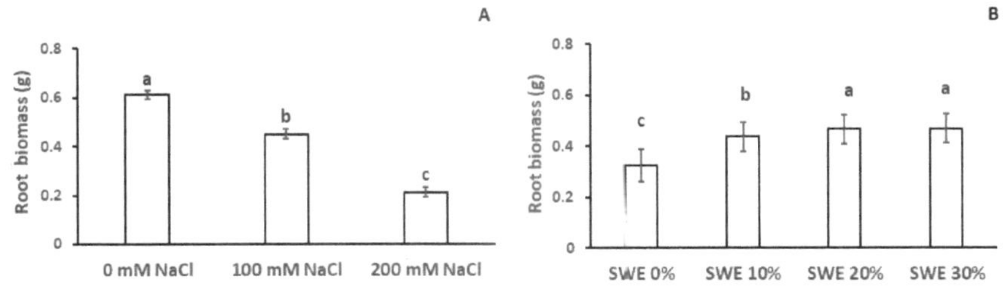


**Figure 4.** Root biomass, as affected by: (**A**) salinity and (**B**) seaweed extracts (SWE). The vertical bars represent the standard error of the mean. Letters above the histograms correspond to the ranking of the Fisher's protected test at $p < 0.05$.

### 3.5. Coleoptile Length

The interaction between treatments was found to be significant for this trait (Table 2). As shown in Figure 5, the application of *Codium fragile* extract was greatly effective for the control treatment (+60% in coleoptile length) and at medium salt stress conditions (coleoptile length doubled), while no remarkable benefit was detected under high salt stress conditions. As previously observed for the other traits, the increase in salt stress severely hindered the development of wheat coleoptile.

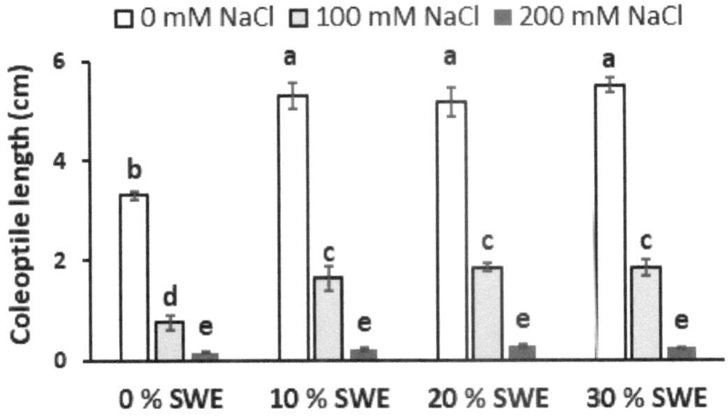


**Figure 5.** Length of the coleoptile as affected by salinity × SWE interaction. The vertical bars represent the standard error of the mean. Letters above the histograms correspond to the ranking of the Fisher's protected test at $p < 0.05$.

### 3.6. Coleoptile Biomass

The interaction between treatments significantly affected coleoptile total biomass (Table 2). The increase in salt concentration strongly reduced the biomass, moving from an average of 0.4 g for the optimal conditions to 0.23 g for 100 mM NaCl and 0.11 g for 200 mM NaCl (Figure 6). The application of SWE increased the performance in all growth conditions. In optimal conditions, all the biostimulant-coated seeds had a higher biomass compared to the control, with a 52% increase. At the medium salinity level, the application of 20% and 30% SWE enhanced the biomass weight compared with the 10% SWE. However, all the SWE-treated seeds performed significantly better than the control. At the high salinity level (200 mM NaCl), all the seed coating treatments enhanced the coleoptile's biomass by 79% as compared to the control.

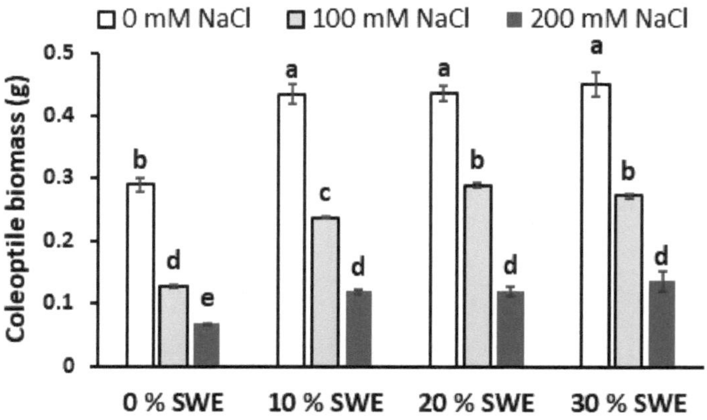


**Figure 6.** Coleoptile biomass as affected by salinity × SWE interaction. The vertical bars represent the standard error of the mean. Letters above the histograms correspond to the ranking of the Fisher's protected test at $p < 0.05$.

## 4. Discussion

According to our first hypothesis, the *Codium fragile* extract significantly enhanced seed germination and seedlings development in each growing condition, especially under salt stress. Conversely, the second hypothesis was not verified since, in most of the measured traits, there is not a clear dose–response effect for the application of *Codium fragile* extract. This means that the SWE is effective also at lower doses.

Furthermore, the results showed that the seed coating can be a viable way to deliver SWE and so it could be applied at an industrial scale too.

According to our findings, seed germination in durum wheat appeared to be stable until medium salt stress (100 mM NaCl), while it was reduced by 30% under high salt stress conditions (200 mM NaCl). This result agrees with other studies found in the literature [67–70]. Even though final germination at 100 mM NaCl did not differ from that of optimal conditions, seeds evidenced a longer median germination time (T50) and a lower coleoptile length. That negative effect will certainly lead to a plant being more susceptible to biotic and abiotic stresses, thus reducing yield and grain quality [22,71–73]. A fast and uniform seed germination, combined with a high seedling vigour, are the fundamental traits used to assess salt tolerance in crops [39,73,74].

Generally, the application of SWE significantly enhanced all the measured traits, except for the coleoptile length under high salt stress. Even though literature on this specific research topic is limited, the positive effect of SWE, and especially of *Codium* spp. extract, in promoting wheat germination (also in controlled salinity conditions) are reported in other studies [51,52]. *Codium* spp., as other macroalgae, appear to be rich in phytohormones such as cytokinin, gibberellins, and auxins [75–77]. These bioactive compounds are certainly involved in the germination process, and their presence can stimulate the metabolic process in wheat seeds [78–80]. Auxins seem to play a major role in promoting seed germination and seedlings development under salt stress conditions. Even though these plant growth regulators are fundamental for the early development of the plants, the stored auxins are not sufficient to guarantee a good stem and roots formation, especially under salt stress [79]. Indeed, plants rapidly decrease the production of auxins during salt stress, triggering down the auxin biosynthesis pathways [81]. Thus, the application of endogenous auxins could fulfil the plants' requirements for wheat germination and seedlings development [79,82]. In addition to that, these phytohormones appear to promote tissue regeneration and limit the uptake and assimilation of toxic mineral elements [79,83]. The cytokinins are also involved in the salt stress regulation, especially in the early stage of the seedling's development [84]. It has been reported that salinity significantly reduces the cytokinins' export from the root to the shoot [85]; therefore, an exogenous application of cytokinins could lead to a better development of the wheat seedlings under salt stress conditions [80]. Additionally, the SWE tested contains amino acids such as proline and glycine betaine [51,52,86]. These two amino acids are well known to enhance crop resistance to drought and salt stress [87–89]. Glycine betaine seems to reduce toxic ions' absorption, protecting ion channels and favourably influencing the properties of the cell membrane. This mechanism greatly favours ion homeostasis under salinity stress [90]. As for proline, this amino acid has proven to limit the negative impact of the salt stress thanks to its role as an osmoprotectant, membrane stabilizer, and scavenger of reactive oxygen species [91,92]. The strong effect of the *Codium fragile* extract is probably due to the combination of these different active compounds that, through different modes of action, confer to durum wheat seedlings a greater tolerance to salt stress. Other studies reported that priming seeds with SWE leads to an improved antioxidant enzyme activity [42]. The seed coating with SWE has also led to a general reduction in germination time, especially under salt stress conditions. Again, this effect could be explained by the presence of phytohormones in the *Codium fragile* extract, as these compounds are able to promote the creation of hydrolytic enzymes and are inhibitors of abscisic acid [93].

As expected, the seedlings' vigour was negatively affected by salinity. In particular, the coleoptile and the root system showed a strong reduction in terms of both length and total dry biomass, even under medium salt stress. This result is consistent with other studies conducted on winter cereals, reporting a different reduction of root length (from 20% to 97%) and dry biomass (from 60% to 70%), depending on the salinity level and the different growing conditions. A similar effect has been shown in shoot length, with reduction rates ranging from 36% to 92% [69,94–96]. However, the application of *Codium fragile* extract, especially at the highest doses, was able to alleviate the salt stress, as evidenced by the improved length and biomass of the coleoptile in SWE-treated seeds. Compared to the control treatment, the application of SWE markedly promoted the development of the root system. A similar positive effect on shoots and roots was observed also in other crops such as tomato, okra, etc. [36,42,97]. It is well known that biostimulants, especially seaweed extracts, can increase crop development, even in stress conditions [33]. This effect is likely due to the presence of phytohormones, as well as the content of macro- and micronutrients. In particular, in *Codium* species, the effect can be due to the presence of elements such as calcium and zinc [86,98]. The exogenous application of these two elements resulted in a general enhancement of the germination percentage and seedling vigour [99–101].

## 5. Conclusions and Future Prospects

Seed treatment with biostimulants is still a relatively under-explored research topic, but it appears extremely interesting. Our study clearly demonstrated that seed coating with a *Codium fragile* extract was effective in enhancing seed germination and early vigour of the durum wheat cultivar Monastir, both under optimal (0 mM NaCl) and salt stress conditions (100 and 200 mM NaCl). A prompt and uniform seedling emergence, combined with a better rooting, can make crop species more tolerant to biotic and abiotic stresses. Moreover, the application of seaweed extracts could be a promising way to decrease the use of synthetic fertilizers in agriculture, without negatively affecting yield potential. The future investigation into all these aspects can greatly improve our understanding of sustainable management of durum wheat. Additional studies will be necessary to gain a deeper understanding of the mechanism of action of this biostimulant. Moreover, since we reported results from a germination trial only, several field experiments will be needed to evaluate the effects on plants during the entire growing cycle, under different soils and climatic conditions. In that way, we will also be able to explore other *Codium fragile* application methods, such as foliar spraying or soil treatments.

**Author Contributions:** Conceptualization, A.R. and F.R.; methodology, R.R. and G.D.M.; formal analysis, A.R., F.R., N.M. and R.R.; investigation, A.R. and G.D.M.; resources, F.R.; data curation, A.R. and R.R.; writing—original draft preparation, A.R. and R.R.; writing—review and editing, F.R. and G.D.M.; visualization, F.R.; supervision, R.R. and F.R. All authors have read and agreed to the published version of the manuscript.

**Funding:** This research was conducted as activity of the PhD course in "Plant and Animal Production Sciences (XXXVIII cycle)" of the University of Tuscia and was co-funded by PNRR, Missione 4, componente 2 "Dalla Ricerca all'Impresa"—Investimento 3.3 and ISLA srl.

**Data Availability Statement:** The data presented in this study are available on request from the corresponding author.

**Acknowledgments:** The authors wish to acknowledge Fakir Mathlouthi (FBSM Nanobiology, Friedrich-Ebert-Anlage 36, 60325 Frankfurt, Germany) for technical support.

**Conflicts of Interest:** The authors declare no conflicts of interest.

## Abbreviations

| | |
|---|---|
| SWE | seaweed extract |
| 0% SWE | seed dressing solution containing 0 g of *Codium fragile* extract |
| 10% SWE | seed dressing solution containing 100 g of *Codium fragile* extract |
| 20% SWE | seed dressing solution containing 200 g of *Codium fragile* extract |
| 30% SWE | seed dressing solution containing 300 g of *Codium fragile* extract |
| T50 | median germination time |

## References


1. Minhas, P.S.; Ramos, T.B.; Ben-Gal, A.; Pereira, L.S. Coping with Salinity in Irrigated Agriculture: Crop Evapotranspiration and Water Management Issues. *Agric. Water Manag.* **2020**, *227*, 105832. [CrossRef]
2. Pimentel, D.; Berger, B.; Filiberto, D.; Newton, M.; Wolfe, B.; Karabinakis, E.; Clark, S.; Poon, E.; Abbett, E.; Nandagopal, S. Water Resources: Agricultural and Environmental Issues. *Bioscience* **2004**, *54*, 909–918. [CrossRef]
3. Jones, A.; Panagos, P.; Erhard, M.; Tóth, G.; Barcelo, S.; Bouraoui, F.; Bosco, C.; Dewitte, O.; Gardi, C.; Hervás, J.; et al. *The State of Soil in Europe: A Contribution of the JRC to the European Environment Agency's Environment State and Outlook Report—SOER 2010*; Publications Office: Luxembourg, 2012; ISBN 9789279228063.
4. Daliakopoulos, I.N.; Tsanis, I.K.; Koutroulis, A.; Kourgialas, N.N.; Varouchakis, A.E.; Karatzas, G.P.; Ritsema, C.J. The Threat of Soil Salinity: A European Scale Review. *Sci. Total Environ.* **2016**, *573*, 727–739. [CrossRef]
5. Corwin, D.L. Climate Change Impacts on Soil Salinity in Agricultural Areas. *Eur. J. Soil. Sci.* **2021**, *72*, 842–862. [CrossRef]
6. Machado, R.M.A.; Serralheiro, R.P. Soil Salinity: Effect on Vegetable Crop Growth. Management Practices to Prevent and Mitigate Soil Salinization. *Horticulturae* **2017**, *3*, 30. [CrossRef]
7. Stolte, J.; Tesfai, M.; Øygarden, L.; Kværnø, S.; Keizer, J.; Verheijen, F.; Panagos, P.; Ballabio, C.; Hessel, R. *Soil Threats in Europe: Status, Methods, Drivers and Effects on Ecosystem Services. A Review Report, Deliverable 2.1 of the RECARE Project*; European Commission DG Joint Research Centre: Luxembourg, 2015.
8. Ferreira, C.S.S.; Seifollahi-Aghmiuni, S.; Destouni, G.; Ghajarnia, N.; Kalantari, Z. Soil Degradation in the European Mediterranean Region: Processes, Status and Consequences. *Sci. Total Environ.* **2022**, *805*, 20. [CrossRef]
9. Bartels, D.; Sunkar, R. Drought and Salt Tolerance in Plants. *CRC Crit. Rev. Plant Sci.* **2005**, *24*, 23–58. [CrossRef]
10. Bernstein, L. Osmotic Adjustment of Plants to Saline Media. *I. Steady State. Am. J. Bot.* **1961**, *48*, 909–918. [CrossRef]
11. Zhao, S.; Zhang, Q.; Liu, M.; Zhou, H.; Ma, C.; Wang, P. Regulation of Plant Responses to Salt Stress. *Int. J. Mol. Sci.* **2021**, *22*, 4609. [CrossRef] [PubMed]
12. Cramer, W.; Guiot, J.; Fader, M.; Garrabou, J.; Gattuso, J.P.; Iglesias, A.; Lange, M.A.; Lionello, P.; Llasat, M.C.; Paz, S.; et al. Climate Change and Interconnected Risks to Sustainable Development in the Mediterranean. *Nat. Clim. Chang.* **2018**, *8*, 972–980. [CrossRef]
13. USDA World Agricultural Production. Available online: https://www.fas.usda.gov/data/world-agricultural-production (accessed on 13 September 2023).
14. FAO World Food Situation. Available online: https://www.fao.org/worldfoodsituation/csdb/en/ (accessed on 13 September 2023).
15. Xynias, I.N.; Mylonas, I.; Korpetis, E.G.; Ninou, E.; Tsaballa, A.; Avdikos, I.D.; Mavromatis, A.G. Durum Wheat Breeding in the Mediterranean Region: Current Status and Future Prospects. *Agronomy* **2020**, *10*, 432. [CrossRef]
16. Giambalvo, D.; Amato, G.; Badagliacca, G.; Ingraffia, R.; Di Miceli, G.; Frenda, A.S.; Plaia, A.; Venezia, G.; Ruisi, P. Switching from Conventional Tillage to No-Tillage: Soil N Availability, N Uptake, 15N Fertilizer Recovery, and Grain Yield of Durum Wheat. *Field Crops Res.* **2018**, *218*, 171–181. [CrossRef]

17. Shafi, M.; Javed Khan, M.; Bakht, J.; Khan, M.A. Response of Wheat Genotypes to Salinity under Field Environment. *Pak. J. Bot.* **2013**, *45*, 787–794.
18. Munns, R.; James, R.A. Screening Methods for Salinity Tolerance: A Case Study with Tetraploid Wheat. *Plant Soil.* **2003**, *253*, 201–218. [CrossRef]
19. Foolad, M.R. Recent Advances in Genetics of Salt Tolerance in Tomato. *Plant Cell Tissue Organ Cult.* **2004**, *76*, 101–119. [CrossRef]
20. Alaoui, M.M.; El Jourmi, L.; Ouarzane, A.; Lazar, S.; El Antri, S.; Zahouily, M.; Hmyene, A. Effet Du Stress Salin Sur La Germination et La Croissance de Six Variétés Marocaines de Blé (Effect of Salt Stress on Germination and Growth of Six Moroccan Wheat Varieties). *J. Mater. Environ. Sci.* **2013**, *4*, 997–1004.
21. Hmissi, M.; Chaieb, M.; Krouma, A. Differences in the Physiological Indicators of Seed Germination and Seedling Establishment of Durum Wheat (*Triticum durum* Desf.) Cultivars Subjected to Salinity Stress. *Agronomy* **2023**, *13*, 1718. [CrossRef]
22. Borrelli, G.; Ficco, D.; Giuzio, L.; Pompa, M.; Cattivelli, L.; Flagella, Z. Durum Wheat Salt Tolerance in Relation to Physiological, Yield and Quality Characters. *Cereal Res. Commun.* **2011**, *39*, 525–534. [CrossRef]
23. Mbarki, S.; Skalicky, M.; Vachova, P.; Hajihashemi, S.; Jouini, L.; Zivcak, M.; Tlustos, P.; Brestic, M.; Hejnak, V.; Khelil, A.Z. Comparing Salt Tolerance at Seedling and Germination Stages in Local Populations of *Medicago ciliaris* L. to *Medicago intertexta* L. and *Medicago scutellata* L. *Plants* **2020**, *9*, 526. [CrossRef]
24. Jacobsen, S.E.; Jensen, C.R.; Liu, F. Improving Crop Production in the Arid Mediterranean Climate. *Field Crops Res.* **2012**, *128*, 34–47. [CrossRef]
25. Tomaz, A.; Palma, P.; Alvarenga, P.; Gonçalves, M.C. Soil Salinity Risk in a Climate Change Scenario and Its Effect on Crop Yield. In *Climate Change and Soil Interactions*; Elsevier: Amsterdam, The Netherlands, 2020; pp. 351–396. [CrossRef]
26. Ronga, D.; Biazzi, E.; Parati, K.; Carminati, D.; Carminati, E.; Tava, A. Microalgal Biostimulants and Biofertilisers in Crop Productions. *Agronomy* **2019**, *9*, 192. [CrossRef]
27. Arora, N.K. Impact of Climate Change on Agriculture Production and Its Sustainable Solutions. *Environ. Sustain.* **2019**, *2*, 95–96. [CrossRef]
28. Mathlouthi, F.; Ruggeri, R.; Rossini, F. Alternative Solution to Synthetic Fertilizers for the Starter Fertilization of Bread Wheat under Mediterranean Climatic Conditions. *Agronomy* **2022**, *12*, 511. [CrossRef]
29. Mathlouthi, F.; Ruggeri, R.; Rossini, A.; Rossini, F. A New Fertilization Approach for Bread Wheat in the Mediterranean Environment: Effects on Yield and Grain Protein Content. *Agronomy* **2022**, *12*, 2152. [CrossRef]
30. Di Miceli, G.; Vultaggio, L.; Sabatino, L.; De Pasquale, C.; La Bella, S.; Consentino, B.B. Synergistic Effect of a Plant-Derived Protein Hydrolysate and Arbuscular Mycorrhizal Fungi on Eggplant Grown in Open Fields: A Two-Year Study. *Horticulturae* **2023**, *9*, 592. [CrossRef]
31. Bulgari, R.; Cocetta, G.; Trivellini, A.; Vernieri, P.; Ferrante, A. Biostimulants and Crop Responses: A Review. *Biol. Agric. Hortic.* **2015**, *31*, 1–17. [CrossRef]
32. Van Oosten, M.J.; Pepe, O.; De Pascale, S.; Silletti, S.; Maggio, A. The Role of Biostimulants and Bioeffectors as Alleviators of Abiotic Stress in Crop Plants. *Chem. Biol. Technol. Agric.* **2017**, *4*, 5. [CrossRef]
33. Rouphael, Y.; Colla, G. Editorial: Biostimulants in Agriculture. *Front. Plant Sci.* **2020**, *11*, 40. [CrossRef] [PubMed]
34. Altındal, D. Effects of Seaweed Extract (SE) Applications on Seed Germination Characteristics of Wheat in Salinity Conditions. *Int. J. Agric. For. Life Sci.* **2019**, *3*, 115–120.
35. Latique, S.; Aymen, E.M.; Halima, C.; Chérif, H.; Mimoun, E.K. Alleviation of Salt Stress in Durum Wheat (*Triticum durum* L.) Seedlings Through the Application of Liquid Seaweed Extracts of Fucus spiralis. *Commun. Soil. Sci. Plant Anal.* **2017**, *48*, 2582–2593. [CrossRef]
36. Hernández-Herrera, R.M.; Santacruz-Ruvalcaba, F.; Ruiz-López, M.A.; Norrie, J.; Hernández-Carmona, G. Effect of Liquid Seaweed Extracts on Growth of Tomato Seedlings (*Solanum lycopersicum* L.). *J. Appl. Phycol.* **2014**, *26*, 619–628. [CrossRef]
37. Makhaye, G.; Aremu, A.O.; Gerrano, A.S.; Tesfay, S.; Du Plooy, C.P.; Amoo, S.O. Biopriming with Seaweed Extract and Microbial-Based Commercial Biostimulants Influences Seed Germination of Five Abelmoschus esculentus Genotypes. *Plants* **2021**, *10*, 1327. [CrossRef]
38. Eugenia, M.; Carvalho, A.; Roberto, P.; Dionisia, A.; Novembre, L.C.; Carvalho, M.E.A ; Castro, P.R.C.; Novembre, A.D.C.; Chamma, H.M.C.P. Seaweed Extract Improves the Vigor and Provides the Rapid Emergence of Dry Bean Seeds. *J. Agric. Environ. Sci.* **2013**, *13*, 1104–1107. [CrossRef]
39. Chanthini, K.M.P.; Senthil-Nathan, S.; Pavithra, G.S.; Malarvizhi, P.; Murugan, P.; Deva-Andrews, A.; Janaki, M.; Sivanesh, H.; Ramasubramanian, R.; Stanley-Raja, V.; et al. Aqueous Seaweed Extract Alleviates Salinity-Induced Toxicities in Rice Plants (*Oryza sativa* L.) by Modulating Their Physiology and Biochemistry. *Agriculture* **2022**, *12*, 2049. [CrossRef]
40. Gul, S.; Abideen, Z.; Adnan, M.Y.; Hanif, M.; Nawaz, M.F.; Shumaila; Keblawy, A.E. Seaweed-Derived Bio-Stimulant Improves Growth and Salt Tolerance of Radish Varieties under Saline Conditions. *Biocatal. Agric. Biotechnol.* **2023**, *52*, 102822. [CrossRef]
41. Supraja, K.V.; Behera, B.; Balasubramanian, P. Efficacy of Microalgal Extracts as Biostimulants through Seed Treatment and Foliar Spray for Tomato Cultivation. *Ind. Crops Prod.* **2020**, *151*, 112453. [CrossRef]
42. Radwan, A.M.; Ahmed, E.A.; Donia, A.M.; Mustafa, A.E.; Balah, M.A. Priming of *Citrullus lanatus* Var. colocynthoides Seeds in Seaweed Extract Improved Seed Germination, Plant Growth and Performance under Salinity Conditions. *Sci. Rep.* **2023**, *13*, 11884. [CrossRef]

43. Hussein, M.H.; Eltanahy, E.; Al Bakry, A.F.; Elsafty, N.; Elshamy, M.M. Seaweed Extracts as Prospective Plant Growth Bio-Stimulant and Salinity Stress Alleviator for Vigna sinensis and Zea mays. *J. Appl. Phycol.* **2021**, *33*, 1273–1291. [CrossRef]
44. Nezamdoost, D.; Ghahremani, Z.; Akbari, M.B.; Barzegar, T.; Ranjbar, M.E. Can Seed Priming with Seaweed Extract Neutralize the Effects of Salinity on New Red Fire Leafy Lettuce Characteristics? *Gesunde Pflanz.* **2023**, *75*, 955–969. [CrossRef]
45. Battacharyya, D.; Babgohari, M.Z.; Rathor, P.; Prithiviraj, B. Seaweed Extracts as Biostimulants in Horticulture. *Sci. Hortic.* **2015**, *196*, 39–48. [CrossRef]
46. Makhaye, G.; Mofokeng, M.M.; Tesfay, S.; Aremu, A.O.; Van Staden, J.; Amoo, S.O. Chapter 5—Influence of Plant Biostimulant Application on Seed Germination. In *Biostimulants for Crops from Seed Germination to Plant Development A Practical Approach*; Academic Press: Cambridge, MA, USA, 2021; pp. 109–135. [CrossRef]
47. Taylor, A.G.; Allen, P.S.; Bennett, M.A.; Bradford, K.J.; Burris, J.S.; Misra, M.K. Seed Enhancements. *Seed Sci. Res.* **1998**, *8*, 245–256. [CrossRef]
48. Amirkhani, M.; Mayton, H.S.; Netravali, A.N.; Taylor, A.G. A Seed Coating Delivery System for Bio-Based Biostimulants to Enhance Plant Growth. *Sustainability* **2019**, *11*, 5304. [CrossRef]
49. Sghaier, Y.R.; Zakhama-Sraieb, R.; Mouelhi, S.; Vazquez, M.; Valle, C.; Ramos-Espla, A.A.; Astier, J.M.; Verlaque, M.; Charfi-Cheikhrouha, F. Review of Alien Marine Macrophytes in Tunisia. *Mediterr. Mar. Sci.* **2016**, *17*, 109–123. [CrossRef]
50. Cherif, W.; Ktari, L.; El Bour, M.; Boudabous, A.; Grignon-Dubois, M. Codium fragile Subsp. fragile (Suringar) Hariot in Tunisia: Morphological Data and Status of Knowledge. *Algae* **2016**, *31*, 129–136. [CrossRef]
51. El-Din, S.M.M. Utilization of Seaweed Extracts as Bio-Fertilizers to Stimulate the Growth of Wheat Seedlings. *Egypt. Soc. Exp. Biol.* **2015**, *11*, 31–39.
52. Nessim, A. Mitigation of Lead Stress in Triticum aestivum by Seed Priming in Aqueous Extracts of The Macroalgea Halimeda opuntia and Codium fragile. *Egypt. J. Bot.* **2018**, *58*, 263–274. [CrossRef]
53. Demir, N.; Dural, B.; Yildirim, K. Effect of Seaweed Suspensions on Seed Germination of Tomato, Pepper and Aubergine. *J. Biol. Sci.* **2006**, *6*, 1130–1133. [CrossRef]
54. Righini, H.; Francioso, O.; Di Foggia, M.; Prodi, A.; Quintana, A.M.; Roberti, R. Tomato Seed Biopriming with Water Extracts from Anabaena minutissima, Ecklonia maxima and Jania adhaerens as a New Agro-Ecological Option against Rhizoctonia solani. *Sci. Hortic.* **2021**, *281*, 109921. [CrossRef]
55. Anisimov, M.M.; Klykov, A.G. Metabolites of Terrestrial Plants and Marine Organisms as Potential Regulators of Growth of Agricultural Plants in the Russian Far East. *J. Agric. Sci.* **2014**, *6*, 88. [CrossRef]
56. Royo, A.; Abiò, D. Salt Tolerance in Durum Wheat Cultivars. *Span J. Agric. Res.* **2003**, *1*, 27–35. [CrossRef]
57. Cuong, D.M.; Kwon, S.J.; Van Nguyen, B.; Chun, S.W.; Kim, J.K.; Park, S.U. Effect of Salinity Stress on Phenylpropanoid Genes Expression and Related Gene Expression in Wheat Sprout. *Agronomy* **2020**, *10*, 390. [CrossRef]
58. Balducci, E.; Tini, F.; Beccari, G.; Ricci, G.; Ceron-Bustamante, M.; Orfei, M.; Guiducci, M.; Covarelli, L. A Two-Year Field Experiment for the Integrated Management of Bread and Durum Wheat Fungal Diseases and of Deoxynivalenol Accumulation in the Grain in Central Italy. *Agronomy* **2022**, *12*, 840. [CrossRef]
59. Ministero dell'Agricoltura e delle Foreste. Metodi Ufficiali di Analisi per le Sementi. *Gazz. Uff. Della Repubb. Ital.* 1993.
60. Soltani, E.; Ghaderi-Far, F.; Baskin, C.C.; Baskin, J.M. Problems with Using Mean Germination Time to Calculate Rate of Seed Germination. *Aust. J. Bot.* **2015**, *63*, 631–635. [CrossRef]
61. Scarici, E.; Ruggeri, R.; Provenzano, M.E.; Rossini, F. Germination and Performance of Seven Native Wildflowers in the Mediterranean Landscape Plantings. *Imp. Jpn. Army* **2018**, *13*, 163–171. [CrossRef]
62. Farooq, M.; Basra, S.M.A.; Ahmad, N.; Hafeez, K. Thermal Hardening: A New Seed Vigor Enhancement Tool in Rice. *J. Integr. Plant Biol.* **2005**, *47*, 187–193. [CrossRef]
63. Ritz, C.; Pipper, C.B.; Streibig, J.C. Analysis of Germination Data from Agricultural Experiments. *Eur. J. Agron.* **2013**, *45*, 1–6. [CrossRef]
64. Rossini, F.; Ruggeri, R.; Celli, T.; Rogai, F.M.; Kuzmanović, L.; Richardson, M.D. Cool-Season Grasses for Overseeding Sport Turfs: Germination and Performance under Limiting Environmental Conditions. *HortScience* **2019**, *54*, 555–563. [CrossRef]
65. R Core Team A Language and Environment for Statistical Computing 2021. Available online: https://www.R-project.org/ (accessed on 17 September 2023).
66. Onofri, A.; Mesgaran, M.B.; Ritz, C. A Unified Framework for the Analysis of Germination, Emergence, and Other Time-to-Event Data in Weed Science. *Weed Sci.* **2022**, *70*, 259–271. [CrossRef]
67. Ayed, S.; Rassaa, N.; Chamekh, Z.; Beji, S.; Karoui, F.; Bouzaien, T.; Mrabit, M.; Ben, Y.M. Effect of Salt Stress (Sodium Chloride) on Germination and Seedling Growth of Durum Wheat (*Triticum durum* Desf.) Genotypes. *Int. J. Biodivers. Conserv.* **2014**, *6*, 320–325. [CrossRef]
68. Hussain, S.; Khaliq, A.; Matloob, A.; Wahid, M.A.; Afzal, I. Germination and Growth Response of Three Wheat Cultivars to NaCl Salinity. *Soil. Environ.* **2013**, *32*, 36–43.
69. Latique, S.; Aymen Elouaer, M.; Chernane, H.; Hannachi, C.; Elkaoua, M. Effect of Seaweed Liquid Extract of Sargassum vulgare on Growth of Durum Wheat Seedlings (*Triticum durum* L.) under Salt Stress. *Int. J. Innov. Appl. Stud.* **2014**, *7*, 1430–1435.
70. Feghhenabi, F.; Hadi, H.; Khodaverdiloo, H.; van Genuchten, M.T. Seed Priming Alleviated Salinity Stress during Germination and Emergence of Wheat (*Triticum aestivum* L.). *Agric. Water Manag.* **2020**, *231*, 106022. [CrossRef]

71. Bonomelli, C.; Celis, V.; Lombardi, G.; Mártiz, J. Salt Stress Effects on Avocado (*Persea americana* Mill.) Plants with and without Seaweed Extract (*Ascophyllum nodosum*) Application. *Agronomy* **2018**, *8*, 64. [CrossRef]
72. De Santis, M.A.; Soccio, M.; Laus, M.N.; Flagella, Z. Influence of Drought and Salt Stress on Durum Wheat Grain Quality and Composition: A Review. *Plants* **2021**, *10*, 2599. [CrossRef]
73. Pastuszak, J.; Dziurka, M.; Hornyák, M.; Szczerba, A.; Kopeć, P.; Płażek, A. Physiological and Biochemical Parameters of Salinity Resistance of Three Durum Wheat Genotypes. *Int. J. Mol. Sci.* **2022**, *23*, 8397. [CrossRef]
74. Pessarakli, M. *Handbook of Plant and Crop Stress*; Dekker, M., Ed.; CRC Press: Boca Raton, FL, USA, 1999; ISBN 0824719484.
75. Hamed, S.M.; Abd El-Rhman, A.A.; Abdel-Raouf, N.; Ibraheem, I.B.M. Role of Marine Macroalgae in Plant Protection & Improvement for Sustainable Agriculture Technology. *Beni. Suef. Univ. J. Basic. Appl. Sci.* **2018**, *7*, 104–110. [CrossRef]
76. Sadak, M.; El-Bassiouny, H.; Mahfouz, S.; El-Enany, M.; Elewa, T. Use of Thiamine, Pyridoxine and Biostimulant for Better Yield of Wheat Plants under Water Stress: Growth, Osmoregulations, Antioxidantive Defense and Protein Pattern. *Egypt. J. Chem.* **2022**, *66*, 407–424. [CrossRef]
77. Stirk, W.A.; Van Staden, J. *Comparison of Cytokinin- and Auxin-like Activity in Some Commercially Used Seaweed Extracts*; Kluwer Academic Publishers: New York, NY, USA, 1997; Volume 8.
78. Iqbal, M.; Ashraf, M. Presowing Seed Treatment with Cytokinins and Its Effect on Growth, Photosynthetic Rate, Ionic Levels and Yield of Two Wheat Cultivars Differing in Salt Tolerance. *J. Integr. Plant Biol.* **2005**, *47*, 1315–1325. [CrossRef]
79. Iqbal, M.; Ashraf, M. Seed Treatment with Auxins Modulates Growth and Ion Partitioning in Salt-Stressed Wheat Plants. *J. Integr. Plant Biol.* **2007**, *49*, 1003–1015. [CrossRef]
80. Iqbal, M.; Ashraf, M.; Jamil, A. Seed Enhancement with Cytokinins: Changes in Growth and Grain Yield in Salt Stressed Wheat Plants. *Plant Growth Regul.* **2006**, *50*, 29–39. [CrossRef]
81. Ribba, T.; Garrido-Vargas, F.; O'Brien, J.A. Auxin-Mediated Responses under Salt Stress: From Developmental Regulation to Biotechnological Applications. *J. Exp. Bot.* **2020**, *71*, 3843–3853. [CrossRef] [PubMed]
82. Ashraf, M.; Foolad, M.R. Pre-Sowing Seed Treatment—A Shotgun Approach to Improve Germination, Plant Growth, and Crop Yield Under Saline and Non-Saline Conditions. *Adv. Agron.* **2005**, *88*, 223–271. [CrossRef]
83. Lutts, S.; Majerus, V.; Kinet, J.M. NaCl Effects on Proline Metabolism in Rice (*Oryza sativa*) Seedlings. *Physiol. Plant* **1999**, *105*, 450–458. [CrossRef]
84. Javid, M.G.; Sorooshzadeh, A.; Moradi, F.; Ali, S.; Sanavy, M.M.; Allahdadi, I. The Role of Phytohormones in Alleviating Salt Stress in Crop Plants. *Aust. J. Crop Sci.* **2011**, *5*, 726–734.
85. Kuiper, D.; Schuit, J.; Kuiper, P.J.C. Actual Cytokinin Concentrations in Plant Tissue as an Indicator for Salt Resistance in Cereals. *Plant Soil* **1990**, *123*, 243–250. [CrossRef]
86. Ortiz, J.; Uquiche, E.; Robert, P.; Romero, N.; Quitral, V.; Llantén, C. Functional and Nutritional Value of the Chilean Seaweeds Codium fragile, Gracilaria chilensis and Macrocystis pyrifera. *Eur. J. Lipid Sci. Technol.* **2009**, *111*, 320–327. [CrossRef]
87. Salama, K.H.A.; Mansour, M.M.F.; Al-Malawi, H.A. Glycinebetaine Priming Improves Salt Tolerance of Wheat. *Biologia* **2015**, *70*, 1334–1339. [CrossRef]
88. Mansour, M.M.F.; Ali, E.F. Evaluation of Proline Functions in Saline Conditions. *Phytochemistry* **2017**, *140*, 52–68.
89. Ambreen, S.; Athar, H.u.R.; Khan, A.; Zafar, Z.U.; Ayyaz, A.; Kalaji, H.M. Seed Priming with Proline Improved Photosystem II Efficiency and Growth of Wheat (*Triticum aestivum* L.). *BMC Plant Biol.* **2021**, *21*, 502. [CrossRef] [PubMed]
90. Khafagy, M.A.; Arafa, A.A.; El-Banna, M.F. Glycinebetaine and Ascorbic Acid Can Alleviate the Harmful Effects of Nacl Salinity in Sweet Pepper. *Aust. J. Crop Sci.* **2009**, *3*, 257–267.
91. Liu, H.-L.; Sha, H.-j.; Wang, J.-g.; Liu, Y.; Zou, D.-t.; Zhao, H.-w. Effect of Seed Soaking with Exogenous Proline on Seed Germination of Rice Under Salt Stress. *J. Northeast Agric. Univ.* **2014**, *21*, 1–6. [CrossRef]
92. Mahboob, W.; Athar Khan, M.; Shirazi, M.U. Induction of Salt Tolerance in Wheat Triticum aestivum L. Seedlings through Exogenous Application of Proline. *Pak. J. Bot.* **2016**, *48*, 861–867.
93. Yang, Y.; Guo, Y. Elucidating the Molecular Mechanisms Mediating Plant Salt-Stress Responses. *New Phytol.* **2018**, *217*, 523–539. [CrossRef] [PubMed]
94. Quitadamo, F.; De Simone, V.; Beleggia, R.; Trono, D. Chitosan-Induced Activation of the Antioxidant Defense System Counteracts the Adverse Effects of Salinity in Durum Wheat. *Plants* **2021**, *10*, 1365. [CrossRef]
95. Mehmood, S.; Khan, A.A.; Shi, F.; Tahir, M.; Sultan, T.; Munis, M.F.H.; Kaushik, P.; Alyemeni, M.N.; Chaudhary, H.J. Alleviation of Salt Stress in Wheat Seedlings via Multifunctional Bacillus aryabhattai pm34: An in-Vitro Study. *Sustainability* **2021**, *13*, 8030. [CrossRef]
96. Seleiman, M.F.; Aslam, M.T.; Alhammad, B.A.; Hassan, M.U.; Maqbool, R.; Chattha, M.U.; Khan, I.; Gitari, H.I.; Uslu, O.S.; Roy, R.; et al. Salinity Stress in Wheat: Effects, Mechanisms and Management Strategies. *Phyton-Int. J. Exp. Bot.* **2022**, *91*, 667–694. [CrossRef]
97. Papenfus, H.B.; Kulkarni, M.G.; Stirk, W.A.; Finnie, J.F.; Van Staden, J. Effect of a Commercial Seaweed Extract (Kelpak®) and Polyamines on Nutrient-Deprived (N, P and K) Okra Seedlings. *Sci. Hortic.* **2013**, *151*, 142–146. [CrossRef]
98. Ferreira, M.; Salgado, J.M.; Fernandes, H.; Peres, H.; Belo, I. Potential of Red, Green and Brown Seaweeds as Substrates for Solid State Fermentation to Increase Their Nutritional Value and to Produce Enzymes. *Foods* **2022**, *11*, 3864. [CrossRef]
99. Itroutwar, P.D.; Kasivelu, G.; Raguraman, V.; Malaichamy, K.; Sevathapandian, S.K. Effects of Biogenic Zinc Oxide Nanoparticles on Seed Germination and Seedling Vigor of Maize (*Zea mays*). *Biocatal. Agric. Biotechnol.* **2020**, *29*, 101778. [CrossRef]

100. Kamran, M.; Wang, D.; Xie, K.; Lu, Y.; Shi, C.; EL Sabagh, A.; Gu, W.; Xu, P. Pre-Sowing Seed Treatment with Kinetin and Calcium Mitigates Salt Induced Inhibition of Seed Germination and Seedling Growth of Choysum (*Brassica rapa* Var. parachinensis) *Ecotoxicol. Environ. Saf.* **2021**, *227*, 112921. [CrossRef] [PubMed]
101. Yang, W.; Gao, Y.; Wang, X.; Li, S.; Zheng, H.; Chen, Z.; Wu, F.; Du, X.; Sui, N. Exogenous Calcium Application Enhances Salt Tolerance of Sweet Sorghum Seedlings. *J. Agron. Crop Sci.* **2022**, *208*, 441–453. [CrossRef]

**Disclaimer/Publisher's Note:** The statements, opinions and data contained in all publications are solely those of the individual author(s) and contributor(s) and not of MDPI and/or the editor(s). MDPI and/or the editor(s) disclaim responsibility for any injury to people or property resulting from any ideas, methods, instructions or products referred to in the content.

*plants*

MDPI

*Article*

# A Biostimulant Based on Silicon Chelates Enhances Growth and Modulates Physiological Responses of In-Vitro-Derived Strawberry Plants to In Vivo Conditions

**Elena Ambros [1,*], Olga Kotsupiy [1], Evgeniya Karpova [1], Ulyana Panova [1], Alexander Chernonosov [2], Elena Trofimova [3] and Boris Goldenberg [4]**

1. Central Siberian Botanical Garden, Siberian Branch of Russian Academy of Sciences, 101 Zolotodolinskaya Str., Novosibirsk 630090, Russia
2. Institute of Chemical Biology and Fundamental Medicine, Siberian Branch of Russian Academy of Sciences, 8 Akad. Lavrentiev Ave., Novosibirsk 630090, Russia
3. Institute of Solid State Chemistry and Mechanochemistry, Siberian Branch of Russian Academy of Sciences, 18 Kutateladze Str., Novosibirsk 630128, Russia
4. Synchrotron Radiation Facility Siberian Circular Photon Source, Boreskov Institute of Catalysis, Siberian Branch of Russian Academy of Sciences, 1 Nikolsky Ave., Koltsovo 630559, Russia

* Correspondence: ambros_ev@mail.ru

**Abstract:** The purpose was to assess the effects of a biostimulant based on silicon chelates in terms of alleviation of the impact of in vivo conditions on strawberry (*Fragaria* × *ananassa* cv. 'Solnechnaya polyanka') in-vitro-derived plants. As a source of silicon chelates, a mechanocomposite (MC) obtained through mechanochemical processing of rice husks and green tea was used. Root treatment of plants with 0.3 g $L^{-1}$ of MC dissolved in tap water was performed at 2 weeks after planting. Control plants were watered with tap water. The greatest shoot height, number of roots per plant, root length, number of stolons per plant, daughter ramets per stolon, relative water content, cuticle thickness, and root and shoot biomasses were achieved with the MC supplementation. The improved parameters were associated with a higher silicon content of roots and shoots of the MC-treated plants. Leaf concentrations of hydrogen peroxide and abscisic acid were reduced by the MC. This effect was accompanied by enhanced activity of superoxide dismutase and catalase. The phenolic profile showed upregulation of *p*-hydroxybenzoic acid, vanillic acid, gallic acid, syringic acid, and ellagic acid derivative 2, while kaempferol rutinoside and catechins were downregulated. Thus, silicon chelates improve growth and trigger the physiological processes that enhance free-radical-scavenging activity in strawberry plants in vivo.

**Keywords:** strawberry; silicon-chelate-based biostimulant; silicon accumulation; plant growth; cuticle; hydrogen peroxide; antioxidant defense; abscisic acid; in vivo conditions

**Citation:** Ambros, E.; Kotsupiy, O.; Karpova, E.; Panova, U.; Chernonosov, A.; Trofimova, E.; Goldenberg, B. A Biostimulant Based on Silicon Chelates Enhances Growth and Modulates Physiological Responses of In-Vitro-Derived Strawberry Plants to In Vivo Conditions. *Plants* **2023**, *12*, 4193. https://doi.org/10.3390/plants12244193

Academic Editors: Spyridon A. Petropoulos, Petronia Carillo and Mohamad Hesam Shahrajabian

Received: 7 November 2023
Revised: 22 November 2023
Accepted: 14 December 2023
Published: 18 December 2023

**Copyright:** © 2023 by the authors. Licensee MDPI, Basel, Switzerland. This article is an open access article distributed under the terms and conditions of the Creative Commons Attribution (CC BY) license (https://creativecommons.org/licenses/by/4.0/).

## 1. Introduction

Strawberry (*Fragaria* × *ananassa* (Duchesne ex Weston) Duchesne ex Rozier) is one of the most cultivated berries in the world. Due to their rich content of biologically active compounds, strawberry fruits are present in the diet of many people. At the same time, the crop, due to its superficial root system, large leaf area, and high water content of the fruits, is extremely sensitive to growth conditions during the cultivation process. The use of environmentally friendly approaches, such as biostimulants, can help sustainable strawberry agricultural practices. Biostimulants reduce the need for plant synthetic chemicals (fertilizers and pesticides) while contributing to the maintenance of an ideal quality and quantity of crop output [1,2].

At low concentrations, biostimulants trigger a number of physiological processes in a plant that enhance growth and yield and mitigate stress-induced limitations [3]. Plant biostimulants based on biogenic silicon (Si), which is absorbed by plants 2–10 times better than salts, are of great interest [4,5]. Silicon is not an essential element for plants, but its presence in different habitats has been shown to help increase plant resistance to negative environmental factors [6–8]. In many cases, exogenous Si treatments improve plant growth and increase productivity, thus enhancing adaptive reactions associated with the activity of the plant antioxidant system, accumulation of metabolites, changes in phytohormonal status, rate of photosynthesis, and nutrient elements' uptake [4,9–16]. The beneficial effects of Si are attributed to the metabolic transformations caused by a modifying impact on sorption properties of the cell wall (where it accumulates in the form of amorphous silica) and by the formation of complexes with various organic compounds [17]. Nonetheless, the mechanisms and trend of Si absorption are different among plant species [18]. The foregoing is the dominant motive for further studies on the effect of Si on strawberry plants.

A promising source of Si-containing biostimulants is renewable plant raw materials obtained using a solid-phase mechanochemical method (without organic solvents) involving biogenic Si in chelate form. The method is also attracting much attention as a step toward green chemistry by reducing the use of hazardous chemicals. Because of such processing, plant raw materials and reagents remain in a stable solid form, which completely prevents their oxidation and loss of biological activity [19,20]. In addition, the presence of a wide range of plant wastes worldwide, including those containing Si, determines the demand for "green chemistry" as a component of organic farming for growing plants [21]. On the other hand, investigations of the behavior and interactions of silicon chelates in growing plants, the mechanism of their influence, and a variety of the effects during application in agriculture are still at a rudimentary stage.

Our previous studies about the effect of silicon chelates on in-vitro-derived strawberry plants have revealed improved growth and a contribution to changes in physiological characteristics. The physiological role of silicon chelates is thought to decrease contents of hydrogen peroxide and abscisic acid (ABA) and to promote activities of enzymatic antioxidants in the leaves, thereby mediating growth and the development of strawberry plants both under in vitro and ex vitro conditions [22,23]. These results lead us to the hypothesis that silicon chelates can similarly modulate the adaptability of strawberry microclones in vivo. A sustainable increase in stress tolerance of the microclones when transferred to in vivo conditions is important for creating an integrated industrial technological scheme to provide the basis for raising the yield of healthy strawberry planting material.

The leaf chlorophyll content is one of the main parameters indicating the physiological state of a plant [24]. This characteristic is often determined for agriculture monitoring and for evaluation of effects of various growth conditions or stimulants [25,26]. The concentration of photosynthetic pigments reflects plant adaptability and can point out a stressor in the environment [27–29].

Plant propagation starting from an explant in vitro to plant growth in the field is associated with sequential changes in the environment and growth conditions. Stress tolerance during all stages is a crucial factor for boosting productivity. The drastic alteration of the surroundings during the transfer of plantlets from in vitro to ex vitro conditions can cause an increase in the stress level and injuries of tissues. Oxidative damage can be triggered by overproduction of reactive oxygen species (ROS), represented predominantly by the superoxide anion ($O_2^-$), hydrogen peroxide ($H_2O_2$), the hydroxyl radical (•OH), and singlet oxygen ($^1O_2$), which cause the oxidation of cell compounds and suppression of plant growth and development. To protect cellular membranes and organelles, plants launch the antioxidant system that consists of both enzymatic and nonenzymatic mechanisms of detoxification. The nonenzymatic antioxidant system includes water-soluble (ascorbic acid and glutathione) and lipid-soluble antioxidants ($\alpha$-tocopherol and $\beta$-carotene) [30]. Another essential part of the plant antioxidant system is phenolic compounds. Due to their aromatic rings and free hydroxyl groups, flavonoids and phenolic acids act as scavengers of ROS [31]. Superoxide dismutase (SOD), catalase (CAT), and peroxidase (PX) are the main enzymatic antioxidants in plants. They catalyze conversion of ROS via enzymatic defense reactions [32], and an increase in their activity is often registered in response to abiotic stress [33,34].

Plant hormones are involved in the regulation of the metabolic response to stress and are capable of inducing sequences of adaptive reactions. They govern plant growth in a wide range of environments [35,36]. Recently, the key role of ABA in plant responses to environmental stressors was documented [37]. ABA is generally maintained at low levels under non-stressful conditions, and its levels can rise significantly in response to environmental stressors and during seed maturation [38]. Effects of ABA on photosynthesis and ROS metabolism have been documented, too [39,40]. In addition, ABA is involved in the regulation of a plant's water status [41] (through osmotic-stress-induced stomatal closure) as well as salt, drought, and cold tolerance [38]. The essential plant hormones cytokinins and auxins also participate in the stress response by maintaining plant growth and development under normal and stressful conditions [36].

Currently, more and more data are accumulating regarding Si's ability to affect plant–environment interactions and to modify the response of the plant defense system [42]. In terms of stress tolerance, Si is regarded as a "multitalented" quasi-essential element. The manipulation of silicon sources to boost plant metabolic activity is starting a trend for sustainable horticulture [43]. Therefore, the present study was focused on assessing the effect of silicon chelates on the acclimation of strawberry plants to in vivo conditions and identifying changes in the growth, development, antioxidant defense system (enzymatic and nonenzymatic antioxidants), contents of photosynthetic pigments, and levels of phytohormones associated with the accumulation of silicon in shoots and roots of plants. An in-depth understanding of the influence of biogenic Si on plant growth and development will contribute to effective regulation of the crop fertilization system, which guarantees a consistently high yield of agricultural plants.

## 2. Results

### 2.1. The Si Content of In-Vitro-Derived Strawberry Plants

Si mainly accumulated in the shoots of the plants, not only in the group of plants supplemented with a mechanocomposite (MC; a source of silicon chelates) but also in the untreated (−MC) group (Figure 1). The Si content in shoots exceeded the Si content of plant roots by 1.1- and 1.3-fold in the MC-treated and untreated plants. The Si content of the strawberry plants grown without MC treatment (−MC) was lower than that of the MC-supplied plants (+MC). Significant elevation of the Si content was found both in the roots and shoots of the plants. The Si accumulation rose 1.6-fold in roots and 1.4-fold in shoots under the action of the (+MC) treatment compared with the (−MC) group ($p < 0.05$).

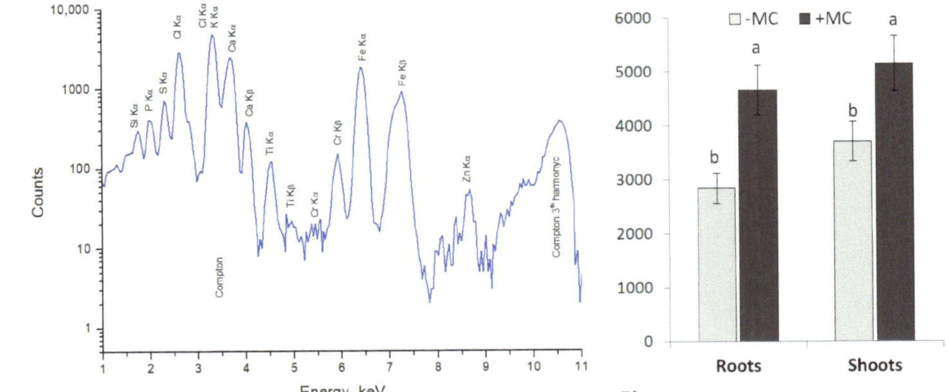


**Figure 1.** Typical X-ray fluorescence analysis with synchrotron radiation (SR-XRF) spectra of roots of in-vitro-derived strawberry plants (**A**). The silicon content of the roots and shoots in strawberry plants

(cv. 'Solnechnaya polyanka') grown without the MC (−MC) or with MC treatment (+MC) in vivo (**B**). The data are means ± standard error. Different lowercase letters above the bars indicate significant differences ($p < 0.05$) between (−MC) and (+MC) treatment groups according to Duncan's multiple-range test. Abbreviations: MC, the mechanocomposite based on silicon chelates; −MC, the mechanocomposite-free group; +MC, mechanocomposite treatment.

### 2.2. Growth Status and Vegetative Productivity of In-Vitro-Derived Strawberry Plants

The growth and development of the roots and above-ground parts of strawberry plants were significantly affected by the MC treatment (Figure 2B). Differences were identified for the root number, root length, and fresh and dry weights of the roots. Mean responses are shown in Table 1. Analysis of variance (ANOVA) indicated that root length in the (+MC) group was greater in comparison to the (−MC) group by 1.2-fold ($p < 0.05$); the root number was greater by 1.5-fold ($p < 0.05$). Root fresh weight and root dry weight were greater by 2.1-fold ($p < 0.05$) and 2.1-fold ($p < 0.01$), respectively.

**Figure 2.** The phenotype of in-vitro-derived strawberry plants (cv. 'Solnechnaya polyanka') grown without the (−MC) (**A**) or with (+MC) treatment (**B**) in vivo. Abbreviations: MC, the mechanocomposite based on silicon chelates; −MC, the mechanocomposite-free group; +MC, mechanocomposite treatment.

**Table 1.** Effects of MC treatment on growth-related parameters of in-vitro-derived strawberry plants (cv. 'Solnechnaya polyanka') during in vivo acclimation.

| Parameter | Treatment Type | |
|---|---|---|
| | (−MC) Group | (+MC) Group |
| root length (cm) | 13.32 ± 0.75 b * | 18.00 ± 1.27 a * |
| roots (number per plant) | 13.30 ± 1.47 b * | 20.35 ± 1.49 a * |
| root fresh weight (g) | 3.18 ± 0.40 b * | 6.52 ± 0.92 a * |
| root dry weight (g) | 0.83 ± 0.08 b ** | 1.72 ± 0.16 a ** |
| shoot height (cm) | 10.88 ± 0.42 b ** | 14.26 ± 0.63 a ** |
| leaves (number per plant) | 10.17 ± 0.66 a | 14.13 ± 1.16 a |
| leaf blade area ($cm^2$) | 39.57 ± 4.34 a | 43.43 ± 3.00 a |
| shoot fresh weight (g) | 14.18 ± 1.54 b * | 26.30 ± 3.11 a * |
| shoot dry weight (g) | 3.33 ± 0.25 b * | 5.35 ± 0.54 a * |

The data are means ± standard error. Different lowercase letters within a row indicate significant differences (* $p < 0.05$; ** $p < 0.01$) between (−MC) and (+MC) treatment groups according to Duncan's multiple-range test. Abbreviations: MC, the mechanocomposite based on silicon chelates; −MC, the mechanocomposite-free group; +MC, mechanocomposite treatment.

A well-developed root system supported the best growth of above-ground parts of strawberry plants (Table 1). For instance, relative to the control (−MC), shoot height in the group with (+MC) was greater by 1.3-fold ($p < 0.01$), the number of leaves was greater by 1.4-fold ($p < 0.05$), and the fresh and dry weights of shoots were greater by 1.9 and 1.6-fold ($p < 0.05$), respectively.

The pronounced effect of MC treatment was observed in stolons and daughter ramet formation (Figure 3). For instance, in the group with (+MC), the number of stolons per mother plant was greater by 2.1-fold ($p < 0.05$). In turn, the yield of daughter ramets per plant amounted to 50 compared to the group without MC, which produced an average of up to 7 ramets per plant (Figure 3A,C).

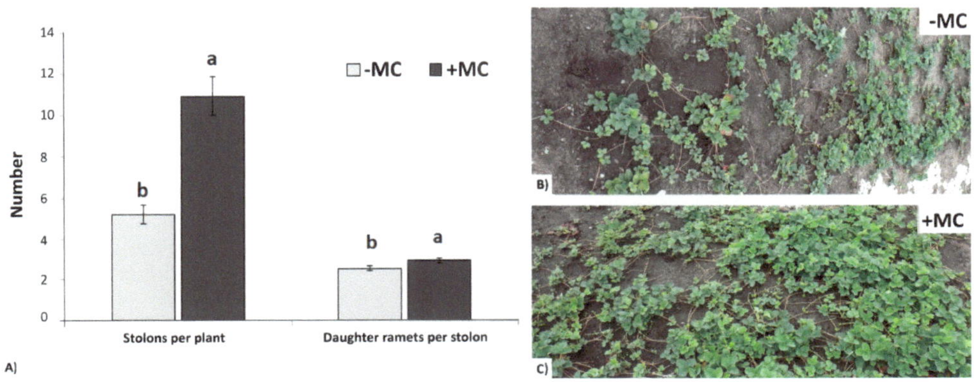


**Figure 3.** Vegetative productivity of in-vitro-derived strawberry plants (cv. 'Solnechnaya polyanka') grown without the MC (−MC) or (+MC) treatment in vivo. Formation of stolons and daughter ramets in the (−MC) group (**A**,**B**) and in the (+MC) group (**A**,**C**). The data are means ± standard error. Different lowercase letters above the bars indicate significant differences ($p < 0.05$) between (−MC) and (+MC) treatment groups according to Duncan's multiple-range test. Abbreviations: MC, the mechanocomposite based on silicon chelates; −MC, mechanocomposite-free group; +MC, mechanocomposite treatment.

## 2.3. Physiological Characteristics of In-Vitro-Derived Strawberry Plants

Changes in essential physiological parameters of the plants are presented in Tables 2 and 3. The relative water content (RWC) of the leaves of the plants was found to be significantly higher with (+MC) treatment (by 1.2-fold) ($p < 0.05$) in comparison to the (−MC) group. There was no noticeable difference in photosynthetic pigment concentrations between the (+MC) treatment group and the (−MC) group. In contrast, a significantly higher chlorophyll a/b ratio ($p < 0.05$) in the leaves of the (+MC) group was observed. In the presence of the MC, the chlorophyll a/b ratio was 1.2-fold higher compared with the control group.

**Table 2.** Effects of MC treatment on physiological parameters in leaves of in-vitro-derived strawberry plants (cv. 'Solnechnaya polyanka') during in vivo acclimation.

| Parameter | Treatment Type | |
|---|---|---|
| | (−MC) Group | (+MC) Group |
| RWC (%) | 55.92 ± 1.71 b | 64.62 ± 2.40 a |
| chlorophyll a (mg $g^{-1}$ FW) | 1.31 ± 0.05 a | 1.38 ± 0.04 a |
| chlorophyll b (mg $g^{-1}$ FW) | 0.33 ± 0.01 a | 0.30 ± 0.01 a |
| total chlorophyll (mg $g^{-1}$ FW) | 1.64 ± 0.07 a | 1.67 ± 0.05 a |
| chlorophyll a/b ratio | 3.94 ± 0.07 b | 4.66 ± 0.13 a |
| carotenoids (mg $g^{-1}$ FW) | 0.44 ± 0.02 a | 0.46 ± 0.02 a |

**Table 2.** *Cont.*

| Parameter | Treatment Type | |
|---|---|---|
| | (−MC) Group | (+MC) Group |
| total chlorophyll/carotenoids | 3.93 ± 0.31 a | 3.66 ± 0.06 a |
| $H_2O_2$ (μmol (g FW)$^{-1}$) | 123.27 ± 24.19 a | 56.08 ± 14.40 b |
| SOD (U (g FW)$^{-1}$) | 98.28 ± 1.21 b | 103.74 ± 0.22 a |
| CAT ((U $H_2O_2$)(g FW)$^{-1}$ min$^{-1}$) | 8.33 ± 2.85 b | 29.51 ± 10.14 a |
| POD (U (g FW)$^{-1}$ min$^{-1}$) | 2.78 ± 1.00 a | 3.13 ± 0.07 a |
| IAA (ng g$^{-1}$ FW) | 6.60 ± 1.01 a | 9.37 ± 0.83 a |
| ABA (ng g$^{-1}$ FW) | 3.57 ± 0.61 a | 1.63 ± 0.10 b |
| t-Z (ng g$^{-1}$ FW) | 2.65 ± 0.49 a | 2.06 ± 0.74 a |

The data are means ± standard error. Different lowercase letters within a row indicate significant differences ($p < 0.05$) between (−MC) and (+MC) treatment groups according to Duncan's multiple-range test. Abbreviations: MC, the mechanocomposite based on silicon chelates; −MC, the mechanocomposite-free group; +MC, mechanocomposite treatment; RWC, the relative water content; FW, fresh weight; $H_2O_2$, hydrogen peroxide; CAT, catalase; POD, peroxidase; SOD, superoxide dismutase; IAA, indoleacetic acid; ABA, abscisic acid; t-Z, trans-zeatin.

**Table 3.** Effects of MC treatment on nonenzymatic antioxidants in leaves of in-vitro-derived strawberry plants (cv. 'Solnechnaya polyanka') during in vivo acclimation.

| Phenolic Compound, mg g−1 DW | $t_R$, min | $\lambda_{max}$, nm | Treatment Type | |
|---|---|---|---|---|
| | | | (−MC) Group | (+MC) Group |
| Hydroxybenzoic acids: | | | | |
| gallic acid | 2.4 | 216, 272 | 0.17 ± 0.01 a | 0.18 ± 0.01 a |
| *p*-hydroxybenzoic acid | 6.8 | 254 | 0.07 ± 0.00 b | 0.12 ± 0.01 a |
| vanillic acid | 9.4 | 260, 290 | 4.03 ± 0.18 b | 5.82 ± 0.28 a |
| syringic acid | 9.7 | 220, 270 | 0.51 ± 0.03 b | 0.69 ± 0.02 a |
| Total | | | 4.78 ± 0.21 b | 6.81 ± 0.26 a |
| Hydroxycinnamic acids: | | | | |
| neochlorogenic acid | 3.8 | 240, 296 sh., 325 | 0.15 ± 0.01 a | 0.17 ± 0.01 a |
| caffeic acid | 8.7 | 218, 240, 298 sh., 324 | 1.19 ± 0.05 a | 1.26 ± 0.06 a |
| *p*-coumaric acid | 10.8 | 226, 310 | 0.11 ± 0.01 a | 0.11 ± 0.05 a |
| ferulic acid | 13.3 | 218, 236, 320 | 0.04 ± 0.00 a | 0.04 ± 0.00 a |
| Total | | | 1.49 ± 0.07 a | 1.58 ± 0.07 a |
| Catechins: | | | | |
| (+)-catechin | 5.3 | 204, 230 sh., 280 | 0.55 ± 0.03 a | 0.49 ± 0.02 a |
| epigallocatechin gallate | 7.7 | 204, 230 sh., 280 | 0.09 ± 0.00 b | 0.06 ± 0.00 a |
| Total | | | 0.64 ± 0.04 a | 0.55 ± 0.02 a |
| Ellagic acid and its derivatives: | | | | |
| ellagic acid | 17.7 | 254, 367 | 1.04 ± 0.05 a | 1.13 ± 0.06 a |
| ellagic acid derivative 1 | 8.0 | 230, 262 | 1.00 ± 0.04 a | 1.13 ± 0.06 a |
| ellagic acid derivative 2 | 11.2 | 230, 276 | 1.16 ± 0.05 b | 1.82 ± 0.05 a |
| ellagic acid derivative 3 | 12.4 | 254, 362 | 0.74 ± 0.03 a | 0.53 ± 0.02 b |
| ellagic acid derivative 4 | 16.3 | 254, 360 | 2.55 ± 0.14 a | 2.26 ± 0.14 a |
| ellagic acid derivative 5 | 16.8 | 254, 360 | 0.75 ± 0.04 a | 0.65 ± 0.04 a |
| Total | | | 7.24 ± 0.35 a | 7.52 ± 0.21 a |
| Flavonol glycosides: | | | | |
| kaempferol rutinoside | 19.2 | 265, 356 | 0.23 ± 0.01 a | 0.13 ± 0.01 b |
| quercetin glycoside 1 | 13.8 | 270, 355 | 4.78 ± 0.27 a | 5.48 ± 0.35 a |
| quercetin glycoside 2 | 14.0 | 270, 355 | 1.54 ± 0.07 a | 1.61 ± 0.07 a |
| Total | | | 6.55 ± 0.36 a | 7.23 ± 0.41 a |
| Total phenolic compounds | | | 20.70 ± 1.03 a | 23.69 ± 0.81 a |

The data are means ± standard error. Different lowercase letters within a row indicate significant differences ($p < 0.05$) between (−MC) and (+MC) treatment groups according to Duncan's multiple-range test. Abbreviations: MC, the mechanocomposite based on silicon chelates; −MC, the mechanocomposite-free group; + MC, mechanocomposite treatment; sh., shoulder; $t_R$, retention time; $\lambda_{max}$, absorption maxima.

The accumulation of $H_2O_2$ in the leaves was lower by 2.2-fold in the presence of the MC. Moreover, the plants grown with the MC exhibited significantly ($p < 0.05$) higher activities of the antioxidant enzymes SOD and CAT. SOD activity was greater by 1.1-fold, and CAT activity was greater by 3.5-fold ($p < 0.05$). POD activity was slightly higher compared with the (−MC) group (by 1.1-fold).

The quantification of phytohormones in the plant leaves showed that the ABA level was significantly lower in the (+MC) group (2.2-fold less than in the (−MC) ($p < 0.05$)). Differences in the indoleacetic acid (IAA) level were weaker. This phytohormone's content grew (by 1.4-fold) during (+MC) treatment. The *trans*-zeatin (t-Z) level did not undergo a significant shift during the MC treatment.

The differences in the phenolic profile between the groups concerned mostly quantitative levels. The total phenolic concentration was higher by 14% compared to the control, and yet the impact on certain compounds was more significant. The most responsive to the treatment were hydroxybenzoic acids (Table 3). *p*-Hydroxybenzoic acid, vanillic acid, and syringic acid showed 1.67-, 1.44-, and 1.3-fold ($p < 0.05$) elevation of their levels in the (+MC) group. The concentration of gallic acid was only slightly higher, and the total hydroxybenzoic acid content was greater by 1.42-fold during the MC treatment.

The majority of hydroxycinnamic acids (neochlorogenic, *p*-coumaric, and ferulic acids) were present in the leaves at low concentrations and did not undergo the effect of the MC. In contrast, caffeic acid showed a substantial level, and it did not undergo significant elevation in the MC treatment group, either. Thus, a significant concentration shift of the total hydroxycinnamic acid content was not revealed during the MC treatment, which was similar to the total concentrations of ellagic acid and its derivatives. Only one compound from the latter group, ellagic acid derivative 2, exhibited substantial (1.57-fold) upregulation comparable to the compounds showing the greatest increase: *p*-hydroxybenzoic and vanillic acids. Similarly to hydroxycinnamic acids and ellagic acid derivatives, flavonol glycosides did not show a significant difference in the total concentration between the plant groups. Meanwhile, concentrations of quercetin glycosides were slightly higher, and the concentration of kaempferol rutinoside was significantly lower in the +MC group. Catechin concentration was slightly lower in the MC's presence.

### 2.4. Cuticle Thickness in In-Vitro-Derived Strawberry Plants

The cuticle of strawberry leaves in the MC-supplied plants (+MC) was significantly thicker than that of plants grown without MC treatment (−MC) (Figure 4). This parameter in the leaves of (+MC) plants was higher by 1.4-fold compared with the (−MC) group ($p < 0.05$).

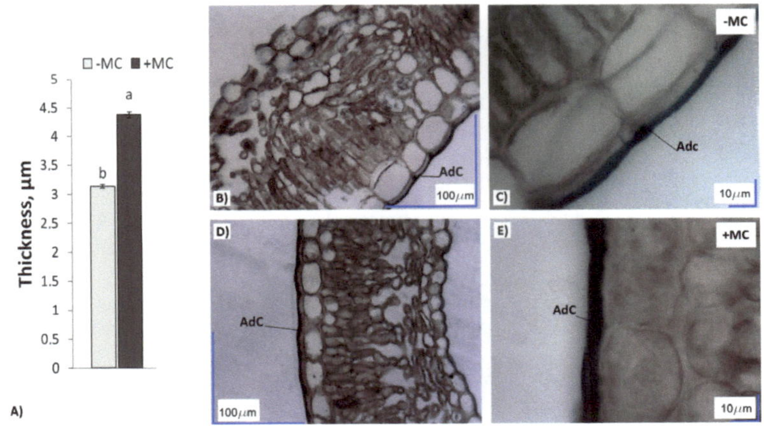


**Figure 4.** Cuticle thickness (**A**) and Sudan black B-stained transverse sections of leaves of in-vitro-derived strawberry plants (cv. 'Solnechnaya polyanka') grown without the MC (−MC) (**B**,**C**) or with

the MC treatment (+MC) (**D**,**E**) in vivo. The data are means ± standard error. Different lowercase letters above the bars indicate significant differences ($p < 0.05$) between (−MC) and (+MC) treatment groups according to Duncan's multiple-range test. Abbreviations: MC, the mechanocomposite based on silicon chelates; −MC, the mechanocomposite-free group; +MC, mechanocomposite treatment; AdC, adaxial cuticle.

## 3. Discussion

Silicon accumulation has been found to be higher in monocotyledonous plants [44]. Most dicotyledons are unable to absorb large amounts of Si from soil. The difference in Si accumulation among species has been attributed to differences in the Si uptake ability of roots [45]. Thus, biochemical treatment of plant roots to enhance their ability to absorb Si and subsequently transfer Si to shoots can help plants overcome a wide range of stressors and improve metabolism [18]. A few studies present conflicting data on the absorption of Si by strawberry plants. There are several articles in which strawberry is classified as a plant that accumulates Si (up to 3% of dry weight) [46–48], whereas the work of M. J. Hodson et al. [49] suggests that strawberry does not absorb Si well. In our study, the elemental analysis confirmed the transfer of Si from the soil to the roots and Si accumulation in the strawberry shoots. Moreover, the strawberry plants absorbed Si from the soil in amounts that benefited the plant. The MC treatment significantly affected the growth and biomass of both roots and shoots of the strawberry plants. This finding is consistent with the results of other studies, indicating Si-mediated enhancement of root and shoot development and also showing active accumulation of Si mainly in organs of cereal and vegetable crops [50]. It is believed that under the influence of Si, the architecture of the cell wall changes and becomes stronger because this element accumulates in rhizodermis cells of the root cap and exudates of root hairs, leading to a total increase in the root surface adsorption [17,51]. After the transport from roots to shoots, Si is accumulated as amorphous silica in cell walls of shoot endodermis, mainly in the transpiration area of shoots, and becomes immobile without relocating anywhere [45,52,53]. These deposits of Si protect plants from multiple abiotic and biotic stresses.

Besides vegetative growth, the MC treatment increased the quantity of ramets from the mother plant. Supposedly, Si can be transported as a nutrient from mother plants to daughter plants. For instance, Y. Li et al. [54] have reported that the foliar spray and root drench of mother plants significantly promote Si accumulation in daughter plants, whereas spraying the runner plants does not increase the Si content of the mother plants. Such changes are undoubtedly useful in the production of strawberry planting material.

Additionally, Si treatment of plants results in the formation a cuticle–Si double layer [55], which helps the plants to hold more water. In our work, the increase in cuticle thickness and in the RWC of the strawberry leaves after the MC treatment confirms other authors' data about the effect of Si on plant water transport processes, including leaf water loss, vessel water transport, and root water uptake [56,57]. In leaves, the deposition of silica phytolith on the cuticle and stomata has been shown to decrease the transpiration loss of water under water deficit stress [58]. It is probable that the increase in the RWC in our experiment was related to both a higher water absorption capacity of plant roots under the influence of the MC and a low rate of its loss by leaves. The increase in the root system's biomass, leading to larger root surface area, proves this supposition. In turn, the low leaf water loss is confirmed by greater silicon accumulation in the shoots of the MC-treated strawberry plants.

Moreover, our results align with the reports describing Si as a trigger that induces physiological mechanisms allowing a strawberry plant to improve resistance to biotic stress by means of defensive compounds and enzymes [59,60]. The main stressors for strawberry plants in the field are suboptimal temperatures, a water deficit or excess, extreme solar radiation, and biotic threats of different natures (bacteria, fungi, or insects). Physiological parameters in our work highlighted a significant effect of the silicon-chelate-based MC on phytohormone and antioxidant systems of strawberry plants. Moreover, the results allow us to correlate some physiological parameters of the plants growing in the field with those of in vitro plantlets [22]. $H_2O_2$ levels and activities of antioxidant enzymes vary strongly throughout the propagation cycle [22], and their concentrations are under a strong influence of the MC. In the absence of the MC, a spike in the $H_2O_2$ level (4.23 and 123.27 °mol $g^{-1}$ during in vitro and in vivo growth, respectively) was registered. Nevertheless, in the +MC treatment group, the increase in $H_2O_2$ during in vivo growth was not so considerable (56.08 °mol $g^{-1}$). At the same time, CAT activity without the MC was modest (10.8 and 8.33 (U $H_2O_2$)(g FW]$^{-1}$ min$^{-1}$) during in vitro multiplication and in vivo growth, respectively), but it was almost 3-fold higher in the MC treatment group (10.4 and 29.51 (U $H_2O_2$)(g FW)$^{-1}$ min$^{-1}$]. SOD and POD activities rose manyfold from the in vitro to in vivo stage. Meanwhile, during the in vivo growth with MC treatment, the increase in SOD activity was not so pronounced (98.28 vs. 103.74 U (g FW)$^{-1}$ in (−MC) and (+MC) groups), and POD activity did not change significantly compared to (−MC) conditions (2.78 vs. 3.13 U (g FW)$^{-1}$) [22].

The level of chlorophyll a substantially went up in the leaves of the plants in the field (1.31 and 1.38 mg $g^{-1}$ FW in the −MC and +MC groups) compared to the in vitro plantlets (0.30 and 0.22 mg $g^{-1}$ FW) [22]. In contrast, the chlorophyll b level did not undergo pronounced changes (0.39 and 0.30 vs. 0.33 and 0.30 mg $g^{-1}$ FW). Obviously, the MC has no significant effect on the chlorophyll content. The carotenoid level also rose throughout plant development (0.14 and 0.10 vs. 0.44 and 0.46 mg $g^{-1}$ FW). Meanwhile, the influence of the MC on this parameter was not substantial either, especially at the in vivo stage.

Our results point out the key role of ABA in the response to Si accumulation. N. Kumar et al. [61] have demonstrated the regulation of the genes responsible for Si accumulation by plant hormones. One of the stress hormones involved in the regulation of Si transport in plants is ABA [62]. Our results are in agreement with this pattern of a stress response, where stress hormones govern Si accumulation after an environmental trigger. Therefore, the lower ABA level in the leaves of the strawberry plants of the MC treatment group may be a consequence of the involvement in Si accumulation. The participation of auxins and cytokinins in Si effects has been poorly investigated [63]. Under in vivo conditions here, these hormones did not exhibit their correlation with the Si effect, yet their potential participation in the implementation of Si's effects may be of interest for our further research. These plant hormones support normal growth and development under stressful conditions. Meanwhile, the precise mechanism of auxin- and cytokinin-mediated stress tolerance is yet to be discovered [36].

Another part of the plant's response to Si application may be a concentration shift of certain phenolic compounds [60]. ABA is reported to take part in the transformation of phenolic metabolism through upregulation or downregulation of the genes related to phenylpropanoid and flavonoid metabolism [64]. G. Hao et al. [65] have demonstrated an enhancement of the activity of phenylalanine ammonia-lyase and chalcone synthase, which are crucial for flavonoid accumulation, in response to the upregulation of ABA. Therefore, the simultaneous decline in the ABA level and the increase in the phenolic compound concentration in the strawberry plants in the MC treatment group may indicate that silicon is a trigger of phenolic compound accumulation [59,60]. This decline in the ABA level may be caused by a reduction in stress-induced signals in MC-treated plants, which possess improved mechanical parameters, and, consequently, a reduced risk of biotic threats and an elevated water content.

In the last few years, data have been accumulating regarding the effects of Si compounds on plants, including, among them, induction of a concentration shift of phenolic compounds as a part of a plant's defense system. Meanwhile, these data are still fragmentary and contradictory. The main reasons for their variation are the differences in the plant genotype, environment, age, physiological state, Si form, concentration in the treatment, and stressor type. I. Vega et al. [66] have revealed a decrease in lipid peroxidation and intensive lignin accumulation induced by Si treatment, mainly with a subsequent reduction in concentrations of caffeoylquinic acid and flavone glycosides in barley under aluminum stress. In an experiment with *Passiflora incarnata*, Si application induced upregulation of total polyphenols and vitexin [67]. A. Carneiro-Carvalho et al. [68] have registered elevation of levels of the major phenolic compounds having antifungal activity in Si-treated chestnut plants up to 2–4-fold compared to untreated ones. The enhancement of production of phenolic compounds was also registered by G. Shahnaz et al. [69] during mitigation using silicon of the negative impact of Al stress on *Borago officinalis*.

The growth of many plants in the field has been shown to be accompanied by high levels of quercetin glycosides, which are effective antioxidants and defensive compounds [70,71]. R. Hajiboland et al. [72] have observed a Si-mediated increase in the leaf concentration of quercetin at three developmental stages of strawberry plants (vegetative, flowering, and fruiting stages). A considerable increase in the concentration of quercetin glycoside rutin (more than 20-fold) induced by Si, with powdery mildew severity reduced, has been revealed in *Rosa hybrida* plants [73]. The strawberry plants in our study similarly contained substantial concentrations of quercetin glycosides, but they were not significantly higher in the +MC group. Consequently, our results do not indicate participation of quercetin glycosides in the response to Si application. In contrast to quercetin glycosides, substantial downregulation of kaempferol rutinoside in the strawberry leaves was revealed in the MC treatment group. Accumulation of kaempferol and/or its glycoside in the cell wall and thickened cuticle of MC-treated plants could be behind this decrease. The essential role of kaempferol glycosides in cuticle formation of some plant species has been demonstrated before [74–76].

Our results are in agreement with R. Hajiboland et al.'s findings [72] about significant upregulation of phenolic acids under the influence of silicon. They showed consistently higher leaf concentrations of gallic, caffeic, and chlorogenic acids in +Si plants at all three developmental stages. Benzoic acids (gallic, *p*-hydroxybenzoic, vanillic, and syringic acids) were the main class of phenolic compounds that ensured the elevation of the total level of phenolic compounds in the strawberry leaves in the field under the influence of Si. Similar phenolic acid upregulations in other plants have also been reported [60,72,73]. These have been associated with the strengthening of plant protection against bacteria, fungi, and herbivores [73,77–80]. It is noteworthy that the decrease in the concentrations of catechins and ellagic acid derivatives in the strawberry leaves treated with Si may be caused by their polymerization and the formation of compounds having a strong defensive effect against various biotic threats [81,82].

In spite of some discrepancies between the literature's data on the responses to Si application, the upregulation of phenolic compounds, including phenolic acids and flavonols, is often recorded. Our findings confirm the involvement of phenolic compounds in the effect of Si on antioxidant status and biotic resistance of strawberry plants and show alterations of the phenolic profile in strawberry leaves under the influence of Si. The downregulation of some compounds under the influence of Si, as observed in our study and some previous papers [66,72,83], can perhaps be attributed to their accumulation in cell-wall-bound fractions [72] and the formation of high-molecular-weight polyphenols [84].

Thus, the results indicate a positive effect of MC treatment on the metabolism and anatomy of cells and tissues of strawberry plants. In the future, to expand the scope of the experiment, we would like to focus on the effect of the MC on soil–nutrient relations. Elemental analysis of the strawberry plants during MC treatment in our greenhouse uncovered a significant increase in the levels of micronutrients (Cu, Zn, Fe, and Mn) (unpublished data). The involvement of the studied biochemical and anatomical alterations in the overall effect on plant nutrition is interesting primarily.

## 4. Materials and Methods

### *4.1. Preparation and Characterization of Silicon Chelates*

As a source of silicon chelates, an MC obtained through mechanochemical processing of plant raw materials was used [20]. The MC was produced from rice husks (*Oryza sativa*, cv. 'Liman', Krasnodar Krai, Russia) and green tea (*Camellia sinensis* L.) (Krasnodar Krai,

Russia) through mechanical activation in a water-cooled flow activator (roller mill RM-20, manufactured at the Institute of Solid State Chemistry and Mechanochemistry, SB RAS, Novosibirsk, Russia). The MC was found to contain 1.4% ± 0.2% of catechins (mean ± SE), 16.7% ± 0.9% of ash, and 15.2% ± 0.7% of silica. In the MC, the mass ratio of rice husks to green tea was 10:1.

### 4.2. Experimental Design

The field experiment was conducted on the experimental plot of the Central Siberian Botanical Garden (CSBG), SB RAS, Novosibirsk, Russia (geographical coordinates 54°49′9.87″ N and 83°6′6.95″ E), during May–August 2022. Dimensions of the experimental plot were 4.0 × 4.0 m. Mechanical properties of the soil in this area are as follows: medium loamy with a large amount of coarse dust fractions, structureless, with poor air and water permeability, rapid settlement and compaction after processing, and a propensity for swimming and crusting. Its pH is slightly acidic at 6.3–6.9. The soil layer from 0 to 20 cm contains 2–4% humus, and at a depth of 50–60 cm, no more than 0.8%. The total natural reserves of nutrients are low; therefore, humus was introduced into the holes during planting.

In-vitro-derived strawberry plants (*F.* × *ananassa* cv. 'Solnechnaya polyanka') were grown in plastic pots containing 250 $cm^3$ of a soil substrate from a mixture of peat with perlite, humus, sand, and a coco substrate (1:1:0.25:0.25, *v*/*v*/*v*/*v*) and were used for planting under field conditions. Single-fruiting cv. 'Solnechnaya polyanka' is characterized by stable high productivity and is recommended for cultivation under Western Siberia conditions.

There were two accounting plots of 4.0 × 1.20 m for the control group and treatment group on the experimental plot. The plants were planted in rows according to a 0.40 × 0.35 m scheme. The row area was 0.4 m × 1.2 m, and 10 plants per row were planted. The experiment was conducted in four biological replicates for each treatment type. One factor was analyzed statistically: the MC application.

Plants were treated by root watering with 0.3 g $L^{-1}$ of MC (the +MC group) dissolved in tap water (2.7 L per plant) at 2 weeks after planting. The control group of plants was watered with tap water (−MC). The experiment consisted of three treatments during the growing season at regular intervals (3 weeks). In between, when the soils dried out, the plants were moistened with water. The plants' above-ground parts were sampled 1 week after the last treatment with the MC to assess the growth and physiological characteristics and Si accumulation.

### 4.3. Analysis of Growth Characteristics

Growth was measured in terms of shoot height (cm), the number of leaves and roots per plant, root length (cm), the number of stolons per plant, the number of daughter ramets per stolon, fresh and dry weights of shoots and roots (g), and leaf blade area ($cm^2$).

The dry weight of the plant material was determined by drying it three times to constant weight at 75 °C for 2 h.

The area of plant leaf blades was calculated using the SIAMS Photolab software with the additional module SIAMS MesoPlant Ekaterinburg, Russia).

### 4.4. Determination of the RWC

This parameter of leaf tissues is used to assess the water status of plants, and it is expressed by: RWC = [(FW − DW)/(TW − DW)] × 100, where FW, DW, and TW are fresh, dry, and turgid weights, respectively, of a plant [85]. To determine the turgid weight, leaf sections (10 mm) were placed in distilled water at 23 ± 2 °C under light for 24 h. The dry mass was measured after drying the leaves at 75 °C for 48 h. For this analysis, the leaves of 20 randomly selected plants from each treatment group were used. The samples for pigment concentrations, analysis of hydrogen peroxide and enzymatic antioxidants, and phytohormone quantification were prepared from combined and homogenized fresh leaves of 20 randomly selected plants from each treatment group. Similarly prepared samples of air-dried leaves were used for analysis of phenolic compounds.

### 4.5. Determination of Pigment Concentrations

Fresh leaf tissues (10 mg) were ground in 96% ethanol (1000 °L). The homogenate was centrifuged using Eppendorf Centrifuge 5430 R (Hamburg, Germany) at 8000 rpm for 10 min. After that, the supernatant's volume was brought to 3 mL with 96% ethanol. The optical density (D) of the solutions was measured immediately at wavelengths of 470, 664, and 649 nm on a UNICO 2100 UV spectrophotometer (Princeton, New Jersey, NJ, USA). The samples containing free 96% ethanol served as a control. The pigment concentration (C) was calculated according to H.K. Lichtenthaler and C. Buschmann [86] using the following formulas:

$$C_a = (13.36 \times A_{664} - 5.19 \times A_{649});$$

$$C_b = (27.43 \times A_{649} - 8.12 \times A_{664});$$

$$C_c = (1000 \times A_{470} - 2.13 \times C_a - 97.64 \times C_b)/209,$$

where $C_a$, $C_b$, and $C_c$ are concentrations of chlorophyll a, chlorophyll b, and carotenoids (mg $mL^{-1}$) and $A_{664}$, $A_{649}$, and $A_{470}$ are the extract's absorbance at 664, 649, and 470 nm.

The concentrations of pigments ($C_P$) in leaf tissue were calculated and expressed in milligrams per gram of fresh weight (FW) (mg $g^{-1}$) as follow: $C_P = C \times V/(1000 \times n)$, where C is pigment concentration in the ethanol extract (mg $mL^{-1}$); V is the volume of the extract (mL); and n is the fresh weight of the sample (g).

### 4.6. Analysis of Hydrogen Peroxide and Enzymatic Antioxidants

The endogenous hydrogen peroxide ($H_2O_2$) level and enzymatic activities were assayed according to previously published protocols [22]. The $H_2O_2$ content was measured based on the peroxide-mediated oxidation of $Fe^{2+}$, followed by the reaction of $Fe^{3+}$ with xylenol orange [87,88]. Absorbance of the $Fe^{3+}$–xylenol orange complex was measured at 560 nm. A standard curve was obtained by diluting 3% $H_2O_2$. The data were expressed in micromoles of $H_2O_2$ per gram of FW (μmol (g $FW)^{-1}$).

Catalase (EC 1.11.1.6) activity was determined according to the decomposition of $H_2O_2$ [89] and measured on the UNICO 2100 UV spectrophotometer (USA) by assessing a decrease in absorbance at 240 nm. The reaction mixture contained 50 mM of Na-K-phosphate buffer (pH 7.0) and 0.03 mL of enzyme extract. CAT activity was expressed in units of $H_2O_2$ decomposed per minute per gram of FW ((U $H_2O_2$)(g $FW)^{-1}min^{-1}$).

Superoxide dismutase (EC 1.15.1.1) activity determination involved ethylenediaminetetraacetic acid (EDTA), a 0.05% (*w*/*v*) nitro blue tetrazolium solution, a 0.025% (*w*/*v*) riboflavin solution, and 0.1 mL of the enzyme extract [90]. The reaction mixture was incubated for 15 min at 25 °C under light with an intensity of 40 °mol $m^{-2}$ $s^{-1}$ (Philips, Pila, Poland), and then absorbance was measured at 560 nm against a nonirradiated blank. Absorbance was expressed in units (U) per gram of FW (U (g $FW)^{-1}$).

Peroxidase (EC 1.11.1.7) activity was measured at 365 nm as a decrease in quercetin concentration [91]. The POD reaction solution contained 0.1 mL of the enzyme extract, 0.2 M of acetate buffer (pH 4.5), 0.1 °M of quercetin, and 10 °M of $H_2O_2$. The reaction mixture contained 50 mM of Na-K-phosphate buffer (pH 7.8) and 0.24% (*w*/*v*) of added $H_2O_2$. The activity was expressed in units of quercetin oxidized per minute per gram of FW (U (g $FW)^{-1}$ $min^{-1}$).

### 4.7. Phytohormone Quantification

Quantification of IAA, ABA, and t-Z was performed using the method of Dobrev et al. [92]. Standards of phytohormones were purchased from Sigma-Aldrich: IAA (Sigma-Aldrich, Chemie GmbH, Buch, Switzerland), ABA (Sigma-Aldrich, Trading Co Ltd, Shanghai, China), and t-Z (Sigma-Aldrich, AG Industriestrasse, Karnataka, India). Internal standards indolepropionic acid (IPA), naphthylacetic acid (NAA), and 6-benzylaminopurine (BA) were acquired from Sigma-Aldrich, CHEMIE GmbH (Shanghai, China). Liquid chromatography/mass spectrometry (LC-MS)-grade acetonitrile was purchased from Biosolve (Dieuze, France). Water for LC-MS was obtained using a Milli-Q purification system from Millipore Corp. (Bedford, MA, USA). Gaseous nitrogen (ultrapure, 99.9%) was produced by means of a Genius NM32LA nitrogen generator (Peak Scientific Instruments Ltd., Scotland, UK). LC-MS-grade acetic acid was acquired from Sigma-Aldrich (St. Louis, MO, USA). LC coupled with tandem mass spectrometry (MS/MS) was performed at the Core Facility of Mass-Spectrometric Analysis of ICBFM SB RAS for quantification of IAA, ABA, and t-Z in an extract. For this analysis, 100 mg of FW of leaves (above-ground parts) of strawberry plants was homogenized in the mixture of methanol, water, and formic acid (15:4:1); internal standards (BA 50 ng and IPA and NAA 500 ng each) were added into the mixture. The homogenate was put into a 2 mL microcentrifuge tube, which was placed into a freezer (–20 °C) for extraction for 1 h. After that, the homogenate was spun in the centrifuge at 20,000× *g* at 4 °C for 20 min. The supernatant was transferred into a new 2 mL tube and evaporated to a volume less than 0.25 mL.

Chromatographic separation was performed on an Agilent 1200 HPLC system (Agilent Technologies, Santa Clara, CA, USA) outfitted with a Prontosil C-18 analytical column (Econova, Novosibirsk, Russia) with an Eclipse XDB-C18 pre-column (Agilent Technologies, Santa Clara, CA, USA). MS detection was conducted on an Agilent 6410 QQQ mass spectrometer (Agilent Technologies, USA) equipped with an electrospray ionization (ESI) source. For IAA and t-Z quantification, detection was carried out in the positive multiple-reaction monitoring (MRM) mode, and for ABA detection, the negative MRM mode was applied.

In the negative mode, transition $m/z$ 263.0 $\rightarrow$ 219.1 was employed for quantification of ABA (collision energy 14 eV, fragmentation voltage 130 V) and transition $m/z$ 185.1 $\rightarrow$ 140.9 was selected for NAA (collision energy 5 eV, fragmentation voltage 130 V). In positive mode, the following transitions were utilized for quantification of analytes: $m/z$ 220.1 $\rightarrow$ 136.1 for t-Z (collision energy 15 eV, fragmentation voltage 110 V), 226.1 $\rightarrow$ 91.1 for BA (collision energy 33 eV, fragmentation voltage 100 V), $m/z$ 176.1 $\rightarrow$ 130.1 for IAA (collision energy 10 eV, fragmentation voltage 70 V), and 190.1 $\rightarrow$ 130.1 for IPA (collision energy 15 eV, fragmentation voltage 130 V). Typical MS conditions were spray voltage at 4000 V, turbo spray temperature of 300 °C, and nitrogen as a nebulizer gas at 40 psi and a 9.2 L $min^{-1}$ flow rate. The duration of a single ion scan was set to 500 ms for t-Z, IAA, AA, and NAA and to 200 ms for BA and IPA. The temperature of the autosampler was adjusted to 6 °C.

For the chromatographic separation of the phytohormones and internal standards, gradient elution with 5 mM of acetic acid in water (eluent A) and 0.5 mM of acetic acid in acetonitrile (eluent B) was conducted. The run with the gradient was started with a solvent A–solvent B mixture at 95:5 ($v/v$) followed by a linear gradient to 50:50 ($v/v$) for the first 15 min, then to 10:90 ($v/v$) from minute 15 to minute 20, and this proportion was constant until minute 22; the mobile phase was returned to 95:5 from minute 22.0 to minute 22.1, and this proportion was constant up to minute 30. The flow rate was set to 0.3 $mL{\cdot}min^{-1}$. Calibration curves for the phytohormones' quantification were built using several concentrations: 1, 5, 10, 30, 50, and 80 $ng{\cdot}mL^{-1}$. Data were processed using the Agilent Technologies Masshunter B03.01 [B2065] software.

### *4.8. High-Performance Liquid Chromatography (HPLC) Analysis of Phenolic Compounds*

For this analysis, air-dried above-ground parts from three randomly selected plant samples were used. Dry weight concentrations in the samples were calculated using the gravimetric method. Then, the samples were ground up, weighed (0.2 g), and extracted three times with an ethanol–water mixture (70:30, $v/v$) in a water bath at 60–70 °C. HPLC analysis of the aqueous ethanol extracts was performed using an Agilent 1200 system with a diode array detector and ChemStation B04.01 SP1[647] software for data processing (Agilent Technologies, Santa Clara, CA, USA). The chromatographic separation was carried out on a Zorbax SB-C18 column (4.6 × 150 mm, 5 °m internal diameter) at 25 °C. The mobile phase consisted of MeOH (solvent A) and 0.1% orthophosphoric acid in water (solvent B).

The separation of compounds in the extracts was performed through gradient 1. The run via this gradient was started with an A–B mixture at 22:78 (*v*/*v*) followed by a linear gradient to 70:30 (*v*/*v*) for the first 30 min, and then to 100:0 (*v*/*v*) from minute 30 to minute 32. The mobile phase was returned to 22:78 (*v*/*v*) from minute 32 to minute 36.

To confirm glycoside composition, flavonoid aglycones were identified after acid hydrolysis of the extracts. The aqueous ethanol extracts of above-ground parts of *F.* × *ananassa* were hydrolyzed with 2N HCl for 1 h in a boiling water bath. The hydrolyzed extracts were separated via gradient 2. The run via this gradient was started with a solvent A–solvent B mixture at 50:50 (*v*/*v*) followed by a linear transition to 52:48 (*v*/*v*) for the first 15 min, then to 100:0 (*v*/*v*) from minute 15 to minute 17, with a return to 50:50 (*v*/*v*) from minute 17 to minute 20. The flow rate in both gradients was set to 1 mL·min$^{-1}$. The sample injection volume was 10 °L. Monitoring of chromatograms was carried out via absorbance at 220, 255, 270, 290, 325, 340, 350, and 370 nm.

Phenolic compounds were quantified using the external standard method. Reference standards of gallic, *p*-hydroxybenzoic, syringic, neochlorogenic, *p*-coumaric, caffeic, and ellagic acids and of (+)-catechin, quercetin, kaempferol, kaempferol rutinoside (Sigma-Aldrich, St. Louis, MO, USA), epigallocatechin gallate (Teavigo, Gevelsberg, Germany), vanillic acid (Serva, Heidelberg, Germany), and ferulic acid (Koch-light Lab Ltd., Haverhill, Suffolk, UK) were used for building calibration curves in the concentration range of 10–100 °g mL$^{-1}$. Ellagic acid derivatives and quercetin glycosides were quantified as ellagic acid and rutin, respectively.

Three biological and three technical replicates were processed in each assay.

### 4.9. X-ray Fluorescence Analysis of Si

Samples of roots and shoots of 20 plants from each treatment group were dried at 50 °C to a constant weight. Then, the material was ground and pressed into tablets weighing 20 mg. The Si content of the samples was determined through SR-XRF at the Siberian Synchrotron and Terahertz Radiation Center (SSTRC) at Budker Institute of Nuclear Physics, SB RAS, Novosibirsk, Russia (BINP SB RAS) at the 1st SR beamline of the VEPP-4M synchrotron [93]. Elemental analysis was carried out using monochromatic exciting synchrotron radiation at 3.53 keV photon energy. The characteristic X-rays emitted from the sample were detected using a Si-PIN diode X-ray detector (Amptek XR-100CR, Bedford, MA, USA; with 165 eV FWHM of MnK$\alpha$, 500 °m Si crystal thickness, and 12.7 °m Be window) positioned at 45° from the normal of the sample surface. The beam size at the sample position was ~1 × 4 mm. For calculation of the Si content, a procedure of the external standard analysis using a standard sample (the Tr1 meadow herbal mix; CRM 8922 2007, produced at Vinogradov Institute of Geochemistry, SB RAS, Irkutsk, Russia) was applied. The concentrations of Si were expressed in ppm (µg g$^{-1}$).

### 4.10. The Cuticle Thickness Measurement

For the analysis, the second leaf was taken from the apical bud of the shoot. Then, 5.0 × 5.0 mm sections were cut out from the middle part of the middle lobe of the trifoliate leaf, avoiding the midrib. The collected leaf materials were fixed in FAA (formalin: glacial acetic acid: 95% ethanol, 7:7:100 (*v*/*v*/*v*)) before storage in 70% ethanol. After that, the leaves were dehydrated using a graded ethanol series. Dehydrated leaf materials were sequentially passed through mixtures made of 95% ethanol/Technovit 7100 resin (Heraeus Kulzer, Germany) combinations (7:3, 15 h; 1:1, 15 h; 3:7, 15 h (*v*/*v*)) and then 100% Technovit for 15 h. The samples were then solidified in Technovit 7100 resin following the manufacturer's protocol. Embedded samples were cut on an HM-325 rotary microtome (Microm,

Germany) into sections of 7 °m thickness, placed onto slide glass, and dried on an electric slide warmer for 12 h. Dried slides were stained with 0.2% Sudan black B (Servicebio Technology Co., Ltd., Wuhan, China) for 60–90 s, rinsed with running water, and again dried by means of the electric slide warmer for more than 6 h to remove water. The stained slides were then mounted with Mowiol® (Sigma-Aldrich, Taufkirchen, Germany) and examined under a VHX-5000 digital light microscope (Keyence, Osaka, Japan). For each type of sample, 300 morphometric measurements were carried out.

### 4.11. Statistical Analysis

For the data analyses, STATISTICA 10.0 software (Statsoft Inc., Tulsa, OK, USA) was used. The results of the number of roots per plant, root length, fresh and dry weights of shoots and roots, and leaf blade area were reported as mean ± standard errors (SE) of 20 individual plants. The results of shoot height, the number of leaves per plant, and the number of stolons per plant were expressed as mean ± standard errors (SE) of 40 individual plants. The results of the biochemical tests and analysis of Si were presented as mean ± SE of 9 (3 biological x 3 technical) replicates. The results were compared using Duncan's multiple-range test to find significant differences among treatment groups. Differences between the means were considered statistically significant at $p < 0.01$ and $p < 0.05$.

## 5. Conclusions

The present results revealed that Si application has beneficial effects on in-vitro-derived strawberry plants during growth in the field. Silicon-chelate-mediated metabolic alterations in the strawberry plants resulted in improvement of protective defense mechanisms in the plants. These strategies involve a complex antioxidant defense system that is composed of antioxidant enzymes and low-molecular-weight compounds (ABA, benzoic acids, and ellagic acid derivative). The activation of the antioxidant defense system promoted growth of the strawberry plants and substantially enhanced their vegetative productivity due to an increase in biomass and in the number of ramets from a mother plant. Cuticle layer thickening in strawberry leaves is proposed as a potential mediator of the modulation of the water balance by silicon chelates. These findings may help produce commercial strawberries worldwide.

**Author Contributions:** Conceptualization, E.A., E.K., O.K. and B.G.; methodology, E.A., E.K., O.K., A.C. and B.G.; validation, E.A., E.K., O.K., A.C. and B.G.; formal analysis, E.A., E.K., O.K., A.C. and B.G.; investigation, E.A., E.K., O.K., U.P. and B.G.; resources, E.A., E.K., O.K., E.T., A.C. and B.G.; data curation, E.A., E.K., O.K., A.C. and B.G.; writing—original draft preparation, E.A., E.K., O.K., A.C. and B.G.; writing—review and editing, E.A., E.K., O.K. and B.G.; visualization, E.A., E.K. and O.K.; supervision, E.A.; project administration, E.A.; funding acquisition, E.A., U.P. and O.K. All authors have read and agreed to the published version of the manuscript.

**Funding:** This research was funded by the Russian Science Foundation and the Government of Novosibirsk Region, grant number 22-26-20061 (https://rscf.ru/project/22-26-20061/ (accessed on 6 November 2023).

**Data Availability Statement:** Raw data from this study are available upon request from the corresponding author.

**Acknowledgments:** In vitro material from the collection of CSBG, SB RAS, was used: unique scientific unit (USU) 440534: "Collection of living plants indoors and outdoors". The determination of the elemental composition using the SR-XRF method was carried out with the equipment at the shared research center SSTRC at the VEPP-4–VEPP-2000 complex at BINP SB RAS.

**Conflicts of Interest:** The authors declare no conflict of interest.

## References

1. Durán-Lara, E.F.; Valderrama, A.; Marican, A. Natural organic compounds for application in organic farming. *Agriculture* **2020**, *10*, 41. [CrossRef]

2. Garza-Alonso, C.A.; Olivares-Sáenz, E.; González-Morales, S.; Cabrera-De la Fuente, M.; Juárez-Maldonado, A.; González-Fuentes, J.A.; Tortella, G.; Valdés-Caballero, M.V.; Benavides-Mendoza, A. Strawberry biostimulation: From mechanisms of action to plant growth and fruit quality. *Plants* **2022**, *11*, 3463. [CrossRef] [PubMed]
3. Rana, V.S.; Sharma, S.; Rana, N.; Sharma, U. Sustainable production through biostimulants under fruit orchards. *CABI Agric. Biosci.* **2022**, *3*, 38. [CrossRef]
4. Savvas, D.; Ntatsi, G. Biostimulant activity of silicon in horticulture. *Sci. Hortic.* **2015**, *196*, 66–81. [CrossRef]
5. Schaller, J.; Cramer, A.; Carminati, A.; Mohsen, Z. Biogenic amorphous silica as main driver for plant available water in soils. *Sci. Rep.* **2020**, *10*, 2424. [CrossRef]
6. Ma, J.F. Role of silicon in enhancing the resistance of plants to biotic and abiotic stresses. *Soil Sci. Plant Nutr.* **2004**, *50*, 11–18. [CrossRef]
7. Zargar, S.M.; Mahajan, R.; Bhat, J.A.; Nazir, M.; Deshmukh, R. Role of silicon in plant stress tolerance: Opportunities to achieve a sustainable cropping system. *3 Biotech* **2019**, *9*, 73. [CrossRef]
8. Verma, K.K.; Song, X.-P.; Tian, D.-D.; Guo, D.-J.; Chen, Z.-L.; Zhong, C.-S.; Nikpay, A.; Singh, M.; Rajput, V.D.; Singh, R.K.; et al. Influence of silicon on biocontrol strategies to manage biotic stress for crop protection, performance, and improvement. *Plants* **2021**, *10*, 2163. [CrossRef]
9. Matichenkov, V.V.; Bocharnikova, E.A. Chapter 13. The relationship between silicon and soil physical and chemical properties. In *Silicon in Agriculture*; Datnoff, L.E., Snyder, G.H., Korndorfer, G.H., Eds.; Elsevier: Amsterdam, The Netherlands, 2001; pp. 209–219. [CrossRef]
10. Vatansever, R.; Ozyigit, I.I.; Filiz, E. Essential and beneficial trace elements in plants, and their transport in roots: A review. *Appl. Biochem. Biotechnol.* **2017**, *181*, 464–482. [CrossRef]
11. Artyszak, A. Effect of silicon fertilization on crop yield quantity and quality—A literature review in Europe. *Plants* **2018**, *7*, 54. [CrossRef]
12. Calero Hurtado, A.; Chiconato, D.A.; de Prado, R.M.; da Sousa Junior, G.S.; Felisberto, G. Silicon attenuates sodium toxicity by improving nutritional efficiency in sorghum and sunflower plants. *Plant Physiol. Biochem.* **2019**, *142*, 224–233. [CrossRef] [PubMed]
13. Coskun, D.; Deshmukh, R.; Sonah, H.; Menzies, J.G.; Reynolds, O.; Ma, J.F.; Kronzucker, H.J.; Belanger, R.R. The controversies of silicon's role in plant biology. *N. Phytol.* **2019**, *221*, 67–85. [CrossRef] [PubMed]
14. Joudmand, A.; Hajiboland, R. Silicon mitigates cold stress in barley plants via modifying the activity of apoplasmic enzymes and concentration of metabolites. *Acta Physiol. Plant.* **2019**, *41*, 29. [CrossRef]
15. Ali, N.; Réthoré, E.; Yvin, J.-C.; Hosseini, S.A. The regulatory role of silicon in mitigating plant nutritional stresses. *Plants* **2020**, *9*, 1779. [CrossRef] [PubMed]
16. Hajiboland, R.; Moradtalab, N.; Aliasgharzad, N.; Eshaghi, Z.; Feizy, J. Silicon influences growth and mycorrhizal responsiveness in strawberry plants. *Physiol. Mol. Biol. Plants* **2018**, *24*, 1103–1115. [CrossRef] [PubMed]
17. Luyckx, M.; Hausman, J.-F.; Lutts, S.; Guerriero, G. Silicon and plants: Current knowledge and technological perspectives. *Front. Plant Sci.* **2017**, *8*, 411. [CrossRef] [PubMed]
18. Sahebi, M.; Hanafi, M.M.; Nor Akmar, A.S.; Rafii, M.Y.; Azizi, P.; Tengoua, F.F.; Azwa, J.N.M.; Shabanimofrad, M. Importance of silicon and mechanisms of biosilica formation in plants. *BioMed Res. Int.* **2015**, *2015*, 396010. [CrossRef]
19. Lomovsky, O.I.; Lomovskiy, I.O.; Orlov, D.V. Mechanochemical solid acid/base reactions for obtaining biologically active preparations and extracting plant materials. *Green Chem. Lett. Rev.* **2017**, *10*, 171–185. [CrossRef]
20. Trofimova, E.G.; Podgorbunskikh, E.M.; Skripkina, T.S.; Bychkov, A.L.; Lomovsky, O.I. Scaling of the mechanochemical process of production of silicon chelates. *Bulg. Chem. Commun.* **2018**, *50*, 45–48.
21. Tolisano, C.; Del Buono, D. Biobased: Biostimulants and biogenic nanoparticles enter the scene. *Sci. Total Environ.* **2023**, *885*, 163912. [CrossRef]
22. Ambros, E.; Karpova, E.; Kotsupiy, O.; Zaytseva, Y.; Trofmova, E.; Novikova, T. Silicon chelates from plant waste promote in vitro shoot production and physiological changes in strawberry plantlets. *Plant Cell Tissue Organ Cult.* **2021**, *145*, 209–221. [CrossRef]
23. Ambros, E.; Karpova, E.; Kotsupiy, O.; Trofimova, E.; Zakabluk, G.; Chernonosov, A.; Koval, V.; Novikova, T. A mechanocomposite based on biogenic silica and green tea flavonoids modulates adaptability of strawberry microclones to in vitro and ex vitro conditions. *J. Soil Sci. Plant Nutr.* **2023**, *23*, 612–627. [CrossRef]
24. Liang, Y.; Urano, D.; Liao, K.L.; Hedrick, T.L.; Gao, Y.; Jones, A.M. A nondestructive method to estimate the chlorophyll content of *Arabidopsis* seedlings. *Plant Methods* **2017**, *13*, 26. [CrossRef] [PubMed]
25. Kurniawan, N.S.H.; Kirana, I.A.P.; Abidin, A.S.; Jupri, A.; Widyastuti, S.; Hernawan, A.; Nikmatullah, A.; Sunarpi, H.; Prasedya, E.S. Analysis of leaf chlorophyll content of paddy plants during vegetative stage grown in soil media containing macroalgae organic fertilizer. *IOP Conf. Ser. Earth Environ. Sci.* **2021**, *913*, 012025. [CrossRef]
26. Qian, X.; Liu, L.; Chen, X.; Zhang, X.; Chen, S.; Sun, Q. The global leaf chlorophyll content dataset over 2003–2012 and 2018–2020 derived from MERIS/OLCI satellite data (GLCC): Algorithm and validation. *Earth Syst. Sci.* **2022**, *in press*. [CrossRef]
27. Borghesi, E.; González-Miret, M.L.; Escudero-Gilete, M.L.; Malorgio, F.; Heredia, F.J.; Meléndez-Martínez, A.J. Effects of salinity stress on carotenoids, anthocyanins, and color of diverse tomato genotypes. *J. Agric. Food Chem.* **2011**, *59*, 11676–11682. [CrossRef]
28. Motyka, O.; Štrbová, K.; Zinicovscaia, I. Chlorophyll content in two medicinal plant species following nano-$TiO_2$ exposure. *Bull. Environ. Contam. Toxicol.* **2020**, *104*, 373–379. [CrossRef]

29. Amri, M.; Abbes, Z.; Trabelsi, I.; Ghanem, M.E.; Mentag, R.; Kharrat, M. Chlorophyll content and fluorescence as physiological parameters for monitoring Orobanche foetida Poir. infection in faba bean. *PLoS ONE* **2021**, *16*, e0241527. [CrossRef]
30. Jiroutova, P.; Kovalikova, Z.; Toman, J.; Dobrovolna, D.; Andrys, R. Complex analysis of antioxidant activity, abscisic acid level, and accumulation of osmotica in apple and cherry in vitro cultures under osmotic stress. *Int. J. Mol. Sci.* **2021**, *22*, 7922. [CrossRef]
31. Vuolo, M.M.; Lima, V.S.; Maróstica Junior, M.R. Phenolic compounds: Structure, classification, and antioxidant power. In *Bioactive Compounds: Health Benefits and Potential Applications Provides*; Campos, M.R.S., Ed.; Elsevier Inc.: Amsterdam, The Netherlands, 2019; pp. 33–50. [CrossRef]
32. Zhang, J.; Kirkham, M.B. Drought-stress-induced changes in activities of superoxide dismutase, catalase, and peroxidase in wheat species. *Plant Cell Physiol.* **1994**, *35*, 785–791. [CrossRef]
33. Leung, D.W.M. Studies of catalase in plants under abiotic stress. In *Antioxidants and Antioxidant Enzymes in Higher Plants*; Gupta, D.K., Palma, J.M., Corpas, F.J., Eds.; Springer International Publishing: Cham, Switzerland, 2018; pp. 27–39.
34. Ranjith, H.V.; Sagar, D.; Kalia, V.K.; Dahuja, A.; Subramanian, S. Differential activities of antioxidant enzymes, superoxide dismutase, peroxidase, and catalase vis-à-vis phosphine resistance in field populations of lesser grain borer (*Rhyzopertha dominica*) from India. *Antioxidants* **2023**, *12*, 270. [CrossRef] [PubMed]
35. Peleg, Z.; Blumwald, E. Hormone balance and abiotic stress tolerance in crop plants. *Curr. Opin. Plant Biol.* **2011**, *14*, 290–295. [CrossRef] [PubMed]
36. Sabagh, A.E.; Islam, M.S.; Hossain, A.; Iqbal, M.A.; Mubeen, M.; Waleed, M.; Reginato, M.; Battaglia, M.; Ahmed, S.; Rehman, A.; et al. Phytohormones as growth regulators during abiotic stress tolerance in plants. *Front. Agron.* **2022**, *4*, 765068. [CrossRef]
37. Lim, C.W.; Baek, W.; Jung, J.; Kim, J.H.; Lee, S.C. Function of ABA in stomatal defense against biotic and drought stresses. *Int. J. Mol. Sci.* **2015**, *16*, 15251–15270. [CrossRef] [PubMed]
38. Xiong, L.; Zhu, J.K. Regulation of abscisic acid biosynthesis. *Plant Physiol.* **2003**, *133*, 29–36. [CrossRef] [PubMed]
39. Jiang, M.Y.; Zhang, J.H. Role of abscisic acid in water stress induced antioxidant defense in leaves of maize seedlings. *Free Radic. Res.* **2002**, *36*, 1001–1015. [CrossRef] [PubMed]
40. Ntatsi, G.; Savvas, D.; Schwarz, D. Role of abscisic acid in the adaptation of grafted tomato to moderately suboptimal temperature stress. *Acta Hortic.* **2012**, *952*, 295–302. [CrossRef]
41. Tuteja, N.; Sopory, S.K. Chemical signaling under abiotic stress environment in plants. *Plant Signal. Behav.* **2008**, *3*, 525–536. [CrossRef]
42. Ahmad, Z.; Abbasi, A.; Sultana, S.; Waraich, E.; Artyszak, A.; Iqbal, M.A.; Barutçular, C. Manipulation of silicon metabolism in plants for stress tolerance. In *Silicon and Nano-Silicon in Environmental Stress Management and Crop Quality Improvement*; Etesami, H., Saeedi, A.H.A., El-Ramady, H., Masayuki, F., Mohammad, P., Anwar, H.M., Eds.; Elsevier: Amsterdam, The Netherlands, 2022; pp. 339–348. [CrossRef]
43. Mir, R.A.; Bhat, B.A.; Yousuf, H.; Islam, S.T.; Raza, A.; Rizvi, M.A.; Charagh, S.; Albaqami, M.; Sofi, P.A.; Zargar, S.M. Multidimensional role of silicon to activate resilient plant growth and to mitigate abiotic stress. *Front. Plant Sci.* **2022**, *13*, 819658. [CrossRef]
44. Currie, H.A.; Perry, C.C. Silica in plants: Biological, biochemical and chemical studies. *Ann. Bot.* **2007**, *7*, 1383–1389. [CrossRef]
45. Ma, J.F.; Takahashi, E. *Soil, Fertilizer, and Plant Silicon Research in Japan*; Elsevier: Amsterdam, The Netherlands, 2002; p. 281. [CrossRef]
46. Miyake, Y.; Takahashi, E. Effect of silicon on the growth and fruit production of strawberry plants in a solution culture. *J. Soil Sci. Plant Nutr.* **1986**, *32*, 321–326. [CrossRef]
47. Ouellette, S.; Goyette, M.H.; Labbé, C.; Laur, J.; Gaudreau, L.; Gosselin, A.; Dorais, M.; Deshmukh, R.K.; Bélanger, R.R. Silicon transporters and effects of silicon amendments in strawberry under high tunnel and field conditions. *Front. Plant Sci.* **2017**, *8*, 949. [CrossRef] [PubMed]
48. Kanto, T.; Miyoshi, A.; Ogawa, T.; Maekawa, K.; Aino, M. Suppressive effect of liquid potassium silicate on powdery mildew of strawberry in soil. *J. Gen. Plant Pathol.* **2006**, *72*, 137–142. [CrossRef]
49. Hodson, M.J.; White, P.J.; Mead, A.; Broadley, M.R. Phylogenetic variation in the silicon composition of plants. *Ann. Bot.* **2005**, *96*, 1027–1046. [CrossRef] [PubMed]
50. Liang, Y.; Nikolic, M.; Bélanger, R.; Gong, H.; Song, A. *Silicon in Agriculture: From Theory to Practice*; Springer: Dordrecht, The Netherlands, 2015; p. 235. [CrossRef]
51. Singh, P.; Kumar, V.; Sharma, A. Interaction of silicon with cell wall components in plants: A review. *J. Appl. Nat. Sci.* **2023**, *15*, 480–497. [CrossRef]
52. Chen, W.; Yao, X.; Cai, K.; Chen, J. Silicon alleviates drought stress of rice plants by improving plant water status, photosynthesis and mineral nutrient absorption. *Biol. Trace Element Res.* **2011**, *142*, 67–76. [CrossRef] [PubMed]
53. Zhu, Y.X.; Gong, H.J.; Yin, J.L. Role of silicon in mediating salt tolerance in plants: A review. *Plants* **2019**, *8*, 147. [CrossRef]
54. Li, Y.; Xiao, J.; Hu, J.; Jeong, B.R. Method of silicon application affects quality of strawberry daughter plants during cutting propagation in hydroponic substrate system. *Agronomy* **2020**, *10*, 1753. [CrossRef]
55. Pozo, J.; Urrestarazu, M.; Morales, I.; Sanchez, J.; Santos, M.; Dianez, F.; Alvaro, J.E. Effects of silicon in the nutrient solution for three horticultural plant families on the vegetative growth, cuticle, and protection against Botrytis cinerea. *HortScience* **2015**, *50*, 1447–1452. [CrossRef]

56. Zhu, Y.; Xu, X.; Hu, Y.; Han, W.; Yin, J.; Li, H.; Gong, H.-J. Silicon improves salt tolerance by increasing root water uptake in *Cucumis sativus* L. *Plant Cell Rep.* **2015**, *34*, 1629–1646. [CrossRef]
57. Chen, D.; Wang, S.; Yin, L.; Deng, X. How does silicon mediate plant water uptake and loss under water deficiency? *Front. Plant Sci.* **2018**, *9*, 281. [CrossRef] [PubMed]
58. Wang, M.; Wang, R.; Mur, L.A.J.; Ruan, J.; Shen, Q.; Guo, S. Functions of silicon in plant drought stress responses. *Hortic. Res.* **2021**, *8*, 254. [CrossRef] [PubMed]
59. Remus-Borel, W.; Menzies, J.G.; Bélanger, R.R. Silicon induces antifungal compounds in powdery mildew-infected wheat. *Physiol. Mol. Plant Pathol.* **2005**, *66*, 108–115. [CrossRef]
60. Rahman, A.; Wallis, C.M.; Uddin, W. Silicon-induced systemic defense responses in perennial ryegrass against infection by Magnaporthe oryzae. *Phytopathology* **2015**, *105*, 748–757. [CrossRef] [PubMed]
61. Kumar, N.; Dubey, A.K.; Upadhyay, A.K.; Gautam, A.; Ranjan, R.; Srikishna, S.; Sahu, N.; Behera, S.K.; Mallick, S. GABA accretion reduces Lsi-1 and Lsi-2 gene expressions and modulates physiological responses in *Oryza sativa* to provide tolerance towards arsenic. *Sci. Rep.* **2017**, *7*, 8786. [CrossRef] [PubMed]
62. Katz, O.; Puppe, D.; Kaczorek, D.; Prakash, N.B.; Schaller, J. Silicon in the soil–plant continuum: Intricate feedback mechanisms within ecosystems. *Plants* **2021**, *10*, 652. [CrossRef]
63. Markovich, O.; Steiner, E.; Kouřil, Š.; Tarkowski, P.; Aharoni, A.; Elbaum, R. Silicon promotes cytokinin biosynthesis and delays senescence in Arabidopsis and Sorghum. *Plant Cell Environ.* **2017**, *40*, 1189–1196. [CrossRef]
64. Gai, Z.; Wang, Y.; Ding, Y.; Qian, W.; Qiu, C.; Xie, H.; Sun, L.; Jiang, Z.; Ma, Q.; Wang, L.; et al. Exogenous abscisic acid induces the lipid and flavonoid metabolism of tea plants under drought stress. *Sci. Rep.* **2020**, *10*, 12275. [CrossRef]
65. Hao, G.; Du, X.; Zhao, F.; Ji, H. Fungal endophytes-induced abscisic acid is required for flavonoid accumulation in suspension cells of Ginkgo biloba. *Biotechnol. Lett.* **2010**, *32*, 305–314. [CrossRef]
66. Vega, I.; Rumpel, C.; Ruíz, A.; Mora, M.D.L.L.; Calderini, D.F.; Cartes, P. Silicon modulates the production and composition of phenols in barley under aluminum stress. *Agronomy* **2020**, *10*, 1138. [CrossRef]
67. Pagassini, J.A.V.; de Godoy, L.J.G.; Campos, F.G.; Barzotto, G.R.; Vieira, M.A.R.; Boaro, C.S.F. Silicon and mechanical damage increase polyphenols and vitexin in *Passiflora incarnata* L. *Sci. Rep.* **2021**, *11*, 22064. [CrossRef] [PubMed]
68. Carneiro-Carvalho, A.; Aires, A.; Anjos, R.; Martins, L.; Pinto, T.; Peixoto, F.; Gomes-Laranjo, J. The role of silicon fertilization in the synthesis of phenolic compounds on chestnut plants infected with *P. cinnamomi* and *C. parasitica*. *J. Plant Dis. Prot.* **2020**, *127*, 211–227. [CrossRef]
69. Shahnaz, G.; Shekoofeh, E.; Kourosh, D.; Moohamadbagher, B. Interactive effects of Silicon and Aluminum on the malondialdehyde (MDA), proline, protein and phenolic compounds in *Borago officinalis* L. *J. Med. Plants Res.* **2011**, *5*, 5818–5827. [CrossRef]
70. Fini, A.; Brunetti, M.D.F.C.; Tattini, M. Stress-induced flavonoid biosynthesis and the antioxidant machinery of plants. *Plant Signal. Behav.* **2011**, *6*, 709–711. [CrossRef] [PubMed]
71. Astashenkov, A.Y.; Karpova, E.A.; Cheryomushkina, V.A. Diversity patterns of life forms and phenolic profiles of endemic Nepeta plants along an aridity gradient of a high-mountain zone in Central Asia. *Taiwania* **2021**, *66*, 541–556. [CrossRef]
72. Hajiboland, R.; Moradtalab, N.; Eshaghi, Z.; Feizy, J. Effect of silicon supplementation on growth and metabolism of strawberry plants at three developmental stages. *N. Z. J. Crop Hortic. Sci.* **2018**, *46*, 144–161. [CrossRef]
73. Shetty, R.; Fretté, X.; Jensen, B.; Shetty, N.P.; Jensen, J.D.; Jorgensen, H.J.L.; Newman, M.; Christensen, L.P. Silicon-induced changes in antifungal phenolic acids, flavonoids, and key phenylpropanoid pathway genes during the interaction between miniature roses and the biotrophic pathogen *Podosphaera pannosa*. *Plant Physiol.* **2011**, *157*, 2194–2205. [CrossRef]
74. Heilemann, J.; Strack, D. Incorporation of kaempferol 3-O-glucoside into the cell walls of Norway spruce needles. *Planta* **1990**, *181*, 599–603. [CrossRef]
75. Adato, A.; Mandel, T.; Mintz-Oron, S.; Venger, I.; Levy, D.; Yativ, M.; Domínguez, E.; Wang, Z.; De Vos, R.C.; Jetter, R.; et al. Fruit-surface flavonoid accumulation in tomato is controlled by a SlMYB12-regulated transcriptional network. *PLoS Genet.* **2009**, *5*, e1000777. [CrossRef]
76. Xu, D.; Qu, S.; Tucker, M.R.; Zhang, D.; Liang, W.; Shi, J. Ostkpr1 functions in anther cuticle development and pollen wall formation in rice. *BMC Plant Biol.* **2019**, *19*, 104. [CrossRef]
77. Cho, J.Y.; Moon, J.H.; Seong, K.Y.; Park, K.H. Antimicrobial activity of 4-hydroxybenzoic acid and trans 4-hydroxycinnamic acid isolated and identified from rice hull. *Biosci. Biotechnol. Biochem.* **1998**, *62*, 2273–2276. [CrossRef] [PubMed]
78. Kępa, M.; Miklasińska-Majdanik, M.; Wojtyczka, R.D.; Idzik, D.; Korzeniowski, K.; Smoleń-Dzirba, J.; Wąsik, T.J. Antimicrobial potential of caffeic acid against *Staphylococcus aureus* clinical strains. *Biomed. Res. Int.* **2018**, *2018*, 7413504. [CrossRef]
79. Li, S.; Pi, J.; Zhu, H.; Yang, L.; Zhang, X.; Ding, W. Caffeic acid in tobacco root exudate defends tobacco plants from infection by *Ralstonia solanacearum*. *Front. Plant Sci.* **2021**, *12*, 690586. [CrossRef] [PubMed]
80. Nehela, Y.; Taha, N.A.; Elzaawely, A.A.; Xuan, T.D.; Amin, M.A.; Ahmed, M.E.; El-Nagar, A. Benzoic acid and its hydroxylated derivatives suppress early blight of tomato (*Alternaria solani*) via the induction of salicylic acid biosynthesis and enzymatic and nonenzymatic antioxidant defense machinery. *J. Fungi.* **2021**, *7*, 663. [CrossRef] [PubMed]
81. Barbehenn, R.V.; Peter Constabel, C. Tannins in plant-herbivore interactions. *Phytochemistry* **2011**, *72*, 1551–1565. [CrossRef] [PubMed]

82. Farha, A.K.; Yang, Q.; Kim, G.; Li, H.; Zhu, F.; Liu, H.; Gan, R.; Corke, H.; Li, K.; Liu, Z.; et al. Tannins as an alternative to antibiotics. *Food Biosci.* **2020**, *38*, 100751. [CrossRef]
83. Kalisz, A.; Huska, D.; Jurkow, R.; Dvořák, M.; Caruso, G.; Sękara, A. Nanoparticles of cerium, iron, and silicon oxides change the metabolism of phenols and flavonoids in butterhead lettuce and sweet pepper seedlings. *Environ. Sci. Nano* **2021**, *8*, 1945–1959. [CrossRef]
84. Weber, F.; Barrantes, A.; Tiainen, H. Silicic acid-mediated formation of tannic acid nanocoatings. *Langmuir* **2019**, *35*, 3327–3336. [CrossRef]
85. Yamasaki, S.; Dillenbur, L.R. Measurements of leaf relative water content in Araucaria angustifolia. *R. Bras. Fisiol. Veg.* **1999**, *11*, 69–75.
86. Lichtenthaler, H.K.; Buschmann, C. Chlorophylls and carotenoids: Measurement and characterization by UV–Vis spectroscopy. *Curr. Protoc. Food Anal. Chem.* **2001**, *1*, F4.2.1–F4.2.6. [CrossRef]
87. Wolff, S.P. Ferrous ion oxidation in presence of ferric iron indicator xylenol orange for measurement of hydroperoxide. *Methods Enzymol.* **1994**, *233*, 182–189. [CrossRef]
88. Bellincampi, D.; Dipperro, N.; Salvi, G.; Cervcone, F.; De Lorenzo, G. Extracellular $H_2O_2$ induced by oligogalacturonides is not involved in the inhibition of the auxin-regulated rolB gene expression in tobacco leaf explants. *Plant Physiol.* **2000**, *122*, 1379–1385. [CrossRef] [PubMed]
89. Mehl, K.A.; Davis, M.; Berger, F.G.; Carson, J.A. Myofber degeneration/regeneration is induced in the cachectic ApsMin/+ mouse. *J. Appl. Physiol.* **2005**, *99*, 2379–2387. [CrossRef] [PubMed]
90. Giannopolitis, C.N.; Ries, S.K. Superoxide dismutase I. occurrence in higher plants. *Plant Physiol.* **1977**, *59*, 309–314. [CrossRef] [PubMed]
91. Zhanaeva, T.A.; Lobanova, I.E.; Kukushkina, T.A. Flavonols and flavonol-oxidizing enzymes during buckwheat ontogenesis. *Biol. Bull. Russ. Acad. Sci.* **1999**, *26*, 105–108.
92. Dobrev, P.I.; Hoyerová, K.; Petrášek, J. Analytical determination of auxins and cytokinins. *Methods Mol. Biol.* **2017**, *1569*, 31–39. [CrossRef]
93. Goldenberg, B.G.; Gusev, I.S.; Zubavichus, Y.V. Synchrotron-radiation technological station at the VEPP-4M storage ring. *J. Surf. Investig.* **2023**, *17*, 1088–1093. [CrossRef]

**Disclaimer/Publisher's Note:** The statements, opinions and data contained in all publications are solely those of the individual author(s) and contributor(s) and not of MDPI and/or the editor(s). MDPI and/or the editor(s) disclaim responsibility for any injury to people or property resulting from any ideas, methods, instructions or products referred to in the content.

plants

MDPI

Article

# Biostimulatory Effects of *Chlorella fusca* CHK0059 on Plant Growth and Fruit Quality of Strawberry

**Young-Nam Kim [1,2,*,†], Jun Hyeok Choi [1,†], Song Yeob Kim [1], Young-Eun Yoon [2], Hyeonji Choe [1], Keum-Ah Lee [2], Vimalraj Kantharaj [2], Min-Jeong Kim [3] and Yong Bok Lee [1,2,*]**

1 Division of Applied Life Science (BK21), Gyeongsang National University, Jinju 52828, Republic of Korea
2 Institute of Agriculture and Life Science (IALS), Gyeongsang National University, Jinju 52828, Republic of Korea
3 Organic Agriculture Division, National Academy of Agriculture Science, Rural Development Administration, Wanju 55365, Republic of Korea
* Correspondence: soilnam@gnu.ac.kr (Y.-N.K.); yblee@gnu.ac.kr (Y.B.L.)
† These authors contributed equally to this work.

**Abstract:** Green algae have been receiving widespread attention for their use as biofertilizers for agricultural production, but more studies are required to increase the efficiency of their use. This study aimed to investigate the effects of different levels of *Chlorella fusca* CHK0059 application on strawberry plant growth and fruit quality. A total of 800 strawberry seedlings were planted in a greenhouse and were grown for seven months under different *Chlorella* application rates: 0 (control), 0.1, 0.2, and 0.4% of the optimal cell density (OCD; $1.0 \times 10^7$ cells $mL^{-1}$). The *Chlorella* application was conducted weekly via an irrigation system, and the characteristics of fruit samples were monitored monthly over a period of five months. The growth (e.g., phenotype, dry weight, and nutrition) and physiological (e.g., Fv/Fm and chlorophylls) parameters of strawberry plants appeared to be enhanced by *Chlorella* application over time, an enhancement which became greater as the application rate increased. Likewise, the hardness and P content of strawberry fruits had a similar trend. Meanwhile, 0.2% OCD treatment induced the highest values of soluble solid content (9.3–12 °Brix) and sucrose content (2.06–2.97 g 100 $g^{-1}$) in the fruits as well as fruit flavor quality indices (e.g., sugars:acids ratio and sweetness index) during the monitoring, whilst control treatment represented the lowest values. In addition, the highest anthocyanin content in fruits was observed in 0.4% OCD treatment, which induced the lowest incidence of grey mold disease (*Botrytis cinerea*) on postharvest fruits for 45 days. Moreover, a high correlation between plants' nutrients and photosynthetic variables and fruits' sucrose and anthocyanin contents was identified through the results of principal component analysis. Overall, *C. fusca* CHK0059 application was found to promote the overall growth and performance of strawberry plants, contributing to the improvement of strawberry quality and yield, especially in 0.2% OCD treatment.

**Keywords:** green algae; biofertilizer; crop performance; fruit flavor; postharvest quality

**Citation:** Kim, Y.-N.; Choi, J.H.; Kim, S.Y.; Yoon, Y.-E.; Choe, H.; Lee, K.-A.; Kantharaj, V.; Kim, M.-J.; Lee, Y.B. Biostimulatory Effects of *Chlorella fusca* CHK0059 on Plant Growth and Fruit Quality of Strawberry. *Plants* **2023**, *12*, 4132. https://doi.org/10.3390/plants12244132

Academic Editors: Spyridon A. Petropoulos and Mohamad Hesam Shahrajabian

Received: 22 November 2023
Revised: 9 December 2023
Accepted: 10 December 2023
Published: 11 December 2023

**Copyright:** © 2023 by the authors. Licensee MDPI, Basel, Switzerland. This article is an open access article distributed under the terms and conditions of the Creative Commons Attribution (CC BY) license (https://creativecommons.org/licenses/by/4.0/).

## 1. Introduction

Over the past century, inorganic fertilizers have been indispensable for improving crop production and quality, thereby ensuring food security in the face of a rapidly growing global population [1]. In addition, the chemical fertilizers used in agricultural lands, along with pesticides and herbicides, are judged to have, to date, met the stability of global food production to some extent [2,3]. However, excessive use of these agrochemicals has caused severe adverse impacts on the natural environments: e.g., soil acidification, mineral depletion, disease incidence, heavy metal contamination, eutrophication, biodiversity loss, and global warming [1,4,5], thereby exacerbating soil quality and health and subsequently reducing crop yield. In turn, these harmful effects, caused by chemical abuse, raise public concerns about food safety and security worldwide [6]. Hence, eco-friendly approaches to the improvement of crop yield and quality as well as efficiency in land management should be explored for sustainable agriculture.

Biofertilizers are substances containing microorganisms (e.g., bacteria, fungi, protozoa, and algae) or organic additives containing natural compounds derived from the microbes that promote plant growth and development by increasing the supply of essential nutrients (e.g., N, P, K, etc.) to the host plants [7]. It has been estimated that the application of biofertilizers may reduce the use of N–P–K fertilizers by up to 60% but increase crop yield by 35% [8]. Additionally, recent studies have demonstrated the superior effects of biofertilizer application on promoting the plant's resistance to pathogens and abiotic stresses [9,10] as well as improving soil structure and water retention capacity [11]. Moreover, the fact that it is eco-friendly and cost-effective is increasing the global interest in biofertilizer development and production, mainly in the USA, Europe, China, and India. Accordingly, as of 2022 the global biofertilizer market was valued at US$2.15 billion and is projected to grow to US$6.83 billion in 2032, at a compound annual growth rate of 12.3% from 2023 to 2032 [12].

The major biofertilizer types include symbiotic (e.g., *Rhizobium*) or non-symbiotic (e.g., *Azotobacter*) N fixers, P solubilizing bacteria (e.g., *Aspergillus*, *Penicillium*), mycorrhizae (e.g., *Glomus*, *Scutellospora*), plant growth promoting bacteria (PGPR; e.g., *Agrobacterium*, *Pseudomonas*), vermicompost, cyanobacteria, etc. [2,4,13,14]. Among these, rhizobia-based biofertilizers account for approximately 80% of the global demand, followed by phosphate-mobilizing biofertilizers (approximately 15%) [15,16]. Albeit less frequent in agricultural use than these microbes, algae containing amino acids, fibers, plant nutrients, vitamins, and antioxidants [7,17] have also been reported for their high potential as biofertilizers. According to the results of many studies, the application of algae extracts considerably increased the yield and quantity of various crops and fruits, including maize, tomato, lettuce, spinach, chives, grape, mango, etc. via foliar or soil applications [18–22]. In addition, it has been confirmed that the algae applications induce enhancement of plant tolerances to pathogen attacks and abiotic stresses (e.g., salinity, drought, etc.), contributing to the promotion of the plant's survival and performance [23].

In South Korea, freshwater algae, i.e., *Chlorella* spp., have been attracting great attention for their application to some leafy vegetables and strawberries (*Fragaria* × *ananassa*) since the mid-2010s. According to a report by the Rural Development Administration (RDA) [24], as of 2020 *Chlorella* farming practices were being applied to 66 crops and to a cultivation area of approximately 3700 ha. Additionally, it was estimated that farm income has increased by about 20% through *Chlorella* application. Kim et al. [25] revealed that the application of the *Chlorella vulgaris* CHK0008 strain to two strawberry cultivars, such as 'Seolhyang' and 'Yukbo', via soil irrigation or foliar treatment led to an increase in soluble solid content and a decrease in the decay rate of strawberry fruits. In addition, the beneficial effects induced by *Chlorella fusca* CHK0059, such as promoting the growth of strawberry plants and suppressing the Fusarium wilt disease of the fruits were reported [26]. Likewise, the role of *Chlorella* spp. as a biofertilizer and biocontrol agent has also been demonstrated in other studies [27–29]. As such, and based on these positive effects, *Chlorella* fertilization is widely used in the strawberry industry in Korea, though there is a lack of mechanism studies on the effects of *Chlorella* application on strawberry growth and fruit quality through monitoring experiments. Here, we aimed to investigate the application effects of *Chlorella fusca* CHK0059 on the morphological and physiological traits of strawberry plants and the quality and antifungal ability of the fruits. Then, through a comprehensive result analysis, we attempted to identify key factors in the beneficial effects of *Chlorella* fertilization and reveal their interactions.

## 2. Results

### 2.1. Plant Characteristics

#### 2.1.1. Plant Biomass and Nutrient Content

The growth of strawberry plants varied greatly with *Chlorella* application rates, as shown in Figure 1A. The stem length and leaf size appeared to be greater as the application rate increased. Consistent with this trend, the highest dry weight (DW) of strawberry plants was observed in 0.4% of the optimal cell density (OCD; $1.0 \times 10^7$ cells $mL^{-1}$) treatment (15.5 g $plant^{-1}$; Figure 1B). The DW values were significantly greater in 0.2% and 0.4% OCD treatments (1.57 and 1.72 times, respectively) than control ($p < 0.05$).

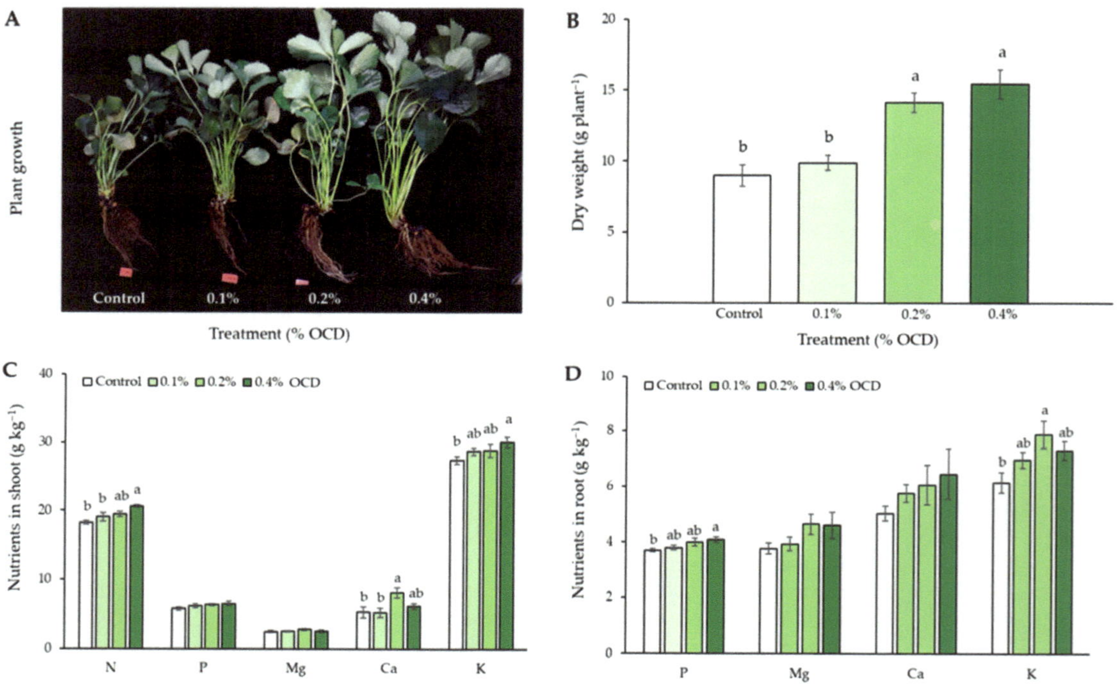


**Figure 1.** Phenotype (**A**), dry weight (**B**), and nutrient content in shoot (**C**) and root (**D**) of strawberry plants grown under different *Chlorella* application rates: 0, 0.1, 0.2, and 0.4% of the optimal cell density (OCD). Data are mean ± standard errors ($n = 6$). Different letters indicate significant difference among the treatments (Tukey's HSD, $p < 0.05$).

It was found that the higher the *Chlorella* application rate, the higher the nutritional level of strawberry plants. Among the nutrients, N, P, and K in the plant shoots were highest at 0.4% OCD (21, 6.4, and 30 g $kg^{-1}$, respectively), while 0.2% OCD represented the highest concentrations of Ca (8.1 g $kg^{-1}$), significantly higher than those in control ($p < 0.05$; Figure 1C). Additionally, the highest values of nutrients in the plant roots were observed in 0.2% and 0.4% OCD treatments for K and P, respectively (Figure 1D).

#### 2.1.2. Photosynthetic Rate and Pigments

Table 1 shows the difference in the photosynthetic properties of plant leaves according to the *Chlorella* application rates. The maximum photosystem II quantum yield (Fv/Fm) was significantly higher in 0.2% and 0.4% OCD than in control ($p < 0.05$). In addition, chlorophyll (Chl) parameters, such as Chl a, Chl b, and total Chl, showed significantly higher values in 0.4% OCD than in control ($p < 0.05$). The carotenoid contents in the three OCD treatments (2.26–2.41 mg 100 $g^{-1}$) were significantly higher than that in control (1.92 mg 100 $g^{-1}$) ($p < 0.05$).

**Table 1.** The maximum photosystem II quantum yield (Fv/Fm) and photosynthetic pigments in leaves of strawberry plants grown under different *Chlorella* application rates.

| Treatment (% OCD) | Fv/Fm | Chlorophyll a (mg 100 $g^{-1}$) | Chlorophyll b (mg 100 $g^{-1}$) | Total Chlorophyll (mg 100 $g^{-1}$) | Carotenoids (mg 100 $g^{-1}$) |
|---|---|---|---|---|---|
| Control | 0.813 b | 0.580 b | 0.195 b | 0.777 b | 1.923 b |
| 0.1% | 0.813 b | 0.628 ab | 0.205 ab | 0.843 ab | 2.257 a |
| 0.2% | 0.834 a | 0.656 ab | 0.208 ab | 0.865 ab | 2.406 a |
| 0.4% | 0.840 a | 0.683 a | 0.235 a | 0.920 a | 2.338 a |

Data represent the mean values of six replicates. Different letters indicate significant difference among the treatments (Tukey's HSD, $p < 0.05$). OCD, the optimal cell density of *Chlorella fusca* CHK0059.

### 2.2. Fruit Characteristics during the Monitoring Period

#### 2.2.1. Hardness, Soluble Solids Content, and Biomass

The properties of strawberry fruits monitored during the five months (from December 2021 to April 2022) are shown in Figure 2. Fresh weights (FW) of the fruits sampled over the period were highest in the 0.2% OCD (18.9–21.7 g $plant^{-1}$) and 0.4% OCD (18.8–21.3 g $plant^{-1}$) treatments, significantly greater than those in control (15.7–20.1 g $plant^{-1}$; $p < 0.05$; Figure 2A). For hardness, the greatest values were observed in the fruits sampled in 0.4% OCD (4.8–5.7 *N*) during the cultivation period, excluding February, which were significantly higher than those in control (4.0–4.5 *N*; $p < 0.05$; Figure 2B). In addition, the highest soluble solid content (SSC) values were found mostly in 0.2% OCD (9.3–12.0 °Brix) and 0.4% OCD (9.3–11.9 °Brix), significantly higher than those in control (7.9–10.9 °Brix; $p < 0.05$; Figure 2C). The most significant difference (approximately 1.23 fold) in the SSC values was found between 0.2% OCD and control treatments, particular in the final month.

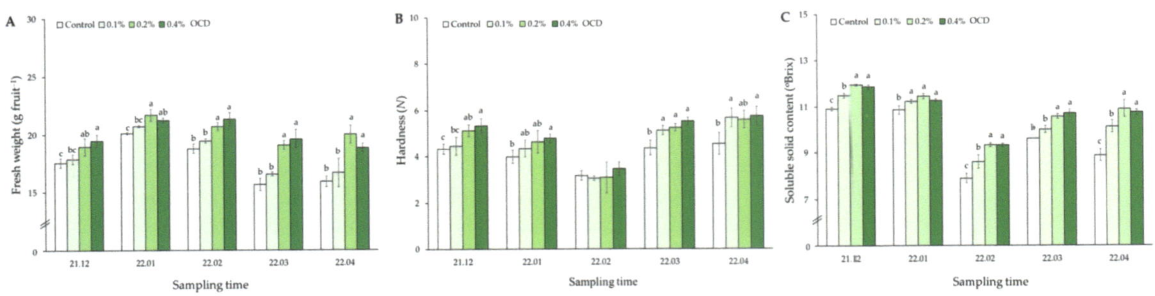


**Figure 2.** The properties of strawberry fruits grown under different *Chlorella* application rates for five months: 0 (control), 0.1, 0.2, and 0.4% of the optimal cell density (OCD). Data are mean ± standard errors ($n$ = 50 for fresh weight and 6 for hardness and soluble solid content). Different letters indicate significant difference among the treatments (Tukey's HSD, $p < 0.05$). (**A**) Fresh weight, (**B**) hardness, (**C**) soluble solid content.

#### 2.2.2. Nutrients

Among the *Chlorella* application rates, 0.4% OCD treatment had the highest concentrations of P (6.97–8.04 g $kg^{-1}$) in fruit samples over the five months (Table 2). Other *Chlorella* treatments induced intermediate concentrations of P (4.74–6.32 g $kg^{-1}$), and control treatment had the lowest P concentration (3.51–5.09 g $kg^{-1}$). Meanwhile, Ca concentration in the fruits showed the opposite trend; the highest value was observed mostly in control while 0.4% OCD had the lowest value. For other nutrients (Mg and K), it seemed that there was no significant difference in the concentrations among the treatments ($p > 0.05$) across the whole cultivation period.

**Table 2.** Concentration of nutrients, including P, Mg, Ca, and K, in strawberry fruits grown under different *Chlorella* application rates for five months.

| Element | Treatment (% OCD) | Sampling Time | | | | |
|---|---|---|---|---|---|---|
| | | 21.12 | 22.01 | 22.02 | 22.03 | 22.04 |
| P ($g\ kg^{-1}$) | Control | 3.61 c | 3.81 c | 3.51 c | 3.70 c | 5.09 b |
| | 0.1% | 5.62 b | 5.57 b | 5.39 b | 4.99 b | 6.14 ab |
| | 0.2% | 6.25 ab | 6.03 b | 4.74 bc | 5.69 b | 6.32 ab |
| | 0.4% | 6.97 a | 8.04 a | 7.44 a | 7.15 a | 7.64 a |
| Mg ($g\ kg^{-1}$) | Control | 1.62 a | 1.45 a | 1.51 ab | 1.62 a | 1.81 a |
| | 0.1% | 1.63 a | 1.50 a | 1.55 ab | 1.57 a | 1.69 a |
| | 0.2% | 1.49 a | 1.41 a | 1.34 b | 1.47 a | 1.56 a |
| | 0.4% | 1.37 a | 1.53 a | 1.65 a | 1.58 a | 1.71 a |
| Ca ($g\ kg^{-1}$) | Control | 3.29 a | 3.70 a | 2.99 a | 3.26 a | 2.61 a |
| | 0.1% | 3.19 a | 2.54 ab | 2.57 a | 3.14 ab | 2.75 a |
| | 0.2% | 2.59 ab | 2.67 b | 2.52 a | 2.35 ab | 1.89 b |
| | 0.4% | 1.85 b | 2.07 b | 2.18 a | 2.26 b | 2.01 b |
| K ($g\ kg^{-1}$) | Control | 15.4 a | 12.4 b | 11.1 b | 12.3 b | 17.0 a |
| | 0.1% | 14.8 a | 13.9 ab | 13.5 ab | 13.2 ab | 16.1 a |
| | 0.2% | 14.2 a | 12.3 b | 11.4 b | 13.4 ab | 15.0 a |
| | 0.4% | 14.1 a | 15.2 a | 16.6 a | 15.3 a | 16.7 a |

Data represent the mean values of six replicates. Different letters indicate significant difference in the concentration of each element among the treatments at each sampling time (Tukey's HSD, $p < 0.05$). OCD, the optimal cell density of *Chlorella fusca* CHK0059.

2.2.3. Sugars

The highest contents of total sugars (TS) in strawberry fruits sampled in December 2021 and January 2022 were observed in 0.2% OCD (5.6–5.9 g 100 $g^{-1}$), and were significantly higher than those in control (2.8–4.3 g 100 $g^{-1}$) ($p < 0.05$; Figure 3A). On the other hand, in the remaining months, TS content in the fruits sampled in control represented the highest value (6.5 g 100 $g^{-1}$) in March and the intermediate values (5.3–5.8 g 100 $g^{-1}$) in February and April. Over the period, the lowest TS values were observed in 0.4% OCD (3.7–4.6 g 100 $g^{-1}$).

The main types of sugar in strawberry fruits cultivated in control were fructose (41%), glucose (31%), and sucrose (27%). Meanwhile, the fruits harvested in 0.2% and 0.4% OCD treatments contained sucrose (43%), fructose (40%), and glucose (27%) in that order (Figure 3). Among these, sucrose contents were highest in 0.2% OCD across the whole cultivation period, significantly higher (34–66%) than those in control ($p < 0.05$; Figure 3B). Fructose and glucose contents were highest in control from February to April, and were significantly higher than those in 0.4% OCD ($p < 0.05$; Figure 3C,D).

2.2.4. Organic Acids

The highest content of total organic acids (TOA) in strawberry fruits sampled in December 2021 was observed in 0.1% OCD (6.52 mg $g^{-1}$) and was significantly higher than that in control ($p < 0.05$; Figure 4A). Over the next three months, TOA content in the fruits gradually increased in all *Chlorella* treatments, but there were no significant differences among the treatments ($p > 0.05$). TOA content in control then increased considerably (by approximately 23%) in April 2022 relative to the previous month, whereas this phenomenon did not occur in the remaining treatments. At this time, the TOA contents in the fruits were highest in control (9.8 mg $g^{-1}$) and were significantly higher than those in the 0.1%, 0.2%, and 0.4% OCD treatments (6.6–8.0 mg $g^{-1}$; $p < 0.05$).

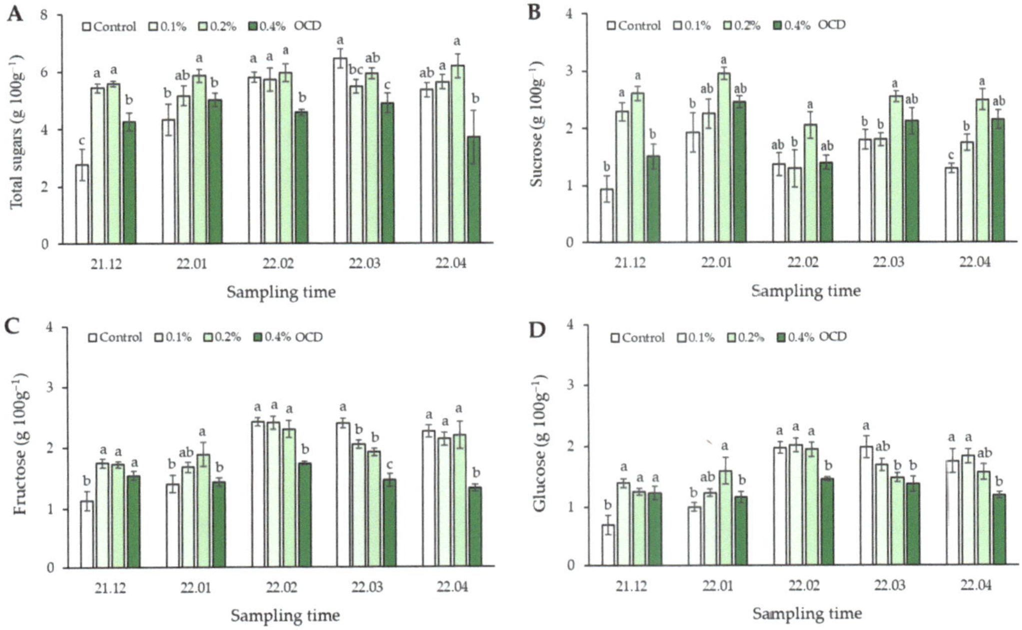

**Figure 3.** Monitoring sugar content and composition in strawberry fruits grown under different *Chlorella* application rates for five months: 0 (control), 0.1, 0.2, and 0.4% of the optimal cell density (OCD). Data are mean ± standard errors ($n = 6$). Different letters indicate significant difference among the treatments (Tukey's HSD, $p < 0.05$). (**A**) Total sugars, (**B**) sucrose, (**C**) fructose, (**D**) glucose.

Overall, the main organic acids in strawberry fruits were found to be citric acid (60–75%), malic acid (25–40%), and acetic acid (<1%). For citric acid, the highest value was observed in the 0.1% OCD treatment in December 2021 while the control treatment represented the highest value in April 2022 (Figure 4B). The remaining organic acids also showed a similar trend (Figure 4C,D). Consistently, these organic acids were found to be significantly lower (23–37%) in the 0.2% and 0.4% OCD treatments than in control ($p < 0.05$). Temporal changes in citric and malic acids appeared to be similar to TOA content, but not for acetic acid, which tended to drop highly in the third month of sampling.

### 2.2.5. Fruit Flavor Quality

The highest sugar–organic acid (S:A) ratio of strawberry fruits was observed in 0.2% OCD (8.47–9.99) during the five month sampling period (Table 3). The lowest S:A ratio was found in control (5.26–7.49) in December, January, and April, while 0.4% OCD represented the lowest S:A ratio (5.77–6.83) in the remaining months. Likewise, the sweetness index (SI) of the fruits also showed a similar trend. The treatment of 0.2% OCD induced the highest SI values (8.76–10.0) in December, January, and April and these were significantly greater than those in control (4.59–8.75; $p < 0.05$).

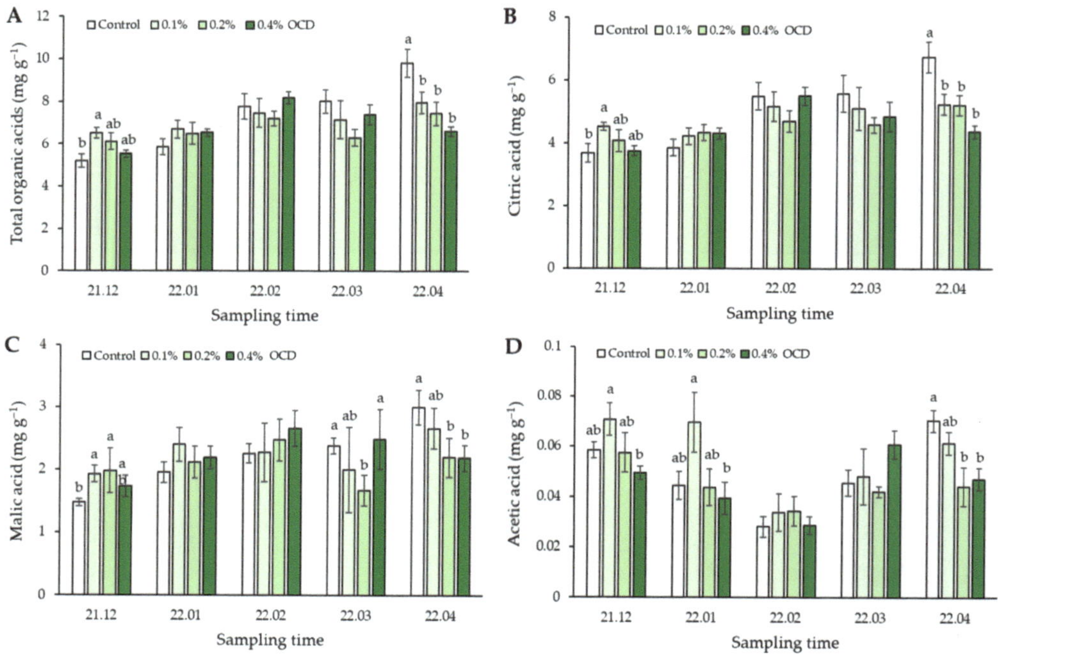


**Figure 4.** Monitoring organic acids content and composition in strawberry fruits grown under different *Chlorella* application rates for five months: 0 (control), 0.1, 0.2, and 0.4% of the optimal cell density (OCD). Data are mean ± standard errors ($n$ = 6). Different letters indicate significant difference among the treatments (Tukey's HSD, $p < 0.05$). (**A**) Total organic acids, (**B**) citric acid, (**C**) malic acid, (**D**) acetic acid.

**Table 3.** Assessment of fruit flavor quality of strawberries grown under different *Chlorella* application rates for five months.

| Quality Index | Treatment (% OCD) | Sampling Time | | | | |
|---|---|---|---|---|---|---|
| | | 21.12 | 22.01 | 22.02 | 22.03 | 22.04 |
| S:A ratio | Control | 5.26 b | 7.49 b | 7.61 ab | 7.92 ab | 5.61 c |
| | 0.1% | 8.38 a | 7.76 b | 7.96 a | 8.39 ab | 7.30 b |
| | 0.2% | 9.36 a | 9.99 a | 8.81 a | 9.58 a | 8.47 a |
| | 0.4% | 7.77 a | 7.76 b | 5.77 b | 6.83 b | 7.12 b |
| SI | Control | 4.59 c | 6.85 b | 9.41 a | 9.97 a | 8.75 b |
| | 0.1% | 8.53 a | 8.16 ab | 9.33 a | 8.90 b | 9.14 ab |
| | 0.2% | 8.76 a | 9.95 a | 10.0 a | 9.37 ab | 10.0 a |
| | 0.4% | 6.83 b | 7.81 b | 7.38 b | 7.64 c | 7.18 c |

Data represent the mean values of six replicates. The fruit quality was evaluated by determining the sugar–organic acid (S:A) ratio and the sweetness index (SI) based on the method of Keutgen and Pawelzik [30]. Different letters indicate significant difference among the treatments (Tukey's HSD, $p < 0.05$). OCD, the optimal cell density of *Chlorella fusca* CHK0059; S:A, total sugars to total organic acids; SI, the sweetness index.

### 2.2.6. Correlations of Fruit Properties

The correlations among the properties of fruits sampled in the second and fifth months, when there was the least damage from climate change and disease outbreaks during strawberry cultivation, are shown in Figure 5. At the early sampling (January 2022; Figure 5A), there were significant positive correlations of the fruits' FW with hardness, P, TS, sucrose, and glucose (r = 0.45, 0.59, 0.54, 0.60, and 0.49, respectively, $p < 0.05$). In addition, SSC (°Brix) had a significant positive correlation with hardness (r = 0.73, $p < 0.05$) and negative correlations with P, Mg, and K (r = −0.49, −0.54, and −0.53, respectively, $p < 0.05$). Among sugar-related parameters, sucrose was significantly correlated to TOA, citric acid, TS, fructose, and glucose (r = 0.41, 0.42, 0.86, 0.60, and 0.54, respectively, $p < 0.05$). As an indicator of fruit flavor quality, the S:A ratio was significantly correlated to FW, TS, sucrose, fructose, and glucose (r = 0.51, 0.77, 0.76, 0.53, and 0.50, respectively, $p < 0.05$).

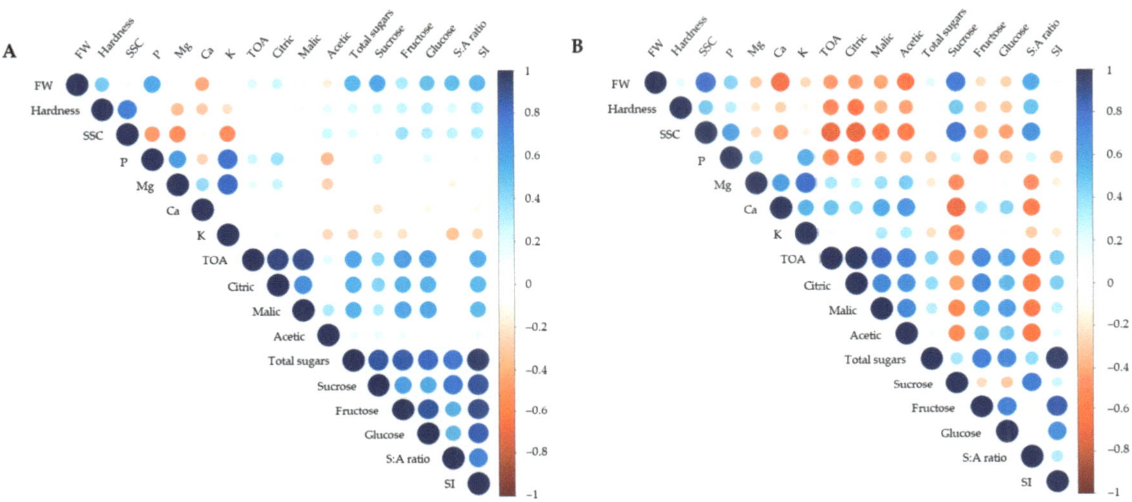


**Figure 5.** Correlation matrix of fruit properties sampled over the two months with the most favorable growing environment conditions during the monitoring period: (**A**) January 2022 and (**B**) April 2022. Positive and negative correlations are displayed in blue and red colors, respectively. Color intensity and the size of the circle are proportional to the Pearson correlation coefficients. The legend color on the right side of the correlogram indicates the correlation coefficient and the corresponding colors.

On the other hand, the correlation trends among the variables of fruits sampled in April 2022 were different (Figure 5B). There was a significant positive correlation between fruit's FW and SSC (r = 0.80, $p < 0.001$), but both variables had significant negative correlations with organic acids (r = −0.43 to −0.73, $p < 0.05$). In addition, the parameters related to sugars showed different correlations with organic acids: positive for fructose and glucose (r = 0.47 to 0.71, $p < 0.05$) and negative for sucrose (r = −0.46 to −0.56, $p < 0.05$). Sucrose was also negatively correlated with Mg, Ca, and K (r = −0.51, −0.73, and −0.50, respectively, $p < 0.05$). Moreover, the S:A ratio was positively correlated with FW, hardness, SSC, TS, and sucrose (r = 0.60, 0.49, 0.68, and 0.75, respectively, $p < 0.01$) but negatively correlated with organic acids (r = −0.60 to −0.66, $p < 0.001$). In contrast, the SI value showed a positive correlation with TOA and citric acid (r = 0.45, $p < 0.05$).

### 2.3. Anthocyanin Content

Figure 6 shows the difference in anthocyanin content in strawberry fruits depending on *Chlorella* application rates. The highest anthocyanin content was observed in 0.4% OCD (27 mg 100 $g^{-1}$), followed by the 0.2% OCD, 0.1% OCD, and control (13 mg 100 $g^{-1}$) treatments. Compared with control, the anthocyanin contents in the fruits of the 0.2% and 0.4% OCD treatments were significantly higher (1.7 and 2.1 times, respectively) than that in control ($p < 0.05$).

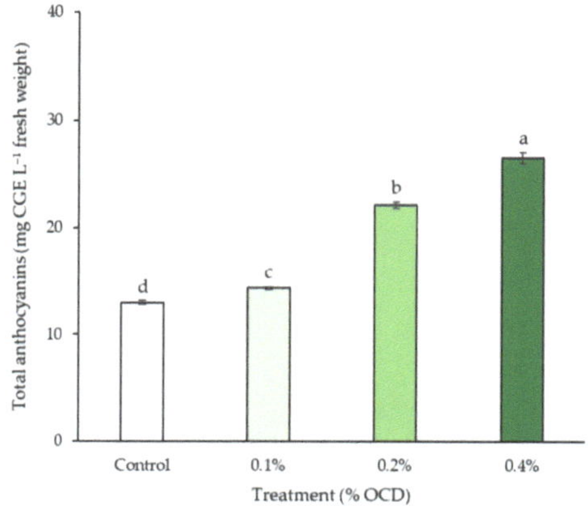


**Figure 6.** Anthocyanin contents in strawberry fruits sampled in different *Chlorella* application rates at the last sampling month: 0 (control), 0.1, 0.2, and 0.4% of the optimal cell density (OCD). Data are mean ± standard errors ($n$ = 6). Different letters indicate significant difference among the treatments (Tukey's HSD, $p < 0.05$).

## 2.4. Postharvest Disease Incidence

As a result of the postharvest fruit storage test, no occurrence of grey disease incidence caused by *Botrytis cinerea* was found in strawberries during the first two weeks (Figure 7). Twenty-five days after the storage, the pathogen was first observed on the surface of fruits in the control and 0.1% OCD treatments, and the disease incidence in both treatments increased sustainably until the last day (Day 45) of the storage test. Over the storage duration, the final incidences of grey disease according to the *Chlorella* application rates were listed in the following order: control (85%), 0.1% OCD (57%), 0.2% OCD (13%), and 0.4% OCD (8%) treatments.

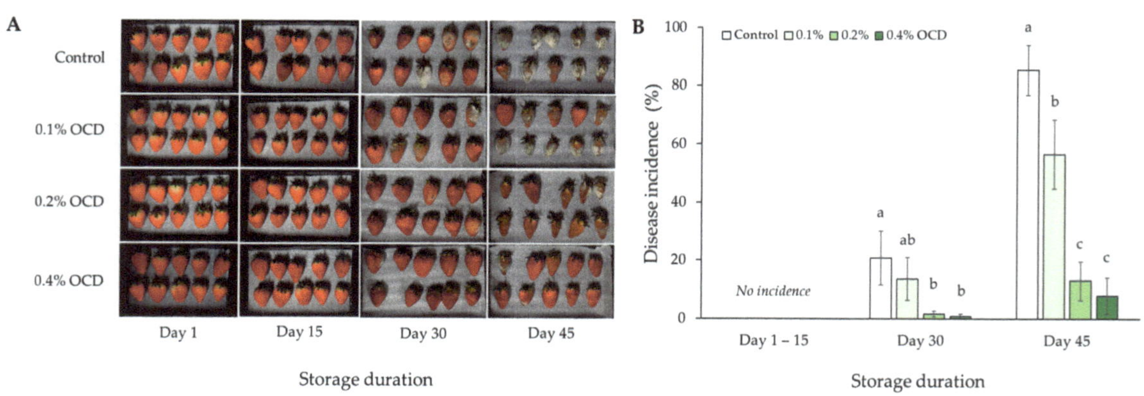


**Figure 7.** Incidence of grey mold disease (*Botrytis cinerea*) on strawberry fruits, sampled in different *Chlorella* application rates including 0 (control), 0.1, 0.2, and 0.4% of the optimal cell density (OCD), during 45 day storage test. (**A**) Temporal changes in the pathogen occurrence and severity on the fruits at 15 day intervals. (**B**) The incidence of grey mold disease is estimated as the proportion of pathogen-caused areas at 15 day intervals. Data are mean ± standard errors ($n$ = 10). Different letters indicate significant difference among the treatments (Tukey's HSD, $p < 0.05$).

## 2.5. Principal Component Analysis

A clear separation of *Chlorella* treatments was confirmed through the principal component analysis (PCA) result of the dataset for plant and fruit properties (Figure 8). Along the

PC1 axis (48% variance), the treatments with different OCD levels were separated primarily due to variables of plant physiology (e.g., DW, chlorophyll, carotenoid, and Fv/Fm) and nutrients (e.g., N and P) as well as fruit quality, such as SSC, FW, hardness, components (e.g., sucrose, organic acids, S:A ratio, and anthocyanins), and disease incidence ($p < 0.01$). Along the PC2 axis (15% variance), the treatments were split by the sweetness index, Mg and Ca in the plants, and fructose and glucose in the fruits ($p < 0.01$). The overall trend of this multivariate analysis indicates that *Chlorella* application at the level of 0.2% OCD induced the most positive impacts on the major variables of plant growth and fruit quality, as well as their significant correlations.

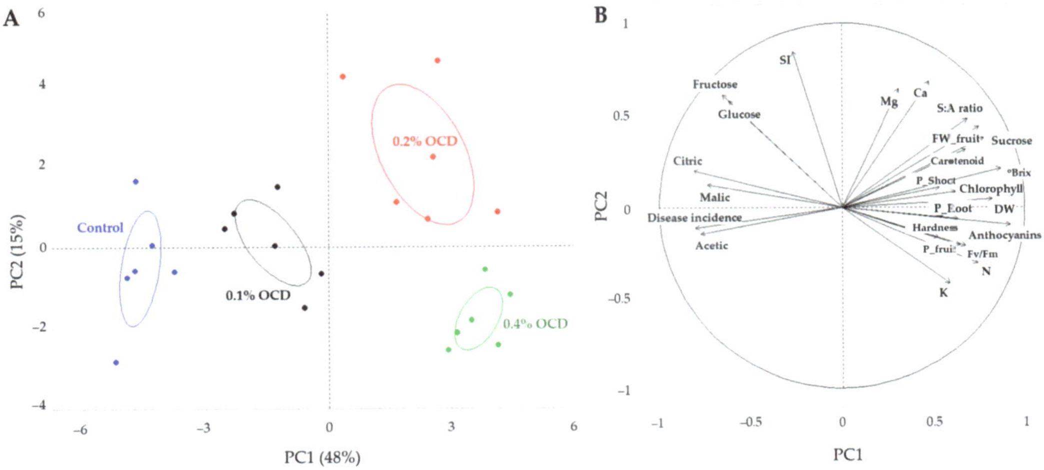


**Figure 8.** Principal component analysis (PCA) of plant and fruit properties in treatments of different *Chlorella* application rates: 0 (control), 0.1, 0.2, and 0.4% of the optimal cell density (OCD). (**A**) Scores for the first two principal components of four different application rates. (**B**) Loading plot of the parameters related to plant physiology and fruit quality.

## 3. Discussion

In modern agriculture, microalgae have been widely used as biofertilizers, biostimulants, and biopesticides because they contain essential sources of nutrients (e.g., N, P, K, etc.), primary and secondary metabolites (e.g., polysaccharides, phenolics, phytohormones, vitamins, etc.), and bioactive compounds to various environmental conditions (e.g., antimicrobial, antioxidant, antiviral or antifungal substances) [31], thereby improving soil fertility and quality and crop yield and performance. As such, in the present study, the improvements in growth performance and nutrient uptake efficiency of strawberry plants by *Chlorella* application were confirmed through increased biomass and nutrient contents (Figure 1) and stimulated photosynthetic activity and pigmentation (Table 1). According to our previous analysis of *C. fusca* CHK0059 cell composition, it contained high levels of N (91.2 g kg$^{-1}$), P (9.38 g kg$^{-1}$), K (2.62 g kg$^{-1}$), Ca (1.05 g kg$^{-1}$), Mg (5.85 mg kg$^{-1}$), etc., inferring that added *Chlorella* is likely to be well absorbed as a nutrient source for strawberry plants. In addition, the improved nutritional status of the plants appeared to positively influence the photosynthetic capacity by regulating their morphological characteristics, such as specific leaf area and leaf mass ratio, that are highly related to the efficiencies of light capture and carbon gain [32,33]. A similar research result has also been reported by Kim et al. [26], who demonstrated significant increases in shoot height, leaf thickness, and leaf pigments of the strawberry cultivar 'Seolhyang' by *C. fusca* CHK0059 application. Moreover, the promoting effect on plant growth performance by microalgae application has been reported in various crops, such as wheat, rice, maize, lettuce, tomato, cabbage, etc., as reviewed in Lee and Ryu [23] and Çakirsoy et al. [34].

Strawberry is one of the most popular fruits in the world because of its sweetness, fragrant flavor, juicy texture, and health-promoting compounds and fibers [35], so it is currently considered a high-value fruit crop that is important for agriculture and local economies, especially in South Korea. With the high demand for better-tasting, longer-lasting, and healthier fruits from consumers, various agricultural technologies including *Chlorella* fertilization, have been applied to strawberry cultivation. In this respect, the application of *C. fusca* CHK0059 (0.2% and 0.4% OCD) in the present study led to considerable increases in the SSC (°Brix), hardness, and biomass of strawberry fruits throughout the entire cultivation (Figure 2), suggesting its superior role in the improvement of fruit quality and yield. Additionally, positive correlations of fruit SSC and hardness with P concentrations in roots, shoots, and fruits may (Figure 8) indicate that highly absorbed P in plant shoots and edible parts following *Chlorella* application also affected the fruit quality parameters. In general, phosphorus is an important macronutrient for photosynthesis and affects sugar partition and transportation as a form of phosphate (Pi) [36]. In this context, the improvement of P status in strawberry roots and shoots following *Chlorella* application seems to significantly stimulate the overall photosynthetic functions (see Table 1), contributing to an increase in sugar accumulation in fruits with high P.

In general, fruit flavor is determined by the quantity and type of sugars, organic acids, and volatile compounds. A good balance of these substances will meet the consumer preference; however, among these, sugars corresponding to the sweetness intensity of fruits are the most significant determinant. In ripe strawberries, the predominant soluble sugars are sucrose, fructose, and glucose, and the content and composition of these have a decisive effect on determining fruit quality [37] but vary with genotypes, development stages, farming practices, and environmental conditions [38]. In this study, the sugar type that responded most positively to *Chlorella* application during strawberry cultivation was sucrose (Figure 3). Sucrose is known to be the ultimate product of photosynthesis. This carbohydrate is converted from all photo-assimilates that are not required for leaf function and transported into sink organs, i.e., roots, shoots, and fruits, through the phloem, leading to coordinating physiological growth and development [39]. The amount of sucrose that can be released from the sources to the receptacle fruit sink via long-distance transport depends on various factors: e.g., photosynthetic activity, partitioning between starch synthesis within the chloroplast and triose-Pi moved out of the chloroplast for sucrose synthesis, and temporary storage of sucrose in the vacuole [37,40,41]. In our findings, a significant correlation of sucrose in fruits with Fv/Fm, chlorophylls, carotenoids, and P in leaves (Figure 8) suggests that the application of *C. fusca* CHK0059 could act as a promoter for fruit flavor as well as for plant growth by increasing photosynthetic efficiency and Pi availability, which are highly related to the sucrose synthesis. Additionally, increased N, Mg, and K contents in the strawberry plants treated with *C. fusca* (Figure 1) seem to promote overall metabolic processes, including photosynthesis, pigmentation, and carbohydrate accumulation in the leaves, subsequently facilitating the sucrose generation and transport into the sink organ [41]. On the other hand, fructose and glucose contents in the fruits showed similar responses to technological (i.e., *Chlorella* application rate) and temporal changes, but these were negatively correlated to sucrose content in the fruits (Figures 5 and 8). A similar result has also been observed by Schwieterman et al. [38]. In addition, the S:A ratio as a taste index was positively correlated with sucrose but not with other sugars in the fruits (Figure 5B), suggesting that the increased sucrose content following *Chlorella* application led to an improvement in the overall strawberry flavor intensity. Furthermore, Schwieterman et al. [38] have reported a high correlation between sucrose and total volatiles, relevant to fruit aroma, in 35 strawberry cultivars. In light of this, in the present study we expect that *Chlorella* application probably also led to an improvement in total volatile substances, which will increase consumer choice.

Besides sweetness, sourness is also a crucial factor in determining strawberry fruit flavor associated with organic acids [30]. In general, organic acids can mediate various functions in plant growth and development by supporting photosynthesis and stomata regulation as well as acting as an alternative energy source, in turn, the metabolic processes contribute to fruit development and quality [42,43]. Additionally, they also play an important role in the gelling properties of pectin in fruit [44]. In the present study, we found an increase in TOA contents, mainly citrate and malate, in the strawberry fruits treated with *C. fusca* CHK0059 (Figure 4), inferring that *Chlorella*-induced organic acids induced benefits not only for plant development but also for development, maturation, and ripening of strawberries [42]; however, such an effect was represented during the initial two or three months but not the remaining period. The qualitative and quantitative variations of organic acids are usually dependent on cultivars, genotypes, and maturation stages of strawberry fruits [45]. Additionally, the difference in accumulation of these organic acids in the fruits may also be attributed to environmental fluctuations, including temperature, ambient $CO_2$, and water and mineral supplies, during the cultivation experiment [46]. Despite the biochemical variation in strawberry fruits, and considering that 0.2% OCD of *C. fusca* CHK0059 led to the highest values of two indices, i.e., S:A ratio and SI, during the entire cultivation period (Table 3), it seems that the effect of *Chlorella*-induced sugar production on the fruit flavor quality far outweighed the effect on organic acid production in this study.

It is known well that anthocyanins are responsible for the red color in strawberries and also serve as health-promoting substances with antioxidant, anticancer, and anti-inflammatory properties [47], thus directly affecting fruit appearance and consumer choice. In the present study, like sucrose, the most noticeably improved fruit parameter with the application of *C. fusca* CHK0059 was anthocyanins, which tended to increase with increasing application rate (Figure 6). Similarly, Roussos et al. [48] have reported that the application of brown alga (*Ascophyllum nodosum*) extract induced a higher content of anthocyanins in strawberry cv. Camarosa. As is widely known, the accumulation of anthocyanins in fruits is affected by sucrose because it not only provides the energy source for the synthesis of anthocyanins but also acts as a signaling molecule inducing transcription of key enzymes (e.g., G6PHD, SKDH, PAL, C4H, 4CL, etc.) related to the biosynthesis pathway [49,50]. In addition to that evidence, the stimulatory effects of sucrose on the biosynthesis of anthocyanins have been documented in various plant species, including strawberries [49,50], grapes [51], broccoli sprouts [52], etc. Likewise, in this study, we found a high correlation between anthocyanins and sucrose contents in the strawberry fruits (Figure 8).

Typically, strawberries consisting of soft tissue are highly susceptible to the attacks of pathogens, including viruses, bacteria, fungi, and nematodes, especially during postharvest storage [47,53]. Among pathogens, the ascomycete *B. cinerea* is regarded as the primary disease of strawberries, causing a significant economic loss in the global strawberry industry. In contrast with such fruit disease damage, the application of *C. fusca* CHK0059 used in this study induced a marked antifungal effect on the postharvest strawberry fruits (Figure 7). Considering the PCA result that showed a significant negative correlation between anthocyanins and disease incidence in fruits (Figure 8), increased anthocyanin accumulation in fruits with the application of *C. fusca* CHK0059 appeared to have acted as an antifungal factor that effectively suppressed the occurrence of grey mold disease during the postharvest storage test. Likewise, reduced susceptibility to postharvest fungal pathogens, including *B. cinerea,* has been reported in other fruits enriched in anthocyanins: grapes [54], mango [55], and tomato [56]. In plants, anthocyanins are produced with other classes of flavonoids, e.g., flavones, flavonols, and tannins, via the same biochemical pathway [57]. Among these defensive secondary compounds, tannins can function to protect plants from both abiotic (e.g., cold, light, etc.) and biotic (e.g., pathogens and herbivores) stresses, and they share the same precursors, i.e., leucoanthocyanins, in the biosynthesis pathway with anthocyanins [54,57,58]. Therefore, it can be assumed in this study that the *Chlorella* application resulted in a simultaneous increase in these two metabolites in strawberry fruits, consequently enhancing the defensive capacity of the postharvest fruits to pathogen attacks. In addition to pathogen resistance, anthocyanin-induced tolerance of postharvest fruits to oxidative stress generally caused by wound and senescence may delay the pathogen infection and development [56].

## 4. Materials and Methods

### 4.1. Chlorella Inoculation

For *Chlorella* inoculation, *Chlorella fusca* CHK0059 was provided by the Organic Agriculture division of the National Institute of Agricultural Sciences, RDA in South Korea. The *Chlorella* cells were placed into commercial mineral water (Ca—25 mg $L^{-1}$, Na—5.3 mg $L^{-1}$; K—1.6 mg $L^{-1}$; Mg—3.0 mg $L^{-1}$) and cultured at the level of optimal cell density ($1.0 \times 10^7$ cells $mL^{-1}$) recommended by the RDA [24]. The *Chlorella* inoculation was conducted in a greenhouse of the experimental farm during the strawberry cultivation, and the culturing conditions were as follows: temperature (25 ± 3 °C), natural sunlight-dependent, and continuous air supply.

### 4.2. Plant Cultivation and Chlorella Treatment

A total of 800 seedlings of strawberry (*Fragaria ananassa* Dutch. cv. 'Kuemsil') were planted in raised beds (ca. 50 cm width) containing artificial soil substrate and cultivated from October 2021 to April 2022 in a greenhouse located in Jinju, South Korea (35°13′35″ N, 127°57′32″ E). The planting distance between the seedings was about 15 cm, and they were planted in two rows on the beds. The nutrient solution was supplied four times daily via an irrigation system and its composition was formulated based on the RDA method [59]. The pH of irrigated water was 5.8–6.0, and the electrical conductivity was adjusted according to the growth stage of strawberry plants: 0.8–1.5 dS $m^{-1}$ (vegetative to budding) and 1.2–1.8 dS $m^{-1}$ (flowering to ripening). During the cultivation period, the temperature in the greenhouse was controlled to maintain the following levels: 26 °C for daytime and more than 8 °C for nighttime. Additionally, the management of flower number was conducted as 5–7 flower buds per inflorescence, and misshapen fruits also were cut off.

There were four different application rates of *C. fusca* CHK0059 culture solution to strawberry seedlings: 0% (control), 0.1%, 0.2%, and 0.4% of OCD. The *Chlorella* solutions were applied weekly to the soil substrate through an irrigation system. The appropriate level of *Chlorella* solution recommended by the RDA for strawberry cultivation is 0.2% OCD [60], so the application rates for this study were installed based on this.

### 4.3. Fruit and Plant Collection

Strawberry fruits, with 80–90% of the fruit surface showing a visible red color [59], were sampled weekly for five months, from December 2021 to April 2022. Of those, up to 50 individuals weighing 15 ± 5 g were selected per treatment each month and their FW recorded. Then, analysis for other fruit characteristics was performed using six fruit samples: e.g., hardness, SSC, total sugars and components (sucrose, fructose, and glucose), total organic acids and components (citric acid, malic acid, and acetic acid), nutrients (P, Mg, Ca, and K), and anthocyanins.

Plant sampling was performed after the last fruit harvest in April 2022. Six individuals of each *Chlorella* treatment were selected, and their photosynthesis rates were measured in situ. Then, the plant samples were collected and transferred to a laboratory of Gyeongsang National University to determine other plant characteristics, such as morphology, dry weight, photosynthetic pigments (Chl a, Chl b, total Chl, and carotenoid), and nutrients (N, P, Mg, Ca, and K).

### 4.4. Grey Mold Incidence during Postharvest Storage

Strawberry fruits harvested in the last sampling month were stored at 4 °C and 80–90% relative humidity in a cold chamber (JSSC-700C, JS Research Inc., Gongju, Republic of Korea). Afterward, the occurrence of grey mold disease (*Botrytis cinerea*) on the fruit surface was observed for 45 days. Each *Chlorella* treatment had 10 individual fruits. The disease incidence (%) of strawberry fruits was estimated as a proportion of pathogen-caused area to total area, the scale of which was measured using the ImageJ v. 1.51i software (https://imagej.nih.gov/ij/index.html: accessed on 1 March 2023). The postharvest storage assay was conducted in duplicate.

*4.5. Analytical*

The hardness (*N*) and SSC (°Brix) of collected fruits were recorded using a fruit hardness tester (KM-1, Fujiwara Factory, Tokyo, Japan) and a Brix refractometer (Surakan SRS03, Snix Electronics Co., Ltd., Cheongju, Republic of Korea), respectively.

The photosynthesis rate in the leaves of strawberry plants was measured using a portable fluorometer with leaf-clip holders (MINI–PAM–II, Heinz Walz GmbH, Effeltrich, Germany). In addition, the photosynthetic pigments, such as chlorophylls and carotenoids, in the leaves were determined following the extraction method with 80% acetone [61]. The extract was measured at 470, 645, and 663 nm wavelengths with a UV spectrophotometer (UV–160A, Shimadzu, Kyoto, Japan) and the pigments of Chl a, Chl b, total Chl (a + b), and carotenoids were estimated with the following formula [1]:

$$\text{Chl a} = 12.7\ \text{A663} - 2.69\ \text{A645}$$

$$\text{Chl b} = 22.9\ \text{A645} - 4.68\ \text{A663}$$

$$\text{Total Chl (a + b)} = 20.29\ \text{A645} + 8.02\ \text{A663}$$

$$\text{Carotenoids} = (1000\ \text{A480} - 1.82\ \text{Chl a} - 85.02\ \text{Chl b})/198$$

where A is the absorbance value at the appropriate wavelength.

The N concentration in plant samples was determined by a Kjeldahl distiller (K–355, BÜCHI, Flawil, Switzerland) following wet digestion with $H_2SO_4$–$HClO_4$ solution [62]. To determine the concentrations of nutrients, including P, K, Ca, and Mg, in plants and fruits, the samples were digested with $HNO_3$–$H_2O_2$ solution and then determined using ICP–OES (iCAP Pro XP Duo, Thermo Scientific, Waltham, MA, USA) [62].

For carbohydrate analyses, dried fruit samples (100 mg) were extracted with 10 mL of distilled water heated for four hours in a water bath at 60 °C and then centrifuged for 15 min at 3000 rpm. The supernatant was filtered with a 0.45-µm syringe filter and the contents of organic acids, such as citric, malic, and acetic acids, were determined using ultra-high performance liquid chromatography (UHPLC) (Dionex Ultimate 3000, Thermo Fisher Scientific Inc., Dreieich, Germany) by injection into a SunFire TM C18 column (4.6 mm × 250 mm, 5 µm, Waters, Milford, MA, USA). The mobile phase used was 0.02 N $H_2SO_4$ and the flow rate was 1 mL $min^{-1}$. The soluble organic acids were detected with a variable wavelength detector (VWD-3100, Thermo Fisher, Crawley, UK). The sum of these three organic acids was considered as TOA.

The supernatant extracted from the fruit samples above was also used to determine soluble sugars according to a method of Keutgen and Pawelzik [30]. Contents of fructose, glucose, and sucrose in the extracts were determined using an Agilent 1200 series HPLC (Agilent, Santa Clara, CA, USA) equipped with a refractive index detector (RID) following injection of the fruit extracts (20 µL) into an Agilent Zorbax carbohydrate column at 30 °C (4.6 mm × 150 mm, I.d., 5 µm). The mobile phase used was 75% acetonitrile and the flow rate was 1.4 mL $min^{-1}$. The sum of fructose, glucose, and sucrose was considered TS.

Anthocyanin content in strawberry fruit was determined with the pH differential method [63]. The frozen sample was mixed thoroughly with 0.025 M potassium chloride buffer (pH 1.0) and 0.4 M sodium acetate buffer (pH 4.5) and then incubated for 90 min at room temperature. After centrifugation, the absorbance of the supernatant was measured at 520 and 700 nm wavelengths with a UV spectrophotometer (UV–160A, Shimadzu, Japan). The following formula was used to calculate the anthocyanin content expressed as cyanidin-3-glucoside equivalents (CGE):

$$\text{Anthocyanin}\left(\text{mg CGE L}^{-1}\ \text{FW}\right) = \frac{\text{A} \times \text{MW} \times \text{DF} \times 10^3}{\varepsilon \times 1}$$

where A: $(A_{520nm} - A_{700nm})_{pH1.0} - (A_{520nm} - A_{700nm})_{pH4.5}$; MW (molecular weight, g $mol^{-1}$) = 449.2; DF = dilution factor; $10^3$ = conversion factor from g to mg; ε = 26,900 (molar extinction coefficient for cyanidin-3-glucoside, $L^{-1}$ $mol^{-1}$ $cm^{-1}$); 1 = pathlength (cm).

### 4.6. Fruit Quality Index

To assess the quality of fruit flavor, i.e., sweetness, data for sugar and organic acids in strawberries were used. To characterize it, the S:A ratio [64] was calculated with the following formula: TS content (sucrose + fructose + glucose)/TOA content (citric + malic + acetic acids). In addition, the sweetness of strawberries was evaluated by a method of Keutgen and Pawelzik [30]. The sweetness index of fruits was calculated based on the degree to which each carbohydrate contributes to overall sweetness as follows: SI = (1.0 × glucose) + (2.3 × fructose) + (1.35 × sucrose).

### 4.7. Statistical Analysis

To compare differences in variables of plant and fruit characteristics among different *Chlorella* treatments, one-way ANOVA with Tukey's honestly significant difference (HSD) post hoc test at the 0.05 probability level was conducted ($n$ = 50 for FW and 6 for others). Pearson correlation analysis was conducted to investigate the correlations among the properties of fruits sampled in the second and last sampling months. Principal component analysis was performed to investigate patterns of variation in the dataset, focusing on the degree of relationships among key parameters related to plant performance and fruit quality throughout different *Chlorella*-level treatments. All statistical analyses in the present study were performed using the R program (version 3.3.3).

## 5. Conclusions

Microalgae are known to be of great utility for agricultural crop production, but their efficiency may vary depending on the species and treatment levels. Our findings obtained from the current fieldwork demonstrate that *C. fusca* CHK0059 has a high value as a biofertilizer resource for strawberry cultivation. Application of *C. fusca* CHK0059 stimulated the nutrient uptake efficiency (e.g., N, P, K, Mg, etc.) and physiological characteristics (e.g., photosynthesis, pigments, and biomass) of strawberry plants, leading to improvements in the fruit characteristics (e.g., SSC and hardness) and flavor quality (e.g., S:A ratio and SI). Among the fruit-related parameters, sucrose and anthocyanin contents tended to show the greatest response to *Chlorella* application, indicating that these substances may be key factors in improving the quality of strawberry fruits during the whole cultivation period. Additionally, the elevated content of anthocyanins in fruits following *Chlorella* application appeared to serve as antifungal agents for postharvest strawberries. Overall, the application of *C. fusca* CHK0059 could promote the overall growth and performance of strawberry plants, contributing to the improvement of strawberry quality and yield as well as postharvest storage capacity, especially in 0.2% OCD treatment. Further studies are needed to specifically identify the mechanisms of the effects of *Chlorella* fertilization through analysis of the interactions among various factors that determine the quality of strawberry fruits.

**Author Contributions:** Conceptualization, Y.-N.K., J.H.C. and Y.B.L.; methodology, Y.-N.K., J.H.C., K.-A.L., H.C. and M.-J.K.; validation, Y.-N.K., K.-A.L.,Y.-E.Y. and V.K.; formal analysis, Y.-N.K., J.H.C., S.Y.K., Y.-E.Y., H.C. and V.K.; investigation, Y.-N.K., J.H.C. and S.Y.K.; resources, M.-J.K.; data curation, J.H.C., K.-A.L. and V.K.; writing—original draft preparation, Y.-N.K., J.H.C. and K.-A.L.; writing—review and editing, Y.-N.K. and Y.B.L.; visualization, Y.-N.K., Y.-E.Y. and V.K.; supervision, Y.B.L.; funding acquisition, Y.-N.K. All authors have read and agreed to the published version of the manuscript.

**Funding:** This work was carried out with the support of "Cooperative Research Program for Agriculture Science and Technology Development (Project No. PJ015641022023 or RS-2021-RD009112)" Rural Development Administration, Republic of Korea.

**Data Availability Statement:** Data are contained within the article.

**Conflicts of Interest:** The authors declare no conflict of interest.

## References


1. Kim, Y.-N.; Cho, Y.-S.; Lee, J.-H.; Seo, H.-R.; Kim, B.-H.; Lee, D.-B.; Lee, Y.B.; Kim, K.-H. Short-Term Responses of Soil Organic Carbon Pool and Crop Performance to Different Fertilizer Applications. *Agronomy* **2022**, *12*, 1106. [CrossRef]
2. Daniel, A.I.; Fadaka, A.O.; Gokul, A.; Bakare, O.O.; Aina, O.; Fisher, S.; Burt, A.F.; Mavumengwana, V.; Keyster, M.; Klein, A. Biofertilizer: The Future of Food Security and Food Safety. *Microorganisms* **2022**, *10*, 1220. [CrossRef]
3. Kim, Y.-N.; Lee, J.-H.; Seo, H.-R.; Kim, J.-W.; Cho, Y.-S.; Lee, D.; Kim, B.-H.; Yoon, J.-H.; Choe, H.; Lee, Y.B.; et al. Co-Responses of Soil Organic Carbon Pool and Biogeochemistry to Different Long-Term Fertilization Practices in Paddy Fields. *Plants* **2022**, *11*, 3195. [CrossRef] [PubMed]
4. Mącik, M.; Gryta, A.; Frąc, M. Biofertilizers in Agriculture: An Overview on Concepts, Strategies and Effects on Soil Microorganisms. *Adv. Agron.* **2020**, *162*, 31–87. [CrossRef]
5. Savci, S. An Agricultural Pollutant: Chemical Fertilizer. *Int. J. Environ. Sci. Dev.* **2012**, *3*, 73. [CrossRef]
6. Liu, Y.; Pan, X.; Li, J. Current Agricultural Practices Threaten Future Global Food Production. *J. Agric. Environ. Ethics* **2015**, *28*, 203–216. [CrossRef]
7. Kim, Y.-N.; Choi, J.H.; Kim, S.Y.; Choe, H.; Shin, Y.; Yoon, Y.-E.; Lee, K.-A.; Kim, M.-J.; Lee, Y.B. Application Effect of *Chlorella Fusca* CHK0059 as a Biofertilizer for Strawberry Cultivation. *Korean J. Environ. Agric.* **2022**, *41*, 282–287. [CrossRef]
8. Singh, D.; Thapa, S.; Geat, N.; Mehriya, M.L.; Rajawat, M.V.S. Biofertilizers: Mechanisms and Application. In *Biofertilizers*; Rakshit, A., Meena, V.S., Parihar, M., Singh, H.B., Singh, A.K., Eds.; Woodhead Publishing: Cambridge, UK, 2021; Chapter 12; pp. 151–166. [CrossRef]
9. Wu, H.; Zhang, Z.; Wang, J.; Qin, X.; Chen, J.; Wu, L.; Lin, S.; Rensing, C.; Lin, W. Bio-Fertilizer Amendment Alleviates the Replanting Disease under Consecutive Monoculture Regimes by Reshaping Leaf and Root Microbiome. *Microb. Ecol.* **2022**, *84*, 425–464. [CrossRef]
10. Sun, W.; Shahrajabian, M.H.; Petropoulos, S.A.; Shahrajabian, N. Developing Sustainable Agriculture Systems in Medicinal and Aromatic Plant Production by Using Chitosan and Chitin-Based Biostimulants. *Plants* **2023**, *12*, 2469. [CrossRef]
11. Chatterjee, A.; Singh, S.; Agrawal, C.; Yadav, S.; Rai, R.; Rai, L.C. Role of Algae as a Biofertilizer. In *Algal Green Chemistry*; Rastogi, R.P., Madamwar, D., Pandey, A., Eds.; Elsevier: Amsterdam, The Netherlands, 2017; Chapter 10; pp. 189–200. [CrossRef]
12. Precedence Research. Biofertilizers Market Size to Hit USD 9.14 Bn by 2032. Available online: https://www.precedenceresearch.com/biofertilizers-market (accessed on 21 September 2023).
13. Thomas, L.; Singh, I. Microbial Biofertilizers: Types and Applications. In *Biofertilizers for Sustainable Agriculture and Environment*; Giri, B., Prasad, R., Wu, Q.-S., Varma, A., Eds.; Soil Biology; Springer International Publishing: Cham, Switzerland, 2019; Volume 55, pp. 1–19. [CrossRef]
14. Sun, W.; Shahrajabian, M.H. The Application of Arbuscular Mycorrhizal Fungi as Microbial Biostimulant, Sustainable Approaches in Modern Agriculture. *Plants* **2023**, *12*, 3101. [CrossRef]
15. Nosheen, S.; Ajmal, I.; Song, Y. Microbes as Biofertilizers, a Potential Approach for Sustainable Crop Production. *Sustainability* **2021**, *13*, 1868. [CrossRef]
16. Raimi, A.; Roopnarain, A.; Adeleke, R. Biofertilizer Production in Africa: Current Status, Factors Impeding Adoption and Strategies for Success. *Sci. Afr.* **2021**, *11*, e00694. [CrossRef]
17. Ann, S.-W.; Lee, J.-M.; Cho, Y.-K. Perilla Leaf Fertilization Effect of Fertilizer by Chlorella and Seafood By-product Fermentation. *J. Environ. Sci. Int.* **2020**, *29*, 423–434. [CrossRef]
18. Dineshkumar, R.; Subramanian, J.; Gopalsamy, J.; Jayasingam, P.; Arumugam, A.; Kannadasan, S.; Sampathkumar, P. The Impact of Using Microalgae as Biofertilizer in Maize (*Zea Mays* L.). *Waste Biomass Valor.* **2019**, *10*, 1101–1110. [CrossRef]
19. El-Sharony, T.F.; El-Gioushy, S.F.; Amin, O.A. Effect of Foliar Application with Algae and Plant Extracts on Growth, Yield and Fruit Quality of Fruitful Mango Trees Cv. *Fagri Kalan. J. Hortic.* **2015**, *2*, 1000162. [CrossRef]
20. Faheed, F.A.; Fattah, Z.A. Effect of *Chlorella Vulgaris* as Bio-Fertilizer on Growth Parameters and Metabolic Aspects of Lettuce Plant. *J. Agric. Soc. Sci.* **2008**, *4*, 165–169.
21. Garcia-Gonzalez, J.; Sommerfeld, M. Biofertilizer and Biostimulant Properties of the Microalga *Acutodesmus Dimorphus. J. Appl. Phycol.* **2016**, *28*, 1051–1061. [CrossRef]
22. Kim, M.-J.; Shim, C.-K.; Kim, Y.-K.; Ko, B.-G.; Park, J.-H.; Hwang, S.-G.; Kim, B.-H. Effect of Biostimulator *Chlorella Fusca* on Improving Growth and Qualities of Chinese Chives and Spinach in Organic Farm. *Plant Pathol. J.* **2018**, *34*, 567–574. [CrossRef]
23. Lee, S.-M.; Ryu, C.-M. Algae as New Kids in the Beneficial Plant Microbiome. *Front. Plant Sci.* **2021**, *12*, 599742. [CrossRef]
24. RDA. Chlorella, Crops Are Growing Rapidly and Farm Income Is also Rapidly (in Korean). Available online: http://www.rda.go.kr/board/board.do?boardId=farmprmninfo&prgId=day_farmprmninfoEntry&dataNo=100000771576&mode=updateCnt (accessed on 21 November 2023).
25. Kim, M.-J.; Shim, C.-K.; Kim, Y.-K.; Park, J.-H.; Hong, S.-J.; Ji, H.-J.; Han, E.-J.; Yoon, J.-C. Effect of *Chlorella vulgaris* CHK0008 Fertilization on Enhancement of Storage and Freshness in Organic Strawberry and Leaf Vegetables. *Hortic. Sci. Technol.* **2014**, *32*, 872–878. [CrossRef]

26. Kim, M.-J.; Shim, C.-K.; Ko, B.-G.; Kim, J. Effect of the Microalga *Chlorella Fusca* CHK0059 on Strawberry PGPR and Biological Control of Fusarium Wilt Disease in Non-Pesticide Hydroponic Strawberry Cultivation. *J. Microbiol. Biotechnol.* **2020**, *30*, 708–716. [CrossRef] [PubMed]
27. El-Ghanam, A.A.; Farfour, S.; Ragab, S. Bio-Suppression of Strawberry Fruit Rot Disease Caused by *Botrytis Cinerea*. *J. Plant Pathol Microbiol.* **2015**, *S3*, 5. [CrossRef]
28. Liu, M.; Wang, G.; Xiao, L.; Xu, X.; Liu, X.; Xu, P.; Lin, X. Bis(2,3-Dibromo-4,5-Dihydroxybenzyl) Ether, a Marine Algae Derived Bromophenol, Inhibits the Growth of *Botrytis Cinerea* and Interacts with DNA Molecules. *Mar. Drugs* **2014**, *12*, 3838–3851. [CrossRef]
29. Gigliotti, G. Effect of Microalgal Biomass on the Strawberry (*Fragaria x Ananassa*) Yield and Quality. Master's Thesis, Universidade do Algarve, Faro, Portugal, 2022.
30. Keutgen, A.; Pawelzik, E. Modifications of Taste-Relevant Compounds in Strawberry Fruit under NaCl Salinity. *Food Chem.* **2007**, *105*, 1487–1494. [CrossRef]
31. Gonçalves, A.L. The Use of Microalgae and Cyanobacteria in the Improvement of Agricultural Practices: A Review on Their Biofertilising, Biostimulating and Biopesticide Roles. *Appl. Sci.* **2021**, *11*, 871. [CrossRef]
32. Hu, H.-J.; Xu, K.; He, L.-C.; Wang, G.-X. A Model for the Relationship between Plant Biomass and Photosynthetic Rate Based on Nutrient Effects. *Ecosphere* **2021**, *12*, e03678. [CrossRef]
33. Lee, K.-A.; Kim, Y.-N.; Kantharaj, V.; Lee, Y.B.; Woo, S.Y. Seedling Growth and Photosynthetic Response of *Pterocarpus Indicus* L. to Shading Stress. *Plant Signal. Behav.* **2023**, *18*, 2245625. [CrossRef]
34. Çakirsoy, I.; Miyamoto, T.; Ohtake, N. Physiology of Microalgae and Their Application to Sustainable Agriculture: A Mini-Review. *Front. Plant Sci.* **2022**, *13*, 1005991. [CrossRef]
35. Liu, Z.; Liang, T.; Kang, C. Molecular Bases of Strawberry Fruit Quality Traits: Advances, Challenges, and Opportunities. *Plant Physiol.* **2023**, *193*, 900–914. [CrossRef]
36. Cao, F.; Guan, C.; Dai, H.; Li, X.; Zhang, Z. Soluble Solids Content Is Positively Correlated with Phosphorus Content in Ripening Strawberry Fruits. *Sci. Hortic.* **2015**, *195*, 183–187. [CrossRef]
37. Del Olmo, I.; Blanch, M.; Romero, I.; Vazquez-Hernandez, M.; Sanchez-Ballesta, M.T.; Escribano, M.I.; Merodio, C. Involvement of Oligosaccharides and Sucrose-Related Genes on Sucrose Retention in Strawberries from Ripening to Shelf-Life. *Postharvest Biol. Technol.* **2020**, *169*, 111301. [CrossRef]
38. Schwieterman, M.L.; Colquhoun, T.A.; Jaworski, E.A.; Bartoshuk, L.M.; Gilbert, J.L.; Tieman, D.M.; Odabasi, A.Z.; Moskowitz, H.R.; Folta, K.M.; Klee, H.J. Strawberry Flavor: Diverse Chemical Compositions, a Seasonal Influence, and Effects on Sensory Perception. *PLoS ONE* **2014**, *9*, e88446. [CrossRef]
39. Osorio, S.; Ruan, Y.-L.; Fernie, A.R. An Update on Source-to-Sink Carbon Partitioning in Tomato. *Front. Plant Sci.* **2014**, *5*, 516. [CrossRef] [PubMed]
40. Lemoine, R.; Camera, S.L.; Atanassova, R.; Dédaldéchamp, F.; Allario, T.; Pourtau, N.; Bonnemain, J.-L.; Laloi, M.; Coutos-Thévenot, P.; Maurousset, L. Source-to-Sink Transport of Sugar and Regulation by Environmental Factors. *Front. Plant Sci.* **2013**, *4*, 272. [CrossRef]
41. Aluko, O.O.; Li, C.; Wang, Q.; Liu, H. Sucrose Utilization for Improved Crop Yields: A Review Article. *Int. J. Mol. Sci.* **2021**, *22*, 4704. [CrossRef]
42. Batista-Silva, W.; Nascimento, V.L.; Medeiros, D.B.; Nunes-Nesi, A.; Ribeiro, D.M.; Zsögön, A.; Araújo, W.L. Modifications in Organic Acid Profiles during Fruit Development and Ripening: Correlation or Causation? *Front. Plant Sci.* **2018**, *9*, 1689. [CrossRef] [PubMed]
43. Zell, M.B.; Fahnenstich, H.; Maier, A.; Saigo, M.; Voznesenskaya, E.V.; Edwards, G.E.; Andreo, C.; Schleifenbaum, F.; Zell, C.; Drincovich, M.F. Analysis of Arabidopsis with Highly Reduced Levels of Malate and Fumarate Sheds Light on the Role of These Organic Acids as Storage Carbon Molecules. *Plant Physiol.* **2010**, *152*, 1251–1262. [CrossRef] [PubMed]
44. Cordenunsi, B.R.; Oliveira do Nascimento, J.R.; Genovese, M.I.; Lajolo, F.M. Influence of Cultivar on Quality Parameters and Chemical Composition of Strawberry Fruits Grown in Brazil. *J. Agric. Food Chem.* **2002**, *50*, 2581–2586. [CrossRef] [PubMed]
45. Wanpeng, X.; Zheng, Q.; Juanfang, L.; Junping, Q. Comparative Analysis of Three Types of Peaches: Identification of the Key Individual Characteristic Flavor Compounds by Integrating Consumers' Acceptability with Flavor Quality. *Hortic. Plant J.* **2017**, *3*, 1–12. [CrossRef]
46. Etienne, A.; Génard, M.; Lobit, P.; Mbeguié-A-Mbéguié, D.; Bugaud, C. What Controls Fleshy Fruit Acidity? A Review of Malate and Citrate Accumulation in Fruit Cells. *J. Exp. Bot.* **2013**, *64*, 1451–1469. [CrossRef] [PubMed]
47. Sun, Y.; Asghari, M.; Zahedipour-Sheshgelani, P. Foliar Spray with 24-Epibrassinolide Enhanced Strawberry Fruit Quality, Phytochemical Content, and Postharvest Life. *J. Plant Growth Regul.* **2020**, *39*, 920–929. [CrossRef]
48. Roussos, P.; Denaxa, N.; Damvakaris, T. Strawberry Fruit Quality Attributes after Application of Plant Growth Stimulating Compounds. *Sci. Hortic.* **2009**, *119*, 138–146. [CrossRef]
49. Jia, H.; Wang, Y.; Sun, M.; Li, B.; Han, Y.; Zhao, Y.; Li, X.; Ding, N.; Li, C.; Ji, W. Sucrose Functions as a Signal Involved in the Regulation of Strawberry Fruit Development and Ripening. *New Phytol.* **2013**, *198*, 453–465. [CrossRef] [PubMed]
50. Li, Z.; Wei, Y.; Xu, Y.; Han, P.; Jiang, S.; Xu, F.; Wang, H.; Tao, N.; Shao, X. Terpinen-4-Ol Treatment Maintains Quality of Strawberry Fruit during Storage by Regulating Sucrose-Induced Anthocyanin Accumulation. *Postharvest Biol. Technol.* **2021**, *174*, 111461. [CrossRef]

51. Guo, R.; Yuan, G.; Wang, Q. Sucrose Enhances the Accumulation of Anthocyanins and Glucosinolates in Broccoli Sprouts. *Food Chem.* **2011**, *129*, 1080–1087. [CrossRef] [PubMed]
52. Olivares, D.; Contreras, C.; Muñoz, V.; Rivera, S.; González-Agüero, M.; Retamales, J.; Defilippi, B.G. Relationship among Color Development, Anthocyanin and Pigment-Related Gene Expression in 'Crimson Seedless' Grapes Treated with Abscisic Acid and Sucrose. *Plant Physiol. Biochem.* **2017**, *115*, 286–297. [CrossRef] [PubMed]
53. Petrasch, S.; Knapp, S.J.; Van Kan, J.A.; Blanco-Ulate, B. Grey Mould of Strawberry, a Devastating Disease Caused by the Ubiquitous Necrotrophic Fungal Pathogen *Botrytis Cinerea*. *Mol. Plant Pathol.* **2019**, *20*, 877–892. [CrossRef]
54. Schaefer, H.M.; Rentzsch, M.; Breuer, M. Anthocyanins Reduce Fungal Growth in Fruits. *Nat. Prod. Commun.* **2008**, *3*, 1934578X0800300808. [CrossRef]
55. Sivankalyani, V.; Feygenberg, O.; Diskin, S.; Wright, B.; Alkan, N. Increased Anthocyanin and Flavonoids in Mango Fruit Peel Are Associated with Cold and Pathogen Resistance. *Postharvest Biol. Technol.* **2016**, *111*, 132–139. [CrossRef]
56. Zhang, Y.; Butelli, E.; De Stefano, R.; Schoonbeek, H.; Magusin, A.; Pagliarani, C.; Wellner, N.; Hill, L.; Orzaez, D.; Granell, A. Anthocyanins Double the Shelf Life of Tomatoes by Delaying Overripening and Reducing Susceptibility to Gray Mold. *Curr. Biol.* **2013**, *23*, 1094–1100. [CrossRef]
57. Schaefer, H.M.; Rolshausen, G. Plants on Red Alert: Do Insects Pay Attention? *Bioessays* **2006**, *28*, 65–71. [CrossRef] [PubMed]
58. Peters, D.J.; Constabel, C.P. Molecular Analysis of Herbivore-induced Condensed Tannin Synthesis: Cloning and Expression of Dihydroflavonol Reductase from Trembling Aspen (*Populus Tremuloides*). *Plant J.* **2002**, *32*. 701–712. [CrossRef] [PubMed]
59. RDA. *Manual for Strawberry Cultivation*; RDA: Wanju, Republic of Korea, 2019.
60. NAAS. *Chlorella Culture Method and Agricultural Application*; RDA: Wanju, Republic of Korea, 2015.
61. Lichtenthaler, H.K. Chlorophylls and Carotenoids: Pigments of Photosynthetic Biomembranes. *Meth. Enzymol.* **1987**, *148*, 350–382.
62. NIAST. *Methods of Analysis of Soil and Plant*; RDA: Suwon, Republic of Korea, 2000.
63. Lee, J.; Durst, R.W.; Wrolstad, R.E. Collaborators: Determination of Total Monomeric Anthocyanin Pigment Content of Fruit Juices, Beverages, Natural Colorants, and Wines by the pH Differential Method: Collaborative Study. *J. AOAC Int.* **2005**, *88*, 1269–1278. [CrossRef] [PubMed]
64. Montero, T.M.; Mollá, E.M.; Esteban, R.M.; López-Andréu, F.J. Quality Attributes of Strawberry during Ripening. *Sci. Hortic.* **1996**, *65*, 239–250. [CrossRef]

**Disclaimer/Publisher's Note:** The statements, opinions and data contained in all publications are solely those of the individual author(s) and contributor(s) and not of MDPI and/or the editor(s). MDPI and/or the editor(s) disclaim responsibility for any injury to people or property resulting from any ideas, methods, instructions or products referred to in the content.

*plants*

*Article*

# Multifarious Characterization and Efficacy of Three Phosphate-Solubilizing *Aspergillus* Species as Biostimulants in Improving Root Induction of Cassava and Sugarcane Stem Cuttings

**Surapong Khuna [1,2,3], Jaturong Kumla [1,2,3], Sirasit Srinuanpan [1,2,3], Saisamorn Lumyong [1,3,4] and Nakarin Suwannarach [1,2,*]**

[1] Center of Excellence in Microbial Diversity and Sustainable Utilization, Chiang Mai University, Chiang Mai 50200, Thailand; surapongway@gmail.com (S.K.); jaturong_yai@hotmail.com (J.K.); sirasit.s@cmu.ac.th (S.S.); saisamorn.l@cmu.ac.th (S.L.)
[2] Office of Research Administration, Chiang Mai University, Chiang Mai 50200, Thailand
[3] Department of Biology, Faculty of Science, Chiang Mai University, Chiang Mai 50200, Thailand
[4] Academy of Science, The Royal Society of Thailand, Bangkok 10300, Thailand
* Correspondence: suwan.462@gmail.com; Tel.: +66-8-6512-7518

**Abstract:** Several soil fungi significantly contribute to the enhancement of plant development by improving nutrient uptake and producing growth-promoting metabolites. In the present study, three strains of phosphate-solubilizing fungi, namely, *Aspergillus chiangmaiensis* SDBR-CMUI4, *A. pseudopiperis* SDBR-CMUI1, and *A. pseudotubingensis* SDBR-CMUO2, were examined for their plant-growth-promoting capabilities. The findings demonstrated that all fungi showed positive siderophore production, but only *A. pseudopiperis* can produce indole-3-acetic acid. All fungi were able to solubilize insoluble phosphate minerals [$Ca_3(PO_4)_2$ and $FePO_4$] by producing phosphatase enzymes and organic acids (oxalic, tartaric, and succinic acids). These three fungal species were grown at a water activity ranging from 0.837 to 0.998, pH values ranging from 4 to 9, temperatures between 4 and 40 °C, and 16–17% NaCl in order to evaluate their drought, pH, temperature, and salt tolerances, respectively. Moreover, the results indicated that *A. pseudopiperis* and *A. pseudotubingensis* were able to tolerate commercial insecticides (methomyl and propargite) at the recommended dosages for field application. The viability of each fungal strain in the inoculum was higher than 50% at 4 and 20 °C after 3 months of storage. Subsequently, all fungi were characterized as plant-growth-promoting strains by improving the root inductions of cassava (*Manihot esculenta* Crantz) and sugarcane (*Saccharum officinarum* L.) stem cuttings in greenhouse experiments. No symptoms of plant disease were observed with any of the treatments involving fungal inoculation and control. The cassava and sugarcane stem cuttings inoculated with fungal strains and supplemented with $Ca_3(PO_4)_2$ exhibited significantly increased root lengths, shoot and root dry biomasses, chlorophyll concentrations, and cellular inorganic phosphate contents. Therefore, the application of these phosphate-solubilizing fungi is regarded as a new frontier in the induction of roots and the promotion of growth in plants.

**Keywords:** economic plants; fungal inoculant; phosphate solubilization; plant growth promotion; soil fungi

**Citation:** Khuna, S.; Kumla, J.; Srinuanpan, S.; Lumyong, S.; Suwannarach, N. Multifarious Characterization and Efficacy of Three Phosphate-Solubilizing *Aspergillus* Species as Biostimulants in Improving Root Induction of Cassava and Sugarcane Stem Cuttings. *Plants* **2023**, *12*, 3630. https://doi.org/10.3390/plants12203630

Academic Editors: Spyridon A. Petropoulos and Mohamad Hesam Shahrajabian

Received: 26 September 2023
Revised: 15 October 2023
Accepted: 18 October 2023
Published: 20 October 2023

**Copyright:** © 2023 by the authors. Licensee MDPI, Basel, Switzerland. This article is an open access article distributed under the terms and conditions of the Creative Commons Attribution (CC BY) license (https://creativecommons.org/licenses/by/4.0/).

## 1. Introduction

The availability of micro- and macro-elements in agricultural soils has a significant impact on the growth, development, and production of plants [1]. Phosphorus (P) is one of the most important macro-elements and is required for almost all of the major metabolic processes in plants, such as signal transduction, energy transfer reactions, respiration, macromolecular biosynthesis, and photosynthesis [2–4]. It is extensively distributed in soils in both inorganic/insoluble (containing aluminum, calcium, and iron phosphates) and organic (organic matter) forms [5,6]. Many agricultural soils contain a substantial total P in the range of 500 to 800 mg/kg of dry soil [7]. However, only 0.1% of the total P in the soil has been shown to be available for assimilation by plants [3,8,9]. Soils are frequently deficient in phosphorus because the majority of the element is insoluble and inaccessible to plants [10]. Consequently, chemical-based fertilizers are widely utilized in order to increase the concentration of available elements, especially P, in the soil that are accessible for the growth and production of crops [2]. Unfortunately, using chemical-based fertilizers excessively, frequently, and inappropriately has resulted in a number of negative side effects, including a decrease in soil fertility, a reduction in microbial diversity, environmental degradation, a decrease in crop yield, and an increase in crop production costs, and this has also caused harm to the health of farmers and consumers [5,11]. Therefore, the concept of employing phosphate-solubilizing fungi as biostimulant suppliers for readily available soil phosphorus for plants provides a strategically advantageous approach from both an economic and environmental perspective [3,12].

Phosphate-solubilizing fungi (PSF) are a type of advantageous microorganism that have the ability to process insoluble forms of phosphorus into soluble forms that can be assimilated easily by plants [13,14]. These fungi are widely dispersed throughout the soil and frequently related to plant rhizospheres [15]. Earlier research suggests that PSF constitutes approximately 0.1% to 0.5% of the overall fungal populations in soil [3,16,17]. These fungi are able to dissolve insoluble phosphate minerals due to their ability to lower pH levels by secreting organic acids (e.g., citric, gluconic, malic, oxalic, succinic, and tartaric acids) and via chelation, exchange reactions, and mineralization [4,18–21]. In addition to making soluble phosphorus available for plant uptake, PSF can also enhance plant growth, induce systematic resistance, and increase stress tolerance through one of the following mechanisms: the production of phytohormones, antibiotics, and siderophores; the solubilization of various other insoluble minerals (calcium, copper, cobalt, iron, manganese, magnesium, potassium, and zinc); and biocontrol activity [3,22–24]. Numerous prior studies have demonstrated that the inoculation of PSF belonging to the genera *Aspergillus*, *Penicillium*, *Rhizopus*, *Talaromyces*, and *Trichoderma* could enhance root and plant growth for several plants (Arabidopsis, chickpea, chili, haricot beans, maize, mung beans, onions, soybeans, and tomatoes) while also improving soil fertility [23,25–33].

From our previous study, three *Aspergillus* strains, including *Aspergillus chiangmaiensis* SDBR-CMUI4, *A. pseudopiperis* SDBR-CMUI1, and *A. pseudotubingensis* SDBR-CMUO2, can solubilize $Ca_3(PO_4)_2$ and promote the growth of Arabidopsis [*Arabidopsis thaliana* (L.) Heynh] and onion (*Allium cepa* L.) [23]. However, their other plant-growth-promoting abilities and tolerances to drought, pH, temperature, salinity, and agrochemicals have not yet been determined. Therefore, the objective of this research was to investigate their plant-growth-promoting abilities, including the production of indole-3-acetic acid (IAA) and siderophore. The production of phosphatase enzymes and organic acids for the solubilization of different forms of phosphate minerals ($AlPO_4$, $Ca_3(PO_4)_2$, and $FePO_4$) was determined. Subsequently, the capacity of these fungi to tolerate conditions of drought, pH values, temperatures, salinity, and the presence of agrochemicals was examined. The viability of fungi in a granular inoculum was studied. Furthermore, the fungal strains were evaluated in terms of their ability to improve root induction in cassava and sugarcane stem cuttings in a greenhouse experiment. The findings obtained from this study will be utilized for the development of a biostimulant-based inoculum using phosphate-solubilizing *Aspergillus*, potentially serving as an alternative to chemical-based fertilizers and supporting sustainable agricultural practices in the future.

## 2. Results

### *2.1. Characterization of Plant Growth Promotion Properties*

#### 2.1.1. Determination of IAA Production

After one week, only the culture supernatant from *A. pseudopiperis* exhibited a positive reaction for IAA, as evidenced by the development of a red to pink color when tested using Salkowski's reagent. A negative result for IAA detection was obtained in the uninoculated medium and culture supernatants from both *A. chiangmaiensis* and *A. pseudotubingensis* Subsequently, the presence of IAA was verified through high-performance liquid chromatography (HPLC). The examination revealed the existence of IAA generated by *A. pseudopiperis*, aligning with the authentic IAA standard that exhibited a retention time of 10.1 min and a peak absorption at 279 nm. Additionally, the identity of fungal IAA was confirmed by a co-injection with the corresponding reference standard. HPLC was also used to measure the concentration of fungal IAA. It was found that *A. pseudopiperis* produced IAA at a level of 33.37 μg/mL (Table 1).

**Table 1.** Plant growth promotion properties and tolerance of fungi in this study.

| Plant Growth Promotion Properties | *A. chiangmaiensis* SDBR-CMUI4 | *A. pseudopiperis* SDBR-CMUI1 | *A. pseudotubingensis* SDBR-CMUO2 |
|---|---|---|---|
| IAA production (μg/mL) | − | 33.37 | − |
| Siderophore production | + | + | + |
| Drought tolerance ($a_w$) | 0.837–0.998 | 0.859–0.998 | 0.837–0.998 |
| pH tolerance | 4.0–9.0 | 4.0–9.0 | 4.0–9.0 |
| Temperature tolerance (°C) | 4–40 | 4–40 | 4–40 |
| Salinity tolerance (% NaCl) | Up to 17% | Up to 16% | Up to 17% |

"+": positive result; "−": negative result.

#### 2.1.2. Determination of Siderophore Production

The findings revealed that all fungal strains exhibited positive siderophore production, which was evident through the formation of a pink zone around their colonies (Table 1 and Figure 1).

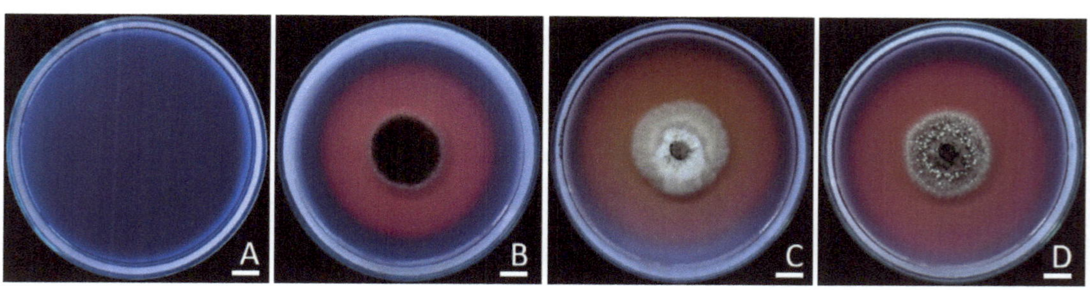


**Figure 1.** The alteration in color zones on chrome azurol S (CAS) agar due to siderophore production in uninoculated medium (**A**), *A. chiangmaiensis* SDBR-CMUI4 (**B**), *A. pseudopiperis* SDBR-CMUI1 (**C**), and *A. pseudotubingensis* SDBR-CMUO2 (**D**). Scale bars: 10 mm.

#### 2.1.3. Solubilization of Phosphate Minerals, Phosphatase Activities, and Organic Acid Production

All fungi could no longer grow in the liquid medium supplemented with $AlPO_4$. During the 15 days of cultivation, the pH values of the culture supernatants from the liquid medium supplemented with $Ca_3(PO_4)_2$ and $FePO_4$ significantly decreased compared to the culture supernatants from the uninoculated treatments (Figure 2A,B). The pH values of the culture supernatants in the solubilizations of $Ca_3(PO_4)_2$ and $FePO_4$ were observed within the ranges of 3.34–5.61 and 2.12–4.84, respectively. The lowest pH values were observed six days after incubation in the solubilizations of both $Ca_3(PO_4)_2$ and $FePO_4$. For the solubilization of $Ca_3(PO_4)_2$, *A. pseudotubingensis* displayed the lowest pH value (pH 3.34), followed by *A. pseudopiperis* (pH 3.49) and *A. chiangmaiensis* (pH 3.72) (Figure 2A). Additionally, the lowest pH value in the $FePO_4$ solubilization was recorded for *A. pseudopiperis* (pH 2.12), followed by *A. pseudotubingensis* (pH 2.16) and *A. chiangmaiensis* (pH 2.40) (Figure 2B). However, the pH values started increasing accordingly with the slight increase in the incubation time.

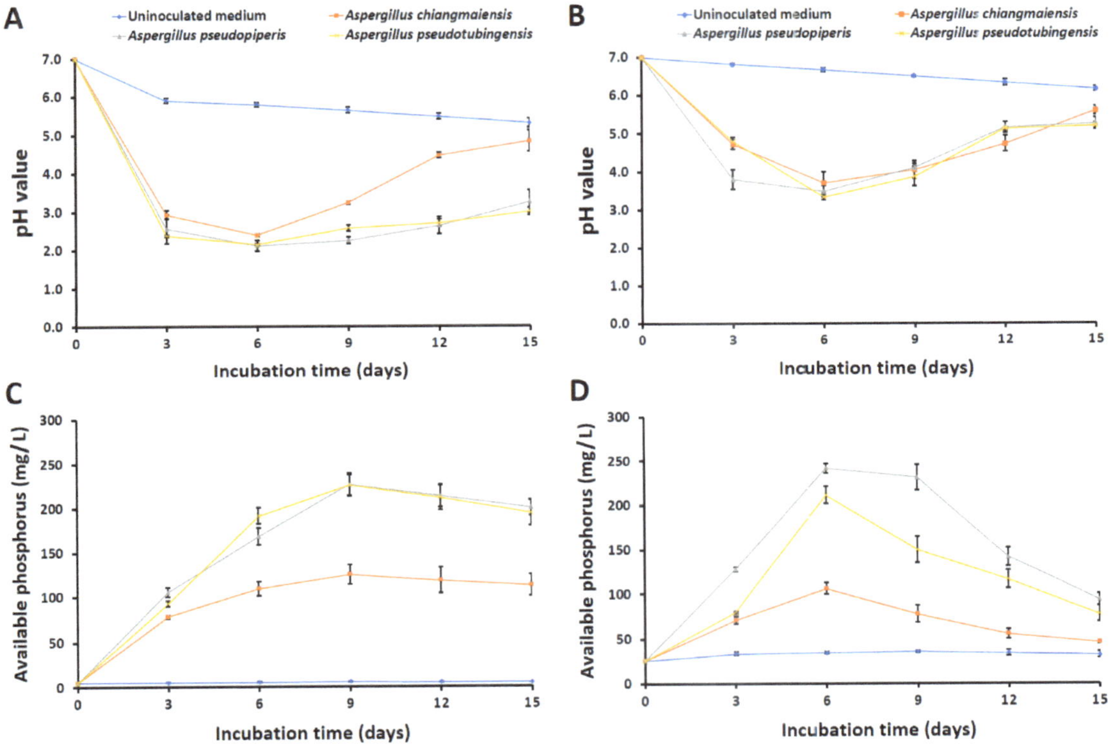


**Figure 2.** Changes in pH and available phosphorus concentration over time during solubilizations of $Ca_3(PO_4)_2$ (**A**,**C**) and $FePO_4$ (**B**,**D**).

During the 15 days of cultivation, the amount of available phosphorus solubilized by the fungi was measured from culture supernatants. The results indicated that the amount of available phosphorus solubilized by each fungal strain was significantly higher than that solubilized by the uninoculated treatments (Figure 2C,D). After 15 days of incubation, the content of available phosphorus from the experiments evaluating $Ca_3(PO_4)_2$ and $FePO_4$ solubilizations ranged from 78.79 to 226.34 mg/L and from 46.50 to 241.48 mg/L, respectively. The maximum available phosphorus of each fungal strain was present in the medium supplemented with $Ca_3(PO_4)_2$ and $FePO_4$ after being incubated for nine and six days, respectively. In terms of the solubilization of $Ca_3(PO_4)_2$, *A. pseudopiperis* had the highest available phosphorus content (226.34 mg/L), followed by *A. pseudotubingensis* (225.78 mg/L) and *A. chiangmaiensis* (125.45 mg/L) after nine days of incubation (Figure 2C). In addition, the highest available phosphorus content was found with *A. pseudopiperis* (241.48 mg/L), followed by *A. pseudotubingensis* (211.73 mg/L) and *A. chiangmaiensis* (106.13 mg/L) after six days of incubation in the solubilization of $FePO_4$ (Figure 2D).

During the solubilizations of $Ca_3(PO_4)_2$ and $FePO_4$, three fungi were found to produce both acid and alkaline phosphatases during the solubilizations of $Ca_3(PO_4)_2$ and $FePO_4$ in the liquid medium. It was found that the acid phosphatase activity was higher than that for the alkaline phosphatase (Figure 3). The values of acid phosphatases in the liquid medium during the solubilizations of $Ca_3(PO_4)_2$ and $FePO_4$ were within the ranges of 31.24–442.67 and 37.14–283.33 μmol *p*NP/mL/h, respectively (Figure 3A,B). *Aspergillus pseudopiperis* showed the highest activity of acid phosphatase in the presence of $Ca_3(PO_4)$ (442.67 μmol *p*NP/mL/h), followed by *A. pseudotubingensis* (274.95 μmol *p*NP/mL/h) and *A. chiangmaiensis* (197.71 μmol *p*NP/mL/h) at day 9 of incubation, respectively (Figure 3A). Furthermore, the highest acid phosphatase in the presence of $FePO_4$ was found with *A. pseudotubingensis* (283.33 μmol *p*NP/mL/h), followed by *A. pseudopiperis* (155.24 μmol *p*NP/mL/h) and *A. chiangmaiensis* (151.33 μmol *p*NP/mL/h) at day 6 of incubation, respectively (Figure 3B).

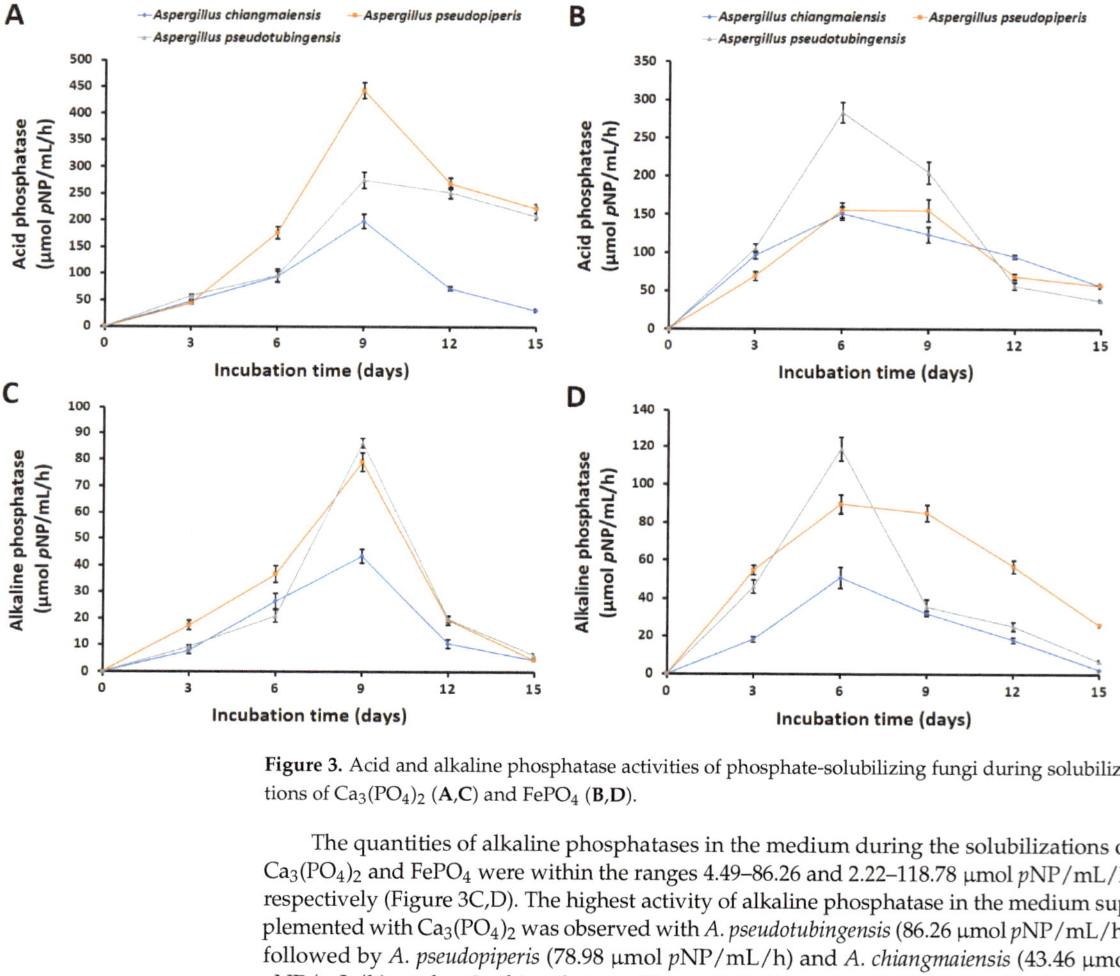


**Figure 3.** Acid and alkaline phosphatase activities of phosphate-solubilizing fungi during solubilizations of $Ca_3(PO_4)_2$ (**A**,**C**) and $FePO_4$ (**B**,**D**).

The quantities of alkaline phosphatases in the medium during the solubilizations of $Ca_3(PO_4)_2$ and $FePO_4$ were within the ranges 4.49–86.26 and 2.22–118.78 μmol *p*NP/mL/h, respectively (Figure 3C,D). The highest activity of alkaline phosphatase in the medium supplemented with $Ca_3(PO_4)_2$ was observed with *A. pseudotubingensis* (86.26 μmol *p*NP/mL/h), followed by *A. pseudopiperis* (78.98 μmol *p*NP/mL/h) and *A. chiangmaiensis* (43.46 μmol *p*NP/mL/h) at day 9 of incubation (Figure 3A). However, the highest alkaline phosphatase in the medium supplemented with $FePO_4$ was obtained with *A. pseudotubingensis* (118.78 μmol *p*NP/mL/h), followed by *A. pseudopiperis* (78.98 μmol *p*NP/mL/h) and *A. chiangmaiensis* (89.75 and 50.80 μmol *p*NP/mL/h) at day 6 of incubation (Figure 3B). The amounts of acid and alkaline phosphatases produced by all fungal strains were found to be correlated with an increase in the amount of available phosphorus. However, the results of the acid and alkaline phosphatase activities were both negative in the uninoculated medium.

The organic acid production from fungi after 15 days of $Ca_3(PO_4)_2$ and $FePO_4$ solubilizations was determined via HPLC. The findings revealed that the organic acid content in the culture supernatants corresponded to the organic acid standards of oxalic acid, tartaric acid, and succinic acid with a relation time of 4.0, 4.5, and 10.7 min, respectively (Figure 4).

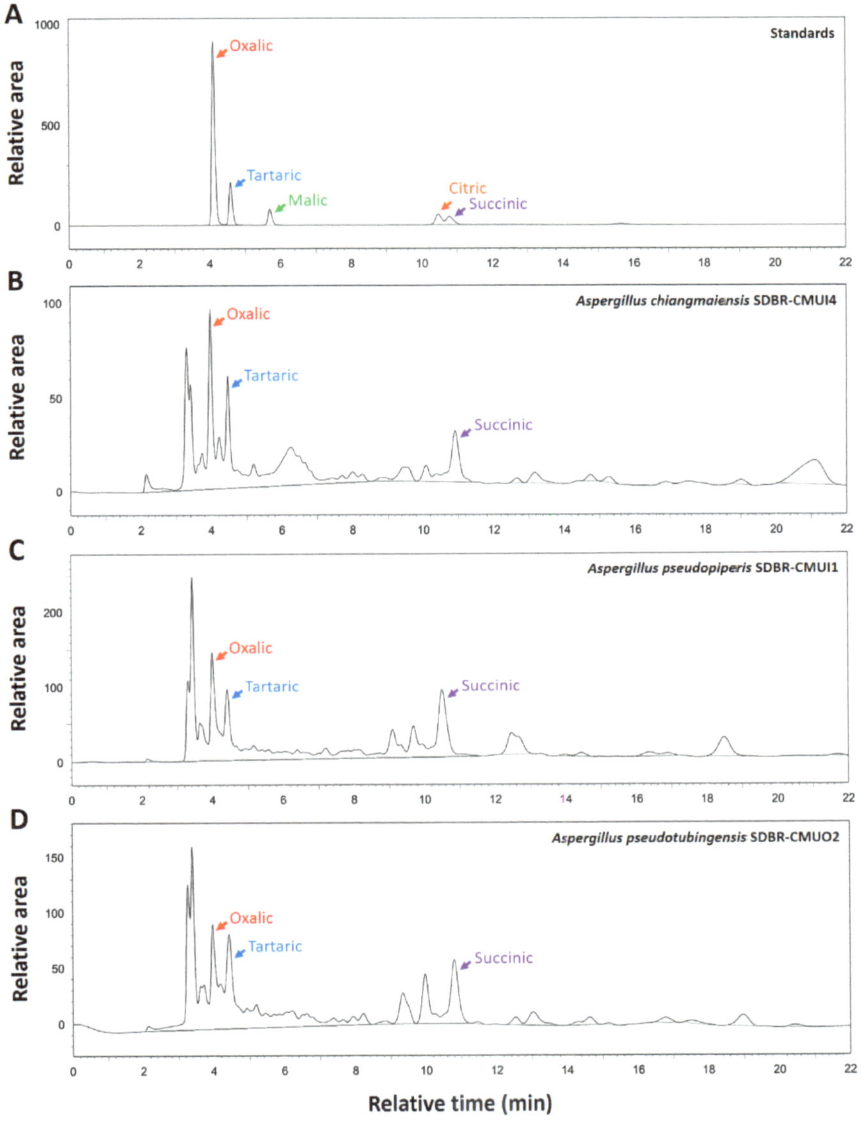


**Figure 4.** HPLC chromatograms of culture supernatant in solubilization of $Ca_3(PO_4)_2$. Organic acid standard (**A**), A. *chiangmaiensis* SDBR-CMUI4 (**B**), A. *pseudopiperis* SDBR-CMUI1 (**C**), and A. *pseudotubingensis* SDBR-CMUO2 (**D**).

The concentration of organic acid in the culture supernatants was also assessed using HPLC and is shown in Figure 5. The obtained values of oxalic acid, tartaric acid, and succinic acid contents of all strains were within the ranges of 4.16–333.65, 5.21–606.88, and 2.79–58.66 μg/mL, respectively. In terms of the solubilization of $FePO_4$, tartaric acid was dominant, followed by oxalic acid and succinic acid. The highest concentrations of tartaric and oxalic acids were observed in the culture supernatants from the liquid medium supplemented with $FePO_4$ and inoculated with each phosphate-solubilizing fungus. Additionally, all strains produced the highest concentration of succinic acid in the presence of $Ca_3(PO_4)_2$ However, organic acids were not detected in the uninoculated treatments.

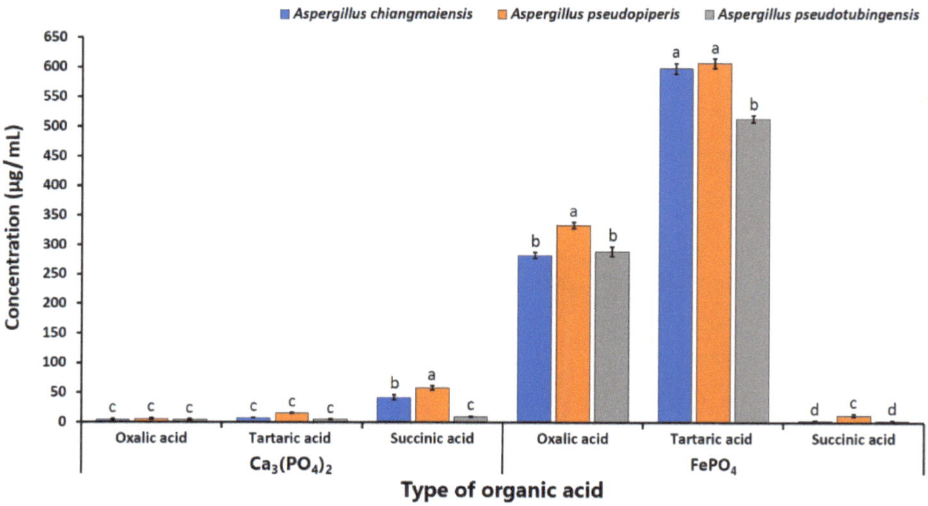


**Figure 5.** Production of organic acids after 15 days of the process of solubilization of $Ca_3(PO_4)_2$ and $FePO_4$ in a liquid medium. The error bars indicate the standard deviation of the average. Distinct letters denote a statistically significant difference ($p \leq 0.05$).

### 2.2. Drought, pH, Temperature, and Salinity Tolerances

Under conditions of drought, all fungal species were capable of growth. Both *A. chiangmaiensis* and *A. pseudotubingensis* demonstrated exceptional drought resistance in vitro, successfully growing in agar with a water availability ($a_w$) of 0.837. However, *A. pseudopiperis* could be grown in up to an $a_w$ value of 0.859. All fungal strains were able to grow on agar within the pH range of 4.0 to 9.0 but did not grow at a pH of 3.0. Additionally, it was found that all fungi could grow at temperatures ranging from 4 to 40 °C. Moreover, *A. chiangmaiensis* and *A. pseudotubingensis* exhibited a salinity tolerance at up to 17% NaCl, whereas *A. pseudopiperis* displayed tolerance to salinity up to 16% NaCl (Table 1).

### 2.3. Agrochemical Tolerance

Three herbicides (glyphosate-isopropylammonium, 2,4-D-dimethylammonium, and paraquat dichloride), two fungicides (metalaxyl and benomyl), and two insecticides (methomyl and propargite) were used in this study. The tolerance of all strains to these agrochemicals is expressed as the TI value, which is shown in Figure 6. The TI value was found to decrease when the concentration of all agrochemical chemicals increased. All of the tested fungicide compounds showed a strong inhibition of the growth of all of the tested fungal species. The outcome showed that paraquat dichloride inhibited the growth of all fungal species (TI value < 50%) at all tested dosages. All if the fungal species were tolerant (TI value > 50%) to glyphosate-isopropylammonium and propargite at half of the recommended dosages. *Aspergillus chiangmaiensis* could tolerate methomyl at half of the recommended dosage (1/2 RD) (Figure 6A). Interestingly, *A. pseudopiperis* showed the highest growth tolerance at half of the recommended dosage of 2,4-D-dimethylammonium and at the recommended dosage (RD) and double the recommended (2RD) dosage of methomyl (Figure 6B). Moreover, *A. pseudotubingensis* could tolerate propargite at the recommended dosages (Figure 6C).

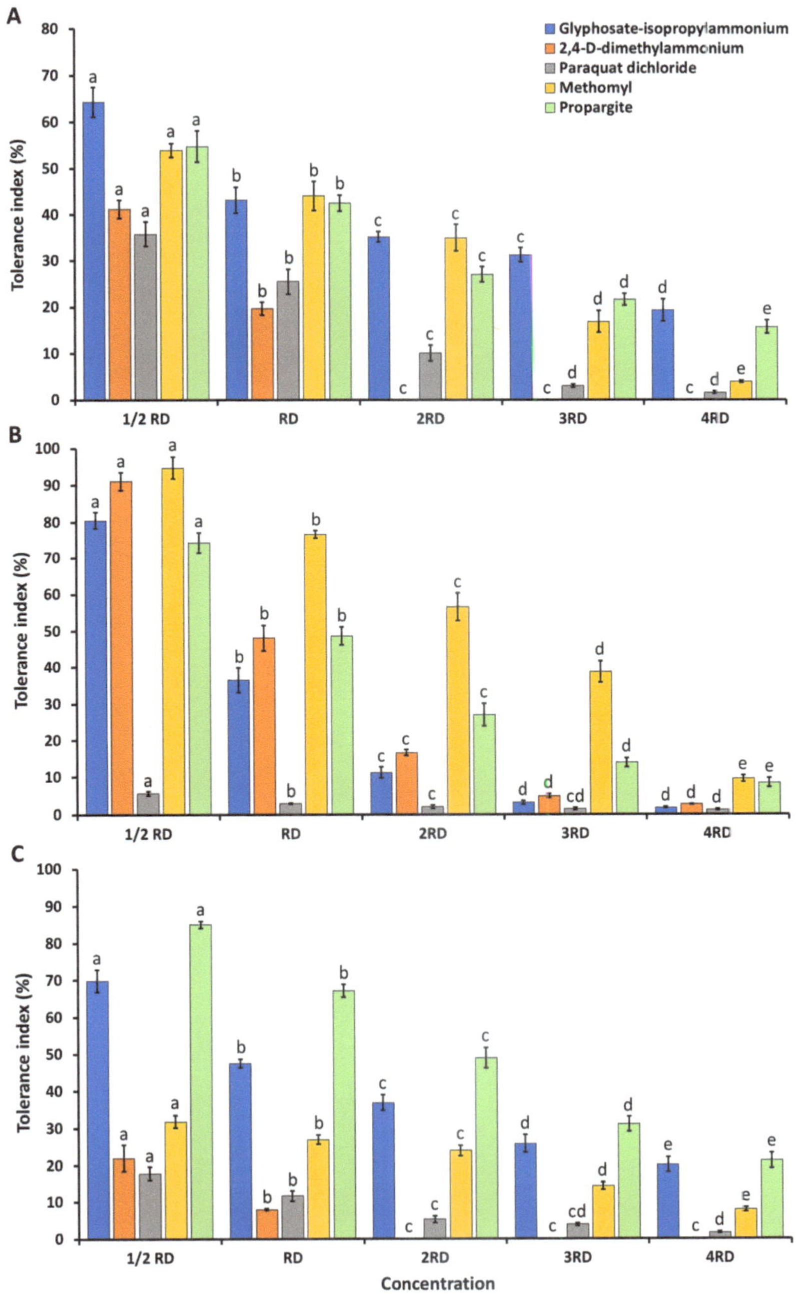


**Figure 6.** Assessment of the tolerance index of three phosphate-solubilizing fungi at various doses of five agrochemicals. *Aspergillus chiangmaiensis* SDBR-CMUI4 (**A**); *A. pseudopiperis* SDBR-CMUI1 (**B**); *A. pseudotubingensis* SDBR-CMUO2 (**C**). The error bars indicate the standard deviation of the average. Distinct letters in each experiment denote a statistically significant difference ($p \leq 0.05$); 1/2 RD, RD, 2RD, 3RD, and 4RD indicate half of the recommended dosage, recommended dosage, double the recommended dosage, triple the recommended dosage, and quadruple the recommended dosage, respectively.

### 2.4. Evaluation of Fungal Viability in a Granular Inoculum

The viability of each fungus in each granular inoculum was determined, and the results are shown in Figure 7. It was found that the viability of each fungus decreased with the increasing storage time and was significantly affected by the storage temperature.

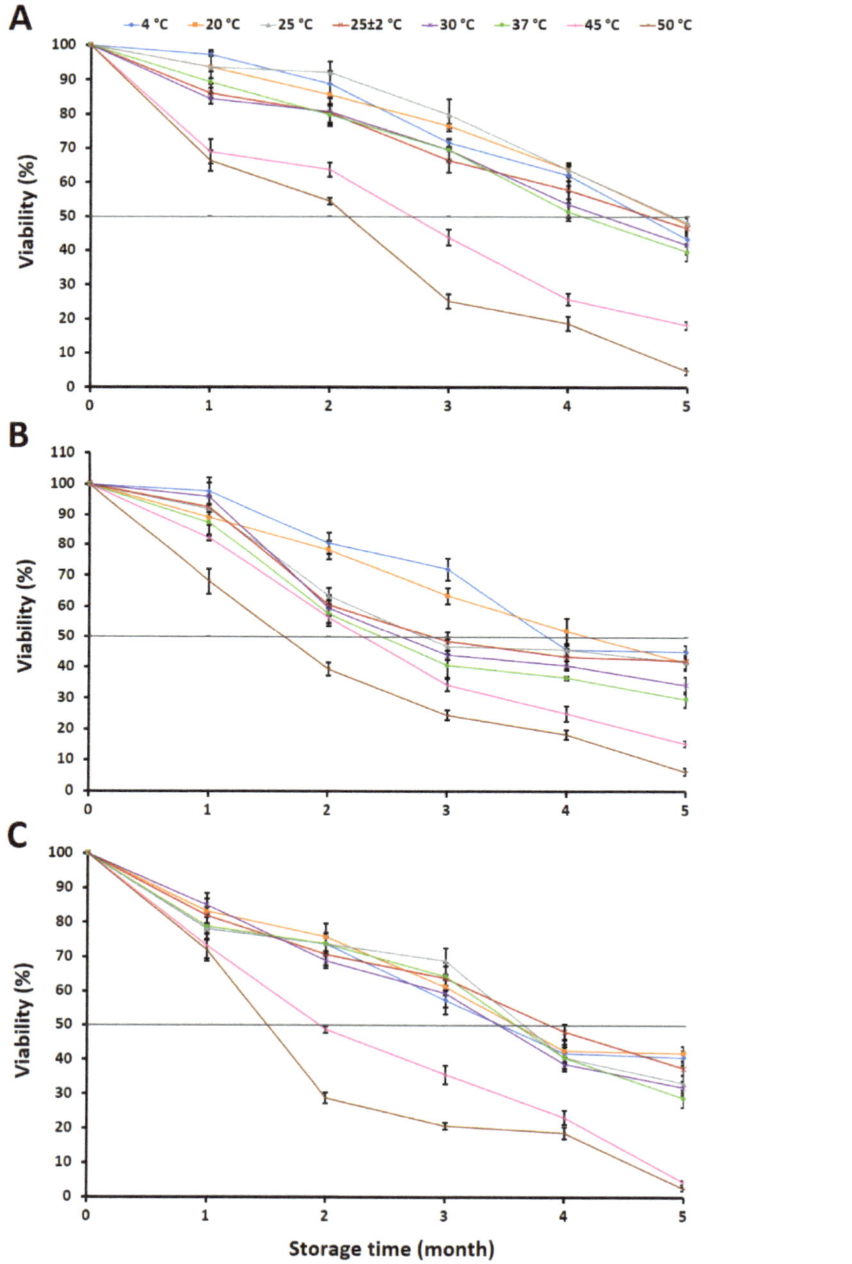


**Figure 7.** Viability of the phosphate-solubilizing fungi in granular inoculum during storage at different temperatures for five months. *Aspergillus chiangmaiensis* SDBR-CMUI4 (**A**); *A. pseudopiperis* SDBR-CMUI1 (**B**); *A. pseudotubingensis* SDBR-CMUO2 (**C**).

The viability of the fungi was reduced to approximately 50% after 3 to 4 months of storage. Among the different storage temperatures, the granular inoculum containing *A. chiangmaiensis* stored at 25 °C retained the maximum viability at 79.68%, followed by 20 °C (76.47%) and 4 °C (71.66%) after 3 months of storage. The minimum viability was observed in the granular inoculum that was stored at 50 °C (25.14%). Nevertheless, the granular inoculum that was stored at 4–37 °C for 4 months showed a viability of this fungus of higher than 50% within the range of 51.34–63.64%. The viability of *A. chiangmaiensis* was reduced to lower than 50% after storage for 5 months at all temperatures (Figure 7A).

The granular inoculum containing *A. pseudopiperis* stored at 4 °C retained the maximum viability (72.00%); this was followed by 20 °C (63.43%) after 3 months of storage. The minimum viability was observed in the granular inoculum that was stored at 50 °C (25.14%). However, the granular inoculum that was stored at 20 °C for 4 months showed the viability of this fungus as being higher than 50% (52.00%). The viability of *A. pseudopiperis* was reduced to less than 50% after storage for 5 months at all temperatures (Figure 7B).

The granular inoculum containing *A. pseudotubingensis* stored at 25 °C retained the maximum viability (68.75%), followed by 37 °C (64.38%) and room temperature (63.75%) after 3 months of storage. However, the minimum viability was observed in the granular inoculum stored at 50 °C (20.63%). The viability of this fungus was reduced to less than 50% after storage for 4 months at all temperatures (Figure 7C).

### 2.5. Root Induction of Cassava and Sugarcane Stem Cuttings under Greenhouse Conditions

#### 2.5.1. Cassava Stem Cuttings

During the planting period, neither the fungal inoculation treatment nor the controls exhibited any indication of plant disease. At 60 days after planting, the findings demonstrate that the cassava stem cuttings inoculated with each fungus along with the addition of $Ca_3(PO_4)_2$ (T6–T8) had significantly higher plant heights, numbers of leaves, root lengths, and dried weights of the stems, leaves, and roots than the plants undergoing the control treatment (T1) (Figure 8A–D,G). Regarding the addition of $Ca_3(PO_4)_2$ and inoculation with *A. pseudopiperis* (T7), the experiment showed a significantly greater root length than the inoculations with *A. chiangmaiensis* (T6) and *A. pseudotubingensis* (T8). This outcome could potentially be attributed to the production of IAA by *A. pseudopiperis*. Additionally, it was observed that the cassava stem cuttings inoculated with each fungus (T3–T5) showed longer roots and higher dry weights for the stems, leaves, and roots when compared to the plants receiving the control treatment (T1).

The concentrations of chlorophyll and cellular inorganic phosphate in the leaves, stems, and roots of the cassava stem cuttings are represented in Figure 8E, and Figure 8F, respectively. The outcomes demonstrated that the inoculation of each fungus in both treatments without additional $Ca_3(PO_4)_2$ (T3–T5) or with additional $Ca_3(PO_4)_2$ (T6–T8) significantly increased the chlorophyll content in the leaves and the cellular inorganic phosphate content in the leaves, stems, and roots of the cassava stem cuttings when compared to the control treatment (T1).

#### 2.5.2. Sugarcane Stem Cuttings

None of the treatments for fungal inoculation and the control exhibited indications of plant disease throughout the planting period of 60 days. The outcomes found that the inoculation of each fungus in both treatments without additional $Ca_3(PO_4)_2$ (T3–T5) or with additional $Ca_3(PO_4)_2$ (T6–T8) could enhance the plant heights, root lengths, and the dried weights of the stems, leaves, and roots of the sugarcane stem cuttings when compared to the control treatment (T1) (Figure 9A–D,G). The highest values of the number of leaves were obtained for the sugarcane stem cuttings that were inoculated with each fungus along with the addition of $Ca_3(PO_4)_2$ (T6–T8). However, the inoculation of *A. pseudopiperis* with the addition of $Ca_3(PO_4)_2$ (T7) resulted in a greater root length compared to *A. chiangmaiensis* (T6) and *A. pseudotubingensis* (T8), while there was no statistically significant difference between these three treatments.

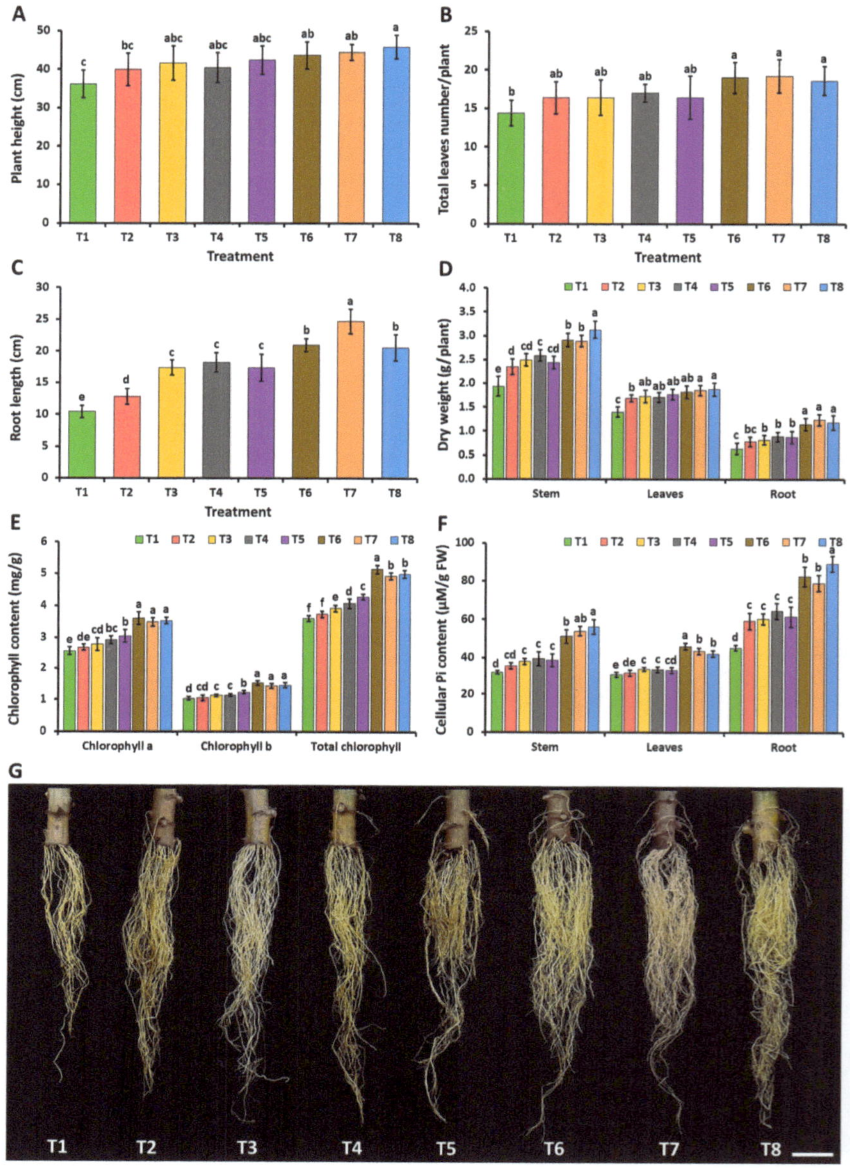


**Figure 8.** Effect of phosphate-solubilizing fungi on the growth and root induction of cassava stem cuttings. Plant height (**A**); total leaf number (**B**); root length (**C**); dry weights of stems, leaves, and roots (**D**); chlorophyll content (**E**); cellular inorganic phosphate content (**F**); root of cassava in each treatment (**G**). The error bars indicate the standard deviation of the average. Distinct letters in each experiment denote a statistically significant difference ($p \leq 0.05$). Scale bar: (**G**) 3 cm.

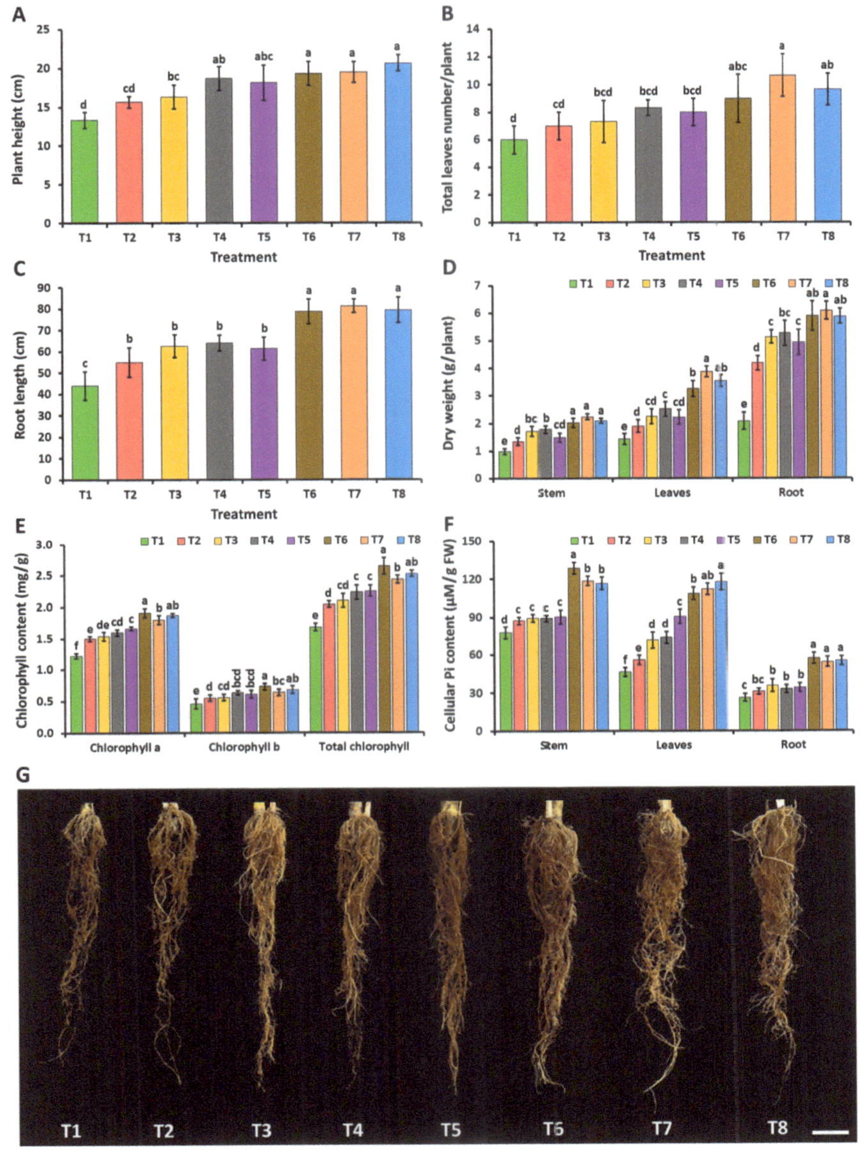


**Figure 9.** Effect of phosphate-solubilizing fungi on the growth and root induction of sugarcane stem cuttings. Plant height (**A**); total leaf number (**B**); root length (**C**); dry weights of stems, leaves, and roots (**D**); chlorophyll content (**E**); cellular inorganic phosphate content (**F**); root of sugarcane in each treatment (**G**). The error bars indicate the standard deviation of the average. Distinct letters in each experiment denote a statistically significant difference ($p \leq 0.05$). Scale bar: (**G**) 10 cm.

The results showed that the inoculation of each fungus with additional $Ca_3(PO_4)_2$ (T6–T8) significantly increased the chlorophyll content in the leaves and the cellular inorganic phosphate content in the leaves, stems, and roots of the sugarcane stem cuttings when compared to the control treatment (T1), the addition of only $Ca_3(PO_4)_2$ (T2), and the inoculation with each fungus (T3–T5) (Figure 9E,F).

## 3. Discussion

In the present study, three PSF, namely, *Aspergillus chiangmaiensis*, *A. pseudopiperis*, and *A. pseudotubingensis*, isolated from soils (loamy sand), were investigated for their plant-growth-promoting capabilities (the production of IAA, siderophores, and the solubilization of phosphorus minerals). IAA is a prevalent type of auxin found in plants and plays a key role in various plant growth reactions, such as elongation, division, the differentiation of cells, and the initiation of roots [34–36]. Several previous studies have found that soil fungi can produce IAA as a part of their metabolism, which plays a significant role in the growth and development of plants [32,33,37]. L-tryptophan (L-trp) served as the main precursor for the synthesis of IAA [38]. Several previous investigations have shown that PSF can produce IAA both with and without L-trp [22,32,39–41]. Our results revealed that *A. pseudopiperis* produced an average of 33.37 μg/mL of IAA when grown in a liquid medium supplemented with L-trp. This outcome is in accordance with previous studies that reported that less than 100 g/mL of IAA was produced from PSF in the genera, *Alternaria*, *Aspergillus*, *Chaetomium*, *Curvularia*, *Fusarium*, *Penicillium*, *Rhizopus*, and *Trichoderma*, isolated from soils [32,39,40,42–46]. Nevertheless, the quantity of IAA produced by PSF in the *Asperigllus* genus differed among various species and strains, with certain strains exhibiting higher levels of IAA production. For example, Li et al. [45] reported that *A. tubingensis* HZ123 could produce IAA at 267.38 μg/mL. Additionally, *A. tubingensis* SFSA1 and *A. fumigatus* SFSA7 showed the highest IAA productions at 260 and 212 μg/mL, respectively [46]. Moreover, numerous previous studies have demonstrated that the IAA production from PSF could improve and increase the root length in various plants [30,39,45,47–49].

Several fungi have demonstrated the ability to synthesize siderophores, which serve to chelate ferric iron in constrained environments. This process aids in the promotion of plant growth and enhances plant nutrition [50,51]. Siderophores also function as potential biocontrol agents against certain plant pathogens by decreasing the available iron from the host environment [52,53]. This study revealed that all strains of fungi exhibited a positive result for siderophore production. This outcome was consistent with prior studies that found that some species of PSF in the genera *Alternaria*, *Aspergillus*, *Curvularia*, *Fusarium*, *Mucor*, *Penicillium*, *Purpureocillium*, *Rhizopus*, *Talaromyces*, and *Trichoderma* could produce siderophores [33,40,44,45,48,54–57]. Additionally, siderophore production has been reported in *Aspergillus* species isolated from soils, including *A. aculeatus*, *A. brunneoviolaceus*, *A. niger*, *A. tubingensis*, and *A. violaceofuscus* [32,40,45,58,59].

The results indicated that three fungal species can solubilize phosphate in the forms of $Ca_3(PO_4)_2$ and $FePO_4$. However, all fungi were found to be unable to solubilize $AlPO_4$. This result is supported by the findings of Hervieux et al. [60], Arriagada et al. [61], and Kolaei et al. [62], who reported that the growth of fungi is significantly inhibited by the presence of aluminum. In this study, *A. chiangmaiensis*, *A. pseudopiperis*, and *A. pseudotubingensis* could produce both acid and alkaline phosphatases and organic acids (oxalic acid, tartaric acid, and succinic acid) for the solubilizations of $Ca_3(PO_4)_2$ and $FePO_4$. These results are consistent with the numerous prior studies that have documented that mechanisms recognized to be responsible for the solubilization of phosphate minerals are the production of different types of organic acids as chelating substances, phosphatase enzymes, and other metabolites [3,14,63–69]. Moreover, several previous studies have reported that the capacity to solubilize phosphatases was contingent on the specific fungal genus, species, and strain [14,40,44,70,71]. Our results are corroborated by previous studies that demonstrated that PSF belonging to the genera *Aspergillus*, *Fusarium*, *Macrophomina*, *Penicillium*, *Talaromyces*, and *Trichoderma* produced phosphatases for solubilizing phosphate minerals [41,72–75]. Additionally, the ability to produce organic acids for solubilizing various forms of insoluble phosphorus sources relies on the specific fungal genus, species, and strain [41,76–78]. In this study, oxalic, tartaric, and succinic acids were identified in the liquid medium for solubilizing $Ca_3(PO_4)_2$ and $FePO_4$ by *A. chiangmaiensis*, *A. pseudopiperis*, and *A. pseudotubingensis*. Tartaric acid was dominant, followed by oxalic acid and succinic acid. This corresponds with the findings of Li et al. [79], who observed that *A. aculeatus* primarily produced tartaric acid for solubilizing $Ca_3(PO_4)_2$, followed by citric and malic acids. According to Wang et al. [70], *A. niger* CS-1 could primarily produce oxalic acid during the solubilization of $Ca_3(PO_4)_2$, which was then followed by tartaric acid and citric acid. Additionally, *A. niger*, *A. awamori*, *A. carbonarius*, and *A. tubingensis* could produce acetic, citric, formic, fumaric, gluconic, malic, propionic, succinic, tartaric, and oxalic acids in the solubilization of inorganic phosphate minerals [25,40,68,78,80–82].

In this study, all three fungi exhibited resistance to the abiotic conditions of drought, pH level, temperature, and salinity in the in vitro experiment. As a result, it was shown that these tolerances were influenced by the fungal species. The outcomes demonstrated that each fungal species had a high level of tolerance to drought, with water availability ranging from 0.837 to 0.859. In previous investigations, *A. aculeatus*, *A. fumigatus* SG-17, *A. violaceofuscus* MH220545, *Penicillium citrinum* 5TAKL-3a, *Pseudeurotium* sp. GRs12, *Talaromyces* sp. GS1, and *Trichoderma* GR21 have been found to be drought-tolerant in vitro, and their beneficial effects on plant growth and the enhancement of drought tolerance have been investigated in several plants, e.g., mulberry, perennial ryegrass, rice, tomato, and wheat [83–87]. All fungi were able to grow in both acidic and alkaline (pH 4–9) conditions. Therefore, these fungi could be cultivated in and endure typical soil conditions with pH levels between 6 and 8, as well as soils that are extremely acidic or alkaline. Our results were in accordance with the results of Rinu and Pandey [88], who reported that *A. candidus*, *A. deflectus*, *A. flavus*, *A. fumigatus*, *A. glaucus*, *A. nidulans*, *A. niger*, *A. parasiticus*, *A. sydowiihad*, and *A. wentii* could tolerate pH ranges from 2 to 12. The range of pH tolerance that *A. niger* and *A. japonicus* displayed was 4 to 11 [89]. In addition, *A. carbonarius* showed the highest growth at pH 4 to 6.5 [90]. In the present study, all three fungi were successfully grown at temperatures ranging from 20 to 40 °C. These results were similar to those of Passamani et al. [90], who found that *A. carbonarius* and *A. niger* could grow at temperatures from 20 to 37 °C. Kurniati et al. [91] indicated that *A. flavus* have readily grown between the temperatures of 25 and 42 °C. *Aspergillus niger* and *A. japonicus* exhibited levels of temperature tolerance ranging from 10 to 45 °C [89]. Furthermore, several *Aspergillus* species have a temperature tolerance ranging from 4 to 42 °C [88]. In this study, all fungi were tolerant to high concentrations of NaCl at 16 to 17%. The ability of the PSF to tolerate salt was demonstrated to depend on the fungal species and strain, as well as NaCl concentration. This result was consistent with the salt tolerance findings of Rinu and Pandey [88], who demonstrated that ten species of phosphate-solubilizing *Aspergillus* displayed varying degrees of salinity tolerance (at 12% up to 15% NaCl). Xiao et al. [89] reported that *A. niger* and *A. japonicus* were tolerant to salinity at up to 3.5% NaCl. Additionally, *A. niger*, *A. fumigatus*, *A. pulverulentus*, *A. parasiticus*, and *A. flavus* could tolerate NaCl concentrations of up to 9% [92]. Tolerance to various stresses is an important factor for the survival and growth of fungi, as is the capacity to solubilize insoluble minerals insoils [2,65,89,93]. Thus, the tolerance to drought, pH, temperature, and salinity of selected plant-growth-promoting microorganisms should be noted. These recommendations are considered to select and apply successful applications [94–96]. Subsequent studies ought to investigate the PSF obtained in this study for their potential to improve plant growth under conditions of drought, varying pH levels, temperature fluctuations, and salinity in both greenhouse and field experiments.

The agrochemical (herbicides, fungicides, and insecticides) tolerance of fungi varied depending on the type and concentration of agrochemicals, as well as the species and strain of fungi [97,98]. The growth and tolerance capacity of each fungal strain could be decreased by an increase in the agrochemical concentration, which was consistent with previous studies [99–101]. In this study, the fungicides (metalaxyl and benomyl) completely inhibited the mycelial growth of all fungi in all concentrations. Each fungal species exhibited varying degrees of tolerance to herbicides (glyphosate-isopropylammonium, 2,4-D-dimethylammonium, and paraquat dichloride) and insecticides (methomyl and propargite), which was contingent upon their species characteristics. It was found that *A. pseudopiperis* and *A. pseudotubingensis* could tolerate methomyl and propargite, respectively, in the field at recommended dosages. These findings are consistent with previous studies that demonstrated that some PSF species of the genera *Aspergillus*, *Chaetomium*, *Clonostachys*, *Grifola*, *Minimedusa*, *Mucor*, *Penicillium*, *Phycomyces*, *Pochonia*, *Purpureocillium*, *Rhizopus*, and *Trichoderma* could tolerate agrochemicals at various dosages depending on the species and strain [101–106]. Notably, Benito et al. [107] found that *A. flavus* AFS 63 and *A. parasiticus* APS 55 were able to tolerate glyphosate (herbicide) at the recommended field dosages. *Aspergillus oryzae* AM 1 exhibited the ability to tolerate chlorpyrifos (insecticide) at both the recommended field dosages and at over double the recommended dosages [108]. Therefore, the knowledge regarding the agrochemical tolerance of the PSF gained from this study could prove valuable insights for practical agricultural purposes, as it might involve the concurrent use of these fungal species with herbicides and insecticides on crops at recommended doses.

Microbial viability is an important feature when developing an inoculum as a biostimulant and plays a crucial role in successful commercialization [109,110]. The survival of microorganisms, carrier characteristics, biological effectiveness, and storage life of the biostimulant inoculum are affected by a variety of factors, including the drying process [111,112] and various storage conditions (e.g., moisture content, temperature, and sunlight intensity) [113–115]. Several researchers have investigated the effect of various temperatures on the shelf life of different microbial inocula [116–118]. Previous studies mentioned that maintaining the viability of microbes that are present in microbial inoculant formulations is challenging, and a loss of viability of less than 50% was considered to be non-effective for application [119,120]. Our results found that the viability of all fungi in a granular inoculum was higher than 50% after storage for 3 months at 4 and 20 °C. The viability of *A. chiangmaiensis* was higher than 50% after storage at 4–37 °C for 4 months, while the viability of *A. pseudopiperis* was higher than 50% at 20 °C after 4 months of storage. Furthermore, the viability of *A. pseudotubingensis* was higher than 50% after storage at 4–37 °C for 3 months. The rapid decrease in viability (<50%) of all fungal species in the granular inoculum was found at high temperatures (45–50 °C). Our findings are corroborated by previous research that found that the survival ability of the microbial species in inoculum depends on the species and strain, a suitable carrier, and the storage temperature [114,118]. Low temperatures are generally recommended for inoculum storage for prolonged microbial viability and high-quality products [49,114,121]. However, the inoculum has a long microbial viability at room temperature, which offers a financial advantage and provides more convenience to users [110,122,123]. The high temperatures have effects on the growth of microbes and decrease conidia germination due to changes in the membrane structure and protein degradation [118]. Prior to this study, Accinelli et al. [124] reported that the viability of *A. flavus* NRRL 30797 on Mater-Bi granules showed a greater decline at 25 °C than at 5 °C during the storage period of six months. The phosphate-solubilizing fungus, *A. awamori* S29, had a viability of 56.4% after six months of storage at room temperature (28 °C) in a free-form-based bio-formulation [125]. A study conducted by Wang et al. [49] indicated that the viability of the phosphate-solubilizing fungus, *A. niger* 1107, in a biofertilizer was more than 50% when stored at 4 °C for up to seven months. Therefore, storage conditions are most important in maintaining the steadiness and quality of an inoculum.

In the current study, the results demonstrated that inoculation stem cuttings of sugarcane and cassava with the granular inocula of *A. chiangmaiensis*, *A. pseudopiperis*, and *A. pseudotubingensis* could significantly increase root induction and plant growth. Moreover, the inoculation of each fungus and the simultaneous addition of an insoluble mineral phosphate, $Ca_3(PO_4)_2$, considerably enhanced the root induction and growth of the sugarcane and cassava stem cuttings. Our results are supported by previous studies that showed that PSF, including species of *Aspergillus* spp., have the ability to dissolve insoluble mineral phosphates in soil, thereby elevating the concentration of accessible phosphorus for enhanced plant growth and productivity [2,3,20,126]. Tariq et al. [127] reported that the inoculation of PSF, *A. versicolor* MF, could significantly increase the root lengths of mint plants (*Mentha viridis*). In greenhouse experiments, the application of PSF, including *A. awamori* S29, *A.*

*awamori* S-19, *A. niger* 1107, *A. niger* FS1, and *A. tubingensis* QF05, along with the addition of an insoluble mineral phosphate, resulted in the enhanced root development and growth of Chinese cabbage (*Brassica rapa*), chickpea (*Cicer arietinum* L.), mung bean (*Vigna radiata*), tomato (*Solanum lycopersicum* L.), and white clover (*Trifolium repens*) [25,30,32,49,126]. Additionally, in the experiments conducted by Kaur and Reddy [27], it was found that the inoculation of PSF, in the form of *A. niger* PSF-7 and *A. tubingensis* PSF-4, could significantly enhance the root lengths, plant growths, and yields of maize (*Zea mays* L.) and wheat (*Triticum aestivum*) in a field experiment. The results of our research revealed that both the sugarcane and cassava stem cuttings had increased chlorophyll and cellular inorganic phosphate concentrations following the inoculation of each fungal strain. Our findings are consistent with previous reports demonstrating that the inoculation of PSF, in the form of *A. awamori* S-19, *A. flavus*, and *A. niger* CSR3, led to a significant increase in the chlorophyll levels of chickpea, maize, and tomato [30,128,129]. Peng et al. [130] reported that the inoculation of six strains of *A. niger* resulted in an increase in the root lengths, plant growths, and chlorophyll contents in peanut plants (*Arachis hypogaea* L.). The phosphorus contents of chickpea, Chinese mustard (*Brassica chinensis* Linn.), and soybeans (*Glycine max*) could be increased via PSF inoculation, including *A. awamori* VHI + VQ2, *A. niger* 6A, and *A. niger* K7 [22,29,131]. Moreover, Shahrajabian et al. [132,133] documented that the application of biostimulants not only has a beneficial effect on plant growth and productivity, but it may also lead to environmentally friendly practices and increase the use efficiency of natural resources for the sustainability of agricultural and horticultural production systems.

## 4. Materials and Methods

### 4.1. Fungal Strains

Three phosphate-solubilizing fungi, namely *A. chiangmaiensis* SDBR-CMUI4, *A. pseudopiperis* SDBR-CMUI1, and *A. pseudotubingensis* SDBR-CMUO2, were isolated from the soil in a longan orchard [23]. All strains of fungi were preserved in the culture collection of the Sustainable Development of Biological Resources, located at the Faculty of Science, Chiang Mai University in Chiang Mai Province, Thailand. The fungi were grown on potato dextrose agar (PDA; Conda, Madrid, Spain) and kept in an incubator set at 25 °C.

### 4.2. Characterization of Plant Growth Promotion Properties

#### 4.2.1. Determination of IAA Production

All fungal strains were cultivated on PDA at a temperature of 30 °C for one week before being used. Five mycelial plugs (5 mm in diameter) taken from the edge of a colony were inoculated into a 125 mL Erlenmeyer flask containing 30 mL of potato dextrose broth (PDB; Conda, Madrid, Spain), with a pH level of 6.0, added to a concentration of 0.2 mg/mL of L-tryptophan (L-trp; Sigma-Aldrich, Steinheim, Germany). The inoculation was carried out in the dark at 25 °C using a reciprocating shaker at 150 rpm. The cultures were centrifuged at 11,000 rpm for 15 min to collect the supernatant after seven days of incubation. The production of IAA was initially assessed using a colorimetric assay by using Salkowski's reagent, following the method described by Tsavkelova et al. [134]. A positive result for IAA production was indicated by a pink to red color. Three replications were made for each treatment. According to the procedure outlined by Kumla et al. [135], the IAA present in the supernatant was extracted and confirmed using high-performance liquid chromatography (HPLC).

#### 4.2.2. Determination of Siderophore Production

Chrome Azurol S (CAS) agar was used to evaluate siderophore production [136]. Colonies were grown in PDA at 30 °C for 7 days; then, mycelial plugs (5 mm in diameter) were taken out from those colonies and placed on CAS agar. The inoculation plates were incubated for three days at 30 °C in the dark. Fungal strains that exhibited a yellow, orange, purplish-red, or purple zone around their colonies were classified as strains capable of producing siderophores [137]. Three replications were set up for each treatment.

#### 4.2.3. Solubilization of Phosphate Minerals, Phosphatase Activities, and Organic Acid Production

Quantitative estimation of soluble phosphate mineral solubilization was performed using basal liquid media (10.0 g glucose, 0.5 g $(NH_4)_2SO_4$, 0.2 g NaCl, 0.1 g $MgSO_4{\cdot}7H_2O$, 0.2 g KCl, 0.5 g yeast extract, 0.002 g $MnSO_4{\cdot}H_2O$, and 1000 mL of deionized water; pH 7.0) with the addition of $AlPO_4$, $Ca_3(PO_4)_2$, and $FePO_4$ to the desired final concentration of 0.5% (*w*/*v*), according to the procedure outlined by Fomina et al. [138]. The starting pH of the medium was set to 7.0 prior to sterilization. Mycelial inocula were formed through culturing on PDA at 30 °C in complete darkness for a period of seven days. Mycelial plugs (with a diameter of 5 mm) taken from the outer edge of the developing colony were utilized for inoculation into the broth media. The control treatment (uninoculated treatment) was run with each experiment without fungi. Three replications were made for each treatment. The flasks were subsequently incubated at 25 ± 2 °C with shaking at 125 rpm. The remaining available phosphorus in the culture supernatant was assessed at intervals of three days over a period of 15 days. The cultures were passed through Whatman number 1 filter paper and then subjected to centrifugation at a speed of 11,000 rpm for a duration of 10 min in order to eliminate any suspended particles and fragments of mycelium. The obtained culture supernatants were used for the determination of pH value, available phosphorus concentration, and phosphatase activities, as well as organic acid production.

A pH meter was used to test the pH of the culture supernatants [139]. Subsequently, the molybdenum blue technique was used to determine the available phosphorus concentration [140,141]. The 500 µL culture supernatant was combined with an equal volume of 10% (*w*/*v*) trichloroacetic acid in a test tube; then, 4 mL of a color reagent (consisting of a 1:1:1:2 ratio of 6 N $H_2SO_4$/2.5% (*w/v*) ammonium molybdate/10% (*w/v*) ascorbic acid/distilled water) was added. The mixture was left to incubate at room temperature (25 ± 2 °C) for 15 min. An uninoculated medium was used as the blank. A spectrophotometer was used to measure the absorbance of the developing blue color at 820 nm. The concentration of available phosphorus was calculated from a calibration curve of $KH_2PO_4$, and the data are expressed as milligrams per liter (mg/L).

The supernatant of each fungal culture from the previous experiment was determined for phosphatase activities. Acid and alkaline phosphatase assays were performed according to the procedures outlined by Gaind and Singh [142] and Adhikari and Pandey [143], respectively. For the acid phosphatase assay, the culture supernatant (0.4 mL) was mixed with 1 mL of 0.1 M citrate buffer (pH 5.0), 0.5 mL of 0.05 M *p*-nitrophenol phosphate (*p*NPP), and 0.1 mL of 1 mM $MgCl_2$. The alkaline phosphatase assay was performed using the same method, but 0.1 M glycine–sodium hydroxide buffer (pH 9.0) was used in place of the citrate buffer. The reaction mixture was incubated at 37 °C in darkness for 60 min. One hundred microliters of 5 M NaOH was added to stop the process. An uninoculated medium was used as the blank. A microplate spectrophotometer was used to measure the intensity of the produced yellow color at 400 nm. The quantity of liberated *p*-nitrophenol (*p*NP) was measured using a *p*-nitrophenol standard as a reference. The phosphatase activity is expressed as 1 µmol *p*NP/mL/h.

The organic acids in the culture supernatant from the broth liquid medium supplemented with different sources of $Ca_3(PO_4)_2$ and $FePO_4$ were analyzed via HPLC. The culture supernatant was filtered using a 0.45 m syringe filter (MILLEX®HA, Merck Millipore Ltd., Tullagreen, Carrigtwohill, County Cork, Ireland) with a volume of one milliliter. The organic acids in the filtrate samples were analyzed using a Shimadzu VP series HPLC system equipped with an SPD-10AVP UV/VIS detector (Shimadzu, Europe, Duisburg, Germany). The temperature was adjusted to 40 °C and the column was an Ultra-Aqueous C18 column (250 × 4.6 mm, 5 µm; Restek Corporation, Benner Circle Bellefonte, PA, USA). The mobile phase included 0.1% phosphoric acid solution flowing at 0.8 µL/min, with a sample volume of 2 µL for each injection. The detection wavelength was 214 nm. A calibration curve was established using five organic acid standards containing varying levels of citric, malic, oxalic, succinic, and tartaric acids. The quantification of organic acids was achieved by comparing the peak areas in the supernatant samples with those on the calibration curve. The results are presented as micrograms of organic acid per milliliter (μg/mL).

### 4.3. Drought, pH, Temperature, and Salinity Tolerances

#### 4.3.1. Drought Tolerance

The technique outlined by Hallsworth et al. [144] was used to assess each fungal strain's resistance to drought using PDA supplemented with sorbitol powder. The sorbitol concentrations were set at 0, 85, 175, 285, 405, 520, 605, 660, and 780 g/L, corresponding to the water activity ($a_w$) levels of 0.998, 0.995, 0.993, 0.963, 0.930, 0.912, 0.886, 0.859, and 0.837, respectively. Three replications were made for each fungal strain.

#### 4.3.2. pH Tolerance

The pH tolerance was determined using PDA. The pH of the medium was modified to the values of 3, 4, 5, 6, 7, 8, and 9 by adding either 1 N HCl or 1 N NaOH before being autoclaved. Mycelial plugs (5 mm in diameter) were placed on the test media. The plates were then incubated at 30 °C after being inoculated. Fungal growth in the tested media was monitored five days later. Three replications were made for each fungal strain.

#### 4.3.3. Temperature Tolerance

PDA was employed to conduct this experiment. A mycelial plug of each fungal strain was inoculated on PDA and incubated at 0, 4, 10, 15, 20, 25, 30, 35, 40, and 45 °C in the dark. After five days of incubation, the growth of fungi in the tested media was measured. All experiments were performed in three replicates.

#### 4.3.4. Salinity Tolerance

The salinity tolerance was determined using PDA supplemented with a graded series of NaCl concentrations (0, 5, 10, 15, 16, 17, 18, 19, and 20%) according to the methods described by Rinu and Pandey [88] and Tresner and Hayes [145]. The pH of all of the tested media was modified using either 1 N HCl or 1 N NaOH before being autoclaved. Five-millimeter-diameter mycelial plugs were placed in the center of the test medium and incubated at 30 °C in the dark. After five days of incubation, the fungal growth of the fungi was observed. Three replications were carried out for each treatment.

### 4.4. Agrochemical Tolerance

Three commercially available herbicides, including glyphosate-isopropylammonium (Glyphosate 48®; Pareto Agro Co., Ltd., Bangkok, Thailand), 2,4-D-dimethylammonium (DMA6®; Zagro (Thailand) Ltd., Pathumthani, Thailand), and paraquat dichloride (Gramm-oxone®; Syngenta Crop Protection Co., Ltd., Bangkok, Thailand); two commercial fungicides, including metalaxyl (Lonsan®; Sahaphol Kemekaset Ltd., Part., Bangkok, Thailand) and benomyl (Belly OD®; Sahaphol Kemekaset Ltd., Part., Bangkok, Thailand); and two commercial insecticides, including methomyl (Garnet®; Alpha Agro Tech Co., Ltd., Samut Prakan, Thailand) and propargite (Omite-20®; Sotus International Co., Ltd., Nonthaburi, Thailand), were tested in this experiment following the methodology outlined by Suwannarach et al. [99]. The recommended dosage for field applications of glyphosate-isopropylammonium, 2,4-D-dimethylammonium, paraquat dichloride, metalaxyl, benomyl, methomyl, and propargite were 7000, 2500, 6250, 2000, 1500, 1750, and 2000 ppm, respectively. Each compound was prepared and introduced into autoclaved PDA in order to achieve final concentrations ranging from half of the recommended dosage to quadruple the recommended dosage. A sterile cellophane disc was used to cover the surface of the test media, and the fungal mycelial plug (with a diameter of 5 mm) was then positioned on the test media [135]. The plates were incubated in the dark at 25 °C. Following a five-day incubation period, the cellophane disc was taken off and subjected to drying at 60 °C for 48 h. Mycelium dry weights were then calculated. The formula provided by Fomina et al. [138]

was used to determine the tolerance Index (TI). TI values of 0% indicated a lethal outcome, while values below 50% signified a growth-inhibiting effect. Three replications of each treatment were performed.

### *4.5. Evaluation of Fungal Viability in a Granular Inoculum*

#### 4.5.1. Preparation of Granular Inoculum

The fungal strain was cultured on PDA for one week at a temperature of 37 °C. A Petri plate containing 5 mL of sterile deionized water was gently scraped to collect conidia. The concentration of conidia in the suspension was determined using a hemocytometer observed through a microscope. Vermiculite, perlite, and peat moss were combined together to produce the carrier material in a ratio of 5:2:3 (*w/w/w*). The combined carrier material was subjected to a drying process at 70 °C for 72 h, and then processed through a blender, filtered using a 2 mm mesh, and finally, sterilized twice in an autoclave set at 121 °C for 30 min. The conidial suspension of each fungus was combined with the sterilized carrier material to achieve a final concentration of approximately $1 \times 10^7$ conidia/g [110]. A tablet pressure machine was used in the granulation process. Subsequently, the granules were dried in an oven at 45 °C for 48 h. Each granular inoculum used in this experiment had an overall diameter of 36 mm and 10 mm length.

#### 4.5.2. Evaluation of Fungal Viability

The viability of fungal conidia in a granular inoculum was investigated. Ten grams of granular inoculum product were packed in 4 oz (4.5 × 9.0 cm) sterilized glass bottles with screwable plastic lids and stored in the dark at different storage temperatures including 4, 20, 25, 25 ± 2 (room temperature), 30, 37, 45, and 50 °C for five months. Three replicates were carried out for each temperature. The fungal viability was monitored at monthly intervals. The viable count was evaluated using the serial dilution spread plate method [146]. A total of 9 mL of the sterile saline solution (0.5% NaCl, *w/v*) was used to suspend 1 g of granules. Suspensions were vigorously shaken using a vortex mixer for 1 min; then, serial dilutions were made. After that, 0.1 mL of each serial dilution was dropped onto PDA and spread. The plates were kept at 30 °C for three days while being kept in the dark. The total number of fungal colonies were counted and recorded. The shelf life of biofertilizer product was calculated and expressed as a percentage (%).

### *4.6. Root Inductions of Cassava and Sugarcane Stem Cuttings under Greenhouse Conditions*

The cassava (*M. esculenta* Crantz) cultivar, "Kaeg Dum", and sugarcane (*S. officinarum* L.) cultivar, "Khon Kaen", were used in this experiment. This experiment employed soil from sugarcane and cassava fields with a pH range of 6.8 to 6.9 as the planting medium. Sterilization of the soil was performed twice at 121 °C for 30 min each. A completely random design (CRD) was used to set up this experiment. The details of each treatment employed in this study are listed in Table 2. The 15 cm long stem cuttings for planting were obtained from the stakes of 10-month-old plants. The stem cuttings were transferred into each plastic pot (19.5 × 13 × 14.5 cm) containing 3 kg of soil in each experiment. Ten replications of each treatment were run twice. The plants were cultivated for a duration of 60 days in a greenhouse situated at the Faculty of Science at Chiang Mai University in Chiang Mai Province, Thailand between September and October 2022. The greenhouse experienced temperature fluctuated between 25 and 35 °C and relative humidity levels ranging from 50% to 85%. The daily peak light intensity varied within the range of 16,000 to 25,000 lux.

#### 4.6.1. Measurement of Plant Growth

The heights of the plants, total number of leaves, leaf lengths, main root lengths, and dry weights of the roots and leaves were measured and recorded for both cassava and sugarcane plants.

**Table 2.** Details of the treatments in this study.

| Treatment Number | Treatment Details |
|---|---|
| T1 | Soil (control) |
| T2 | Soil (3 kg) + $Ca_3(PO_4)_2$ (1.5 g) |
| T3 | Soil (3 kg) + inoculum of *Aspergillus chiangmaiensis* SDBR-CMUI4 (3 g) |
| T4 | Soil (3 kg) + inoculum of *Aspergillus pseudopiperis* SDBR-CMUI1 (3 g) |
| T5 | Soil (3 kg) + inoculum of *Aspergillus pseudotubingensis* SDBR-CMUO2 (3 g) |
| T6 | Soil (3 kg) + $Ca_3(PO_4)_2$ (1.5 g) + inoculum of *A. chiangmaiensis* SDBR-CMUI4 (3 g) |
| T7 | Soil (3 kg) + $Ca_3(PO_4)_2$ (1.5 g) + inoculum of *A. pseudopiperis* SDBR-CMUI1 (3 g) |
| T8 | Soil (3 kg) + $Ca_3(PO_4)_2$ (1.5 g) + inoculum of *A. pseudotubingensis* SDBR-CMUO2 (3 g) |

4.6.2. Determination of Chlorophyll and Cellular Inorganic Phosphate Contents in Plants

The methodologies outlined in the studies by Lichtenthaler and Wellburn [147] and Liang et al. [148] were used to determine the concentrations of chlorophyll (chlorophyll a, chlorophyll b, and total chlorophyll) in the leaves of the cassava and sugarcane plants. A total of 0.2 g of fresh leaf samples was submerged in 8 mL of an 80% (*v/v*) acetone solution, followed by incubation for 24 h at 25 °C in darkness until the tissue became white. A microplate spectrophotometer was employed to measure the supernatant at 645 and 663 nm. Chlorophyll concentration was computed, and the results were represented in milligrams per gram (mg/g).

The methodology outlined by Ames [149] and Wang et al. [150] was used to measure the concentrations of cellular inorganic phosphate. Plant tissues were weighed and soaked in 1 mL of 1% glacial acetate. Eight occurrences of freezing and thawing followed. Subsequently, 100 µL of the extract was combined with 700 µL of a phosphate reaction buffer (consisting of A = 0.42% ammonium molybdate and 2.85% (*v/v*) sulfuric acid and B = 10% (*w/v*) ascorbic acid; A/B (*v/v*) in a ratio of 6:1) and 200 µL of deionized water. A spectrophotometer was used to determine the absorbance at 820 nm after the reaction had been carried out at 37 °C for 60 min. A dipotassium hydrogen phosphate calibration curve was used to determine the concentrations of cellular inorganic phosphate, and the results are expressed as µmol/g fresh weight (µmol/g FW).

### *4.7. Statistical Analysis*

Statistical variances among the treatments were evaluated using one-way analysis of variance (ANOVA) conducted with the SPSS software for Microsoft Windows (version 16). Significant differences at $p \leq 0.05$ were determined using Duncan's multiple range test (DMRT).

## 5. Conclusions

In the present study, three phosphate-solubilizing *Aspergillus* species, *A.* chiangmaiensis, *A. pseudopiperis*, and *A. pseudotubingensis*, could produce siderophores and solubilize the insoluble mineral phosphate minerals ($Ca_3(PO_4)_2$ and $FePO_4$). These three fungal species produce both acid and alkaline phosphatases and organic acids (tartaric, oxalic, and succinic acids) that indicate the mechanisms for solubilizing insoluble phosphate minerals. Moreover, *A. pseudopiperis* could produce IAA. Furthermore, their tolerances to drought, pH value, temperature, salinity, and agrochemicals as well as their fungal viability were characterized for consideration in the field applications. The application of a granular inoculum of each fungal strain as a biostimulant to the cassava and sugarcane stem cuttings enhanced root development and did not cause any disease symptoms. Moreover, the presence of these three fungi along with the addition of $Ca_3(PO_4)_2$ led to significant enhancements in the growths, chlorophyll contents, and cellular inorganic phosphate levels in both the cassava and sugarcane plants. Thus, these three fungi have the potential to serve as biostimulants to promote plant growth. Subsequent research will concentrate on utilizing the biostimulants derived from these three fungi to assess their impacts on the growths and yields of various plant species through field experiments. This endeavor will provide researchers with improved insights into the development of efficient biostimulants that have the potential to replace the detrimental chemical-based fertilizers presently employed in agriculture. To establish the safety of these three fungi, clinical investigations on toxicity will be necessary in further studies.

**Author Contributions:** Conceptualization, S.K., J.K., S.L. and N.S.; methodology, S.K., J.K. and N.S.; software, S.K.; validation, S.K., J.K. and N.S.; formal analysis, S.K., J.K. and N.S.; investigation, S.K., J.K., S.S. and N.S.; resources, S.K., J.K. and N.S.; data curation, S.K., J.K., S.S. and N.S.; writing—original draft preparation, S.K., J.K. and N.S.; writing—review and editing, S.K., J.K., S.S., S.L. and N.S.; supervision, N.S. All authors have read and agreed to the published version of the manuscript.

**Funding:** This work was supported by the Fundamental Fund 2023 (grant number FF66/065) and the Technology to Industry Convergence (grant number 111/2565), Chiang Mai University, Thailand.

**Data Availability Statement:** Not applicable.

**Acknowledgments:** The authors are grateful to Russell Kirk Hollis for his kind help in the English correction.

**Conflicts of Interest:** The authors declare no conflict of interest.

## References

1. Kumar, S.; Kumar, S.; Mohapatra, T. Interaction between macro- and micro-nutrients in plants. *Front. Plant Sci.* **2021**, *12*, 665583. [CrossRef]
2. Khan, M.S.; Zaidi, A.; Ahemad, M.; Ahemad, M.; Wani, P.A. Plant growth promotion by phosphate solubilizing fungi–current perspective. *Arch. Agron. Soil Sci.* **2010**, *56*, 73–98. [CrossRef]
3. Sharma, S.B.; Sayyed, R.Z.; Trivedi, M.H.; Gobi, T.A. Phosphate solubilizing microbes: Sustainable approach for managing phosphorus deficiency in agricultural soils. *SpringerPlus* **2013**, *2*, 587. [CrossRef]
4. Jain, R.; Saxena, J.; Sharma, V. Differential effects of immobilized and free forms of phosphate-solubilizing fungal strains on the growth and phosphorus uptake of mung bean plants. *Ann. Microbiol.* **2014**, *64*, 1523–1534. [CrossRef]
5. Gyaneshwar, P.; Naresh, K.G.; Parekh, L.J.; Poole, P.S. Role of soil microorganisms in improving P nutrition of plants. *Plant Soil* **2002**, *245*, 83–93. [CrossRef]
6. Mendoza-Arroyo, G.E.; Chan-Bacab, M.J.; Aguila-Ramírez, R.N.; Ortega-Morales, B.O.; Solís, R.E.C.; Chab-Ruiz, A.O.; Cob-Rivera, K.I.; Dzib-Castillo, B.; Tun-Che, R.E.; Camacho-Chab, J.C. Inorganic phosphate solubilization by a novel isolated bacterial strain *Enterobacter* sp. ITCB-09 and its application potential as biofertilizer. *Agriculture* **2020**, *10*, 383. [CrossRef]
7. Mengel, K.; Kirkby, E.A.; Kosegarten, H.; Appel, T. Phosphorus. In *Principles of Plant Nutrition*; Mengel, K., Kirkby, E.A., Kosegarten, H., Appel, T., Eds.; Springer: Dordrecht, The Netherlands, 2001; pp. 453–479. [CrossRef]
8. Zhou, K.; Binkley, D.; Doxtader, K.G. A new method for estimating gross phosphorus mineralization and immobilization rates in soils. *Plant Soil* **1992**, *147*, 243–250. [CrossRef]
9. Bechtaoui, N.; Rabiu, M.K.; Raklami, A.; Oufdou, K.; Hafidi, M.; Jemo, M. Phosphate-dependent regulation of growth and stresses management in plants. *Front. Plant Sci.* **2021**, *12*, 679916. [CrossRef]
10. Rengel, Z.; Marschner, P. Nutrient availability and management in the rhizosphere: Exploiting genotypic differences. *New Phytol.* **2005**, *168*, 305–312. [CrossRef]
11. Hasnain, M.; Chen, J.; Ahmed, N.; Memon, S.; Wang, L.; Wang, Y.; Wang, P. The effects of fertilizer type and application time on soil properties, plant traits, yield and quality of tomato. *Sustainability* **2020**, *12*, 9065. [CrossRef]
12. Chang, C.H.; Yang, S.S. Thermo-tolerant phosphate-solubilizing microbes for multi-functional biofertilizer preparation. *Bioresour. Technol.* **2009**, *100*, 1648–1658. [CrossRef] [PubMed]
13. Padmavathi, T. Optimization of phosphate solubilization by *Aspergillus niger* using plackett-burman and response surface methodology. *J. Soil Sci. Plant Nutr.* **2015**, *15*, 781–793. [CrossRef]
14. Kalayu, G. Phosphate solubilizing microorganisms: Promising approach as biofertilizers. *Int. J. Agron.* **2019**, *2019*, 4917256. [CrossRef]
15. Shrivastava, M.; Srivastava, P.C.; D'Souza, S.F. Phosphate-solubilizing microbes: Diversity and phosphates solubilization mechanism. In *Rhizospheric Microbes in Soil*; Meena, V., Ed.; Springer: Singapore, 2018; pp. 137–165. [CrossRef]
16. Kucey, R.M.N. Phosphate solubilizing bacteria and fungi in various cultivated and virgin Alberta soils. *Can. J. Soil Sci.* **1983**, *63*, 671–678. [CrossRef]
17. Kaul, S.; Sharma, S.; Apra; Dhar, M.K. Phosphate-solubilising fungi and their potential role in sustainable agriculture. In *Biofertilizers for Sustainable Agriculture and Environment*; Soil Biology; Giri, B., Prasad, R., Wu, Q.S., Varma, A., Eds.; Springer: Cham, Switzerland, 2019; Volume 55, pp. 371–393. [CrossRef]

18. Mendes, G.O.; Freitas, A.L.M.; Pereira, O.L.; Silva, I.R.; Vassilev, B.; Costa, M.D. Mechanism of phosphate solubilization by fungal isolates when exposed to different P sources. *Ann. Microbiol.* **2014**, *64*, 239–249. [CrossRef]
19. Pawar, V.C.; Thaker, V.S. Acid phosphatase and invertase activities of *Aspergillus niger*. *Mycoscience* **2009**, *50*, 323–330. [CrossRef]
20. Singh, H.; Reddy, M.S. Effect of inoculation with phosphate solubilizing fungus on growth and nutrient uptake of wheat and maize plants fertilized with rock phosphate in alkaline soils. *Eur. J. Soil Biol.* **2011**, *47*, 30–34. [CrossRef]
21. Mayadunna, N.; Karunarathna, S.C.; Asad, S.; Stephenson, S.L.; Elgorban, A.M.; Al-Rejaie, S.; Kumla, J.; Yapa, N.; Suwannarach, N. Isolation of phosphate-solubilizing microorganisms and the formulation of biofertilizer for sustainable processing of phosphate rock. *Life* **2023**, *13*, 782. [CrossRef]
22. Mittal, V.; Singh, O.; Nayyar, H.; Kaur, J.; Tewari, R. Stimulatory effect of phosphate-solubilizing fungal strains (*Aspergillus awamori* and *Penicillium citrinum*) on the yield of chickpea (*Cicer arietinum* L. cv. GPF2). *Soil Biol. Biochem.* **2008**, *40*, 718–727. [CrossRef]
23. Khuna, S.; Suwannarach, N.; Kumla, J.; Frisvad, J.C.; Matsui, K.; Nuangmek, W.; Lumyong, S. Growth enhancement of Arabidopsis (*Arabidopsis thaliana*) and onion (*Allium cepa*) with inoculation of three newly identified mineral-solubilizing fungi in the genus *Aspergillus* section *Nigri*. *Front. Microbiol.* **2021**, *12*, 705896. [CrossRef]
24. Rawat, P.; Das, S.; Shankhdhar, D.; Shankhdhar, S.C. Phosphate-solubilizing microorganisms: Mechanism and their role in phosphate solubilization and uptake. *J. Soil Sci. Plant Nutr.* **2021**, *21*, 49–68. [CrossRef]
25. Jain, R.; Saxena, J.; Sharma, V. Effect of phosphate-solubilizing fungi *Aspergillus awamori* S29 on mungbean (*Vigna radiata* cv. RMG 492) growth. *Folia Microbiol.* **2012**, *57*, 533–541. [CrossRef]
26. Patel, S.; Panchal, B.; Karmakar, N.; Katagi, R.B.; Jha, S. Solubilization of rock phosphate by two *Rhizopus* species isolated from coastal areas of South Gujarat and its effect on chickpea. *Ecol. Environ. Conserv.* **2015**, *21*, 223–231.
27. Kaur, G.; Reddy, M.S. Improvement of crop yield by phosphate-solubilizing *Aspergillus* species in organic farming. *Arch. Agron. Soil Sci.* **2016**, *63*, 24–34. [CrossRef]
28. Saxena, J.; Saini, A.; Ravi, I.; Chandra, S.; Garg, V. Consortium of phosphate-solubilizing bacteria and fungi for promotion of growth and yield of chickpea (*Cicer arietinum*). *J. Crop. Improv.* **2015**, *29*, 353–369. [CrossRef]
29. Saxena, J.; Rawat, J.; Sanwal, P. Enhancement of growth and yield of *Glycine max* plants with inoculation of phosphate solubilizing fungus *Aspergillus niger* K7 and biochar amendment in soil. *Commun. Soil Sci. Plant Anal.* **2016**, *47*, 2334–2347. [CrossRef]
30. Saxena, J.; Saini, A.; Kushwaha, K.; Ariño, A. Synergistic effect of plant growth promoting bacterium *Pseudomonas fluorescens* and phosphate solubilizing fungus *Aspergillus awamori* for growth enhancement of chickpea. *Indian J. Biochem. Biophys.* **2016**, *53*, 135–143.
31. Zhao, L.; Liu, Q.; Zhang, Y.; Cui, Q.; Liang, Y. Effect of acid phosphatase produced by *Trichoderma asperellum* Q1 on growth of *Arabidopsis* under salt stress. *J. Integr. Agric.* **2017**, *16*, 1341–1346. [CrossRef]
32. Zhao, J.; Liu, W.; Liu, D.; Lu, C.; Zhang, D.; Wu, H.; Dong, D.; Meng, L. Identification and evaluation of *Aspergillus tubingensis* as a potential biocontrol agent against grey mould on tomato. *J. Gen. Plant Pathol.* **2018**, *84*, 148–159. [CrossRef]
33. Naziya, B.; Murali, M.; Amruthesh, K.N. Plant growth-promoting fungi (PGPF) Instigate plant growth and induce disease resistance in *Capsicum annuum* L. upon infection with *Colletotrichum capsici* (Syd.) Butler & Bisby. *Biomolecules* **2019**, *10*, 41. [CrossRef]
34. Aloni, R.; Aloni, E.; Langhans, M.; Ullrich, C.I. Role of cytokinin and auxin in shaping root architecture: Regulating vascular differentiation, lateral root initiation, root apical dominance and root gravitropism. *Ann. Bot.* **2006**, *97*, 883–893. [CrossRef]
35. Fu, J.; Wang, S. Insights into auxin signaling in plant–pathogen interactions. *Front. Plant Sci.* **2011**, *2*, 74. [CrossRef]
36. Tian, H.; De Smet, I.; Ding, Z. Shaping a root system: Regulating lateral versus primary root growth. *Trends Plant Sci.* **2014**, *19*, 426–431. [CrossRef]
37. Keswani, C.; Singh, S.P.; Cueto, L.; García-Estrada, C.; Mezaache-Aichour, S.; Glare, T.R.; Borriss, R.; Singh, S.P.; Blázquez, M.A.; Sansinenea, S. Auxins of microbial origin and their use in agriculture. *Appl. Microbiol. Biotechnol.* **2020**, *104*, 8549–8565. [CrossRef]
38. Spaepen, S.; Vanderleyden, J. Auxin and plant-microbe interactions. *Cold Spring Harb. Perspect. Biol.* **2011**, *3*, a001438. [CrossRef]
39. Yadav, J.; Verma, J.P.; Tiwari, K.N. Plant growth promoting activities of fungi and their effect on chickpea plant growth. *Asian J. Biol. Sci.* **2011**, *4*, 291–299. [CrossRef]
40. Gaind, S. Phosphate dissolving fungi: Mechanism and application in alleviation of salt stress in wheat. *Microbiol. Res.* **2016**, *193*, 94–102. [CrossRef]
41. Amrutha, G.; Savalgi, V.P.; Jagadeesh, K.S.; Hebsur, N.S. Isolation screening and selection of phosphate solubilizing fungi from maize rhizosphere. *Int. J. Curr. Microbiol. Appl. Sci.* **2018**, *7*, 988–998. [CrossRef]
42. Srinivasan, R.; Alagawadi, A.R.; Yandigeri, M.S.; Meena, K.K.; Saxena, A.K. Characterization of phosphate-solubilizing microorganisms from salt-affected soils of India and their effect on growth of sorghum plants [*Sorghum bicolor* (L.) Moench]. *Ann. Microbiol.* **2012**, *62*, 93–105. [CrossRef]
43. Bader, A.N.; Salerno, G.L.; Covacevich, F.; Consolo, V.F. Native *Trichoderma harzianum* strains from Argentina produce indole-3-acetic acid and phosphorus solubilization, promote growth and control wilt disease on tomato (*Solanum lycopersicum* L.). *J. King Saud Univ. Sci.* **2020**, *32*, 867–873. [CrossRef]
44. Imran, M.; Abulreesh, H.H.; Monjed, M.K.; Elbanna, K.; Samreen; Ahmad, I. Multifarious functional traits of free-living rhizospheric fungi, with special reference to *Aspergillus* spp. isolated from North Indian soil, and their inoculation effect on plant growth. *Ann. Microbiol.* **2021**, *71*, 31. [CrossRef]

45. Li, X.; Li, D.; Yan, J.; Zhang, Y.; Wang, H.; Zhang, J.; Ahmed, T.; Li, B. Effect of plant-growth-promoting fungi on eggplant (*Solanum melongena* L.) in new reclamation land. *Agriculture* **2021**, *11*, 1036. [CrossRef]
46. Nafaa, M.; Rizk, S.M.; Aly, T.A.-G.A.; Rashed, M.A.-S.; Abd El-Moneim, D.; Ben Bacha, A.; Alonazi, M.; Magdy, M. Screening and identification of the rhizosphere fungal communities associated with land reclamation in Egypt. *Agriculture* **2023**, *13*, 215. [CrossRef]
47. Jain, R.; Saxena, J.; Sharma, V. Solubilization of inorganic phosphates by *Aspergillus awamori* S19 isolated from rhizosphere soil of a semi-arid region. *Ann. Microbiol.* **2012**, *62*, 725–735. [CrossRef]
48. Patel, D.; Patel, A.; Patel, M.; Goswami, D. *Talaromyces pinophilus* strain M13: A portrayal of novel groundbreaking fungal strain for phytointensification. *Environ. Sci. Pollut. Res.* **2021**, *28*, 8758–8769. [CrossRef]
49. Wang, H.; Liu, S.; Zhai, L.; Zhang, J.; Ren, T.; Fan, B.; Liu, H. Preparation and utilization of phosphate biofertilizers using agricultural waste. *J. Integr. Agric.* **2015**, *14*, 158–167. [CrossRef]
50. Kannahi, M.; Senbagam, N. Studies on siderophore production by microbial isolates obtained from rhizosphere soil and its antibacterial activity. *J. Chem. Pharm. Res.* **2014**, *6*, 1142–1145.
51. Saha, M.; Sarkar, S.; Sarkar, B.; Sharma, B.K.; Bhattacharjee, S.; Tribedi, P. Microbial siderophores and their potential applications: A review. *Environ. Sci. Pollut. Res.* **2016**, *23*, 3984–3999. [CrossRef]
52. Crowley, D.E. Microbial siderophores in the plant rhizosphere. In *Iron Nutrition in Plants and Rhizospheric Microorganisms*; Barton, L.L., Abadia, J., Eds.; Springer: Dordrecht, The Netherlands, 2006; pp. 169–198. [CrossRef]
53. Ahmed, E.; Holmström, S.J.M. Siderophores in environmental research: Roles and applications. *Microb. Biotechnol.* **2014**, *7*, 196–208. [CrossRef]
54. Babu, A.G.; Shim, J.; Bang, K.; Shea, P.J.; Oh, B. *Trichoderma virens* PDR-28: A heavy metal-tolerant and plant growth promoting fungus for remediation and bioenergy crop production on mine tailing soil. *J. Environ. Manag.* **2014**, *132*, 129–134. [CrossRef]
55. Babu, A.G.; Kim, S.W.; Yadav, D.R.; Hyum, U.; Adhikari, M.; Lee, Y.S. *Penicillium menonorum*: A novel fungus to promote growth and nutrient management in cucumber plants. *Mycobiology* **2015**, *43*, 49–56. [CrossRef] [PubMed]
56. Banerjee, S.; Dutta, S. Plant growth promoting activities of a fungal strain *Penicillium commune* MCC 1720 and it's effect on growth of black gram. *Pharm. Innov. J.* **2019**, *8*, 121–127.
57. Moreno-Salazar, R.; Sánchez-García, I.; Chan-Cupul, W.; Ruiz-Sánchez, E.; Hernández-Ortega, H.A.; Pineda-Lucatero, J.; Figueroa-Chávez, D. Plant growth, foliar nutritional content and fruit yield of *Capsicum chinense* biofertilized with *Purpureocillium lilacinum* under greenhouse conditions. *Sci. Hortic.* **2020**, *261*, 108950. [CrossRef]
58. Xie, Y.; Han, S.; Li, X.; Amombo, E.; Fu, J. Ameliorates of salt stress on bermudagrass by the fungus *Aspergillus aculeatus*. *Mol. Plant-Microbe Interac.* **2017**, *30*, 245–254. [CrossRef]
59. Muthuraja, R.; Muthukumar, T. Isolation and characterization of potassium solubilizing *Aspergillus* species isolated from saxum habitats and their effect on maize growth in different soil types. *Geomicrobiol. J.* **2021**, *38*, 672–685. [CrossRef]
60. Hervieux, V.; Yaganza, E.S.; Arul, J.; Tweddell, R.J. Effect of organic and inorganic salts on the development of *Helminthosporium solani*, the causal agent of potato silver scurf. *Plant Dis.* **2002**, *86*, 1014–1018. [CrossRef]
61. Arriagada, C.A.; Herrera, M.A.; Borie, F.; Ocampo, J.A. Contribution of arbuscular mycorrhizal and saprobe fungi to the aluminum resistance of *Eucalyptus globulus*. *Water Air Soil Pollut.* **2007**, *182*, 383–394. [CrossRef]
62. Kolaei, E.A.; Cenatus, C.; Tweddell, R.J.; Avis, T.J. Antifungal activity of aluminium-containing salts against the development of carrot cavity spot and potato dry rot. *Ann. Appl. Biol.* **2013**, *136*, 311–317. [CrossRef]
63. Jones, D.J.; Oburger, E. Solubilization of phosphorus by soil microorganisms. In *Phosphorus in Action*; Bünemann, E.K., Oberson, A., Frossard, E., Eds.; Springer: Heidelberg/Berlin, Germany, 2011; pp. 169–198. [CrossRef]
64. Mapelli, F.; Marasco, R.; Balloi, A.; Rolli, E.; Cappitelli, F.; Daffonchio, D.; Borin, S. Mineral-microbe interactions: Biotechnological potential of bioweathering. *J. Biotechnol.* **2012**, *157*, 473–481. [CrossRef]
65. Alori, E.T.; Glick, B.R.; Babalola, O.O. Microbial phosphorus solubilization and its potential for use in sustainable agriculture. *Front. Microbiol.* **2017**, *8*, 971. [CrossRef]
66. Illmer, P.; Barbato, A.; Schinner, F. Solubilization of hardly soluble $AlPO_4$ with P-solubilizing microorganisms. *Soil Biol. Biochem.* **1995**, *27*, 265–270. [CrossRef]
67. Behera, B.C.; Singdevsachan, S.K.; Mishra, R.R.; Dutta, S.K.; Thatoi, H.N. Diversity, mechanism and biotechnology of phosphate solubilising microorganism in mangrove—A review. *Biocatal. Agric. Biotechnol.* **2014**, *3*, 97–110. [CrossRef]
68. Jamshidi, R.; Jalili, B.; Bahmanyar, M.A.; Salek-Gilani, S. Isolation and identification of a phosphate solubilising fungus from soil of a phosphate mine in Chaluse, Iran. *Mycology* **2016**, *7*, 134–142. [CrossRef] [PubMed]
69. Vera-Morales, M.; López Medina, S.E.; Naranjo-Morán, J.; Quevedo, A.; Ratti, M.F. Nematophagous fungi: A review of their phosphorus solubilization potential. *Microorganisms* **2023**, *11*, 137. [CrossRef] [PubMed]
70. Wang, X.; Wang, C.; Sui, J.; Liu, Z.; Li, Q.; Ji, C.; Song, X.; Hu, Y.; Wang, C.; Sa, R.; et al. Isolation and characterization of phosphofungi, and screening of their plant growth-promoting activities. *AMB Express* **2018**, *8*, 63. [CrossRef]
71. Escobar Diaz, P.A.; Gil, O.J.A.; Barbosa, C.H.; Desoignies, N.; Rigobelo, E.C. *Aspergillus* spp. and *Bacillus* spp. as growth promoters in cotton plants under greenhouse conditions. *Front. Sustain. Food Syst.* **2021**, *5*, 709267. [CrossRef]
72. Zhao, L.; Zhang, Y. Effects of phosphate solubilization and phytohormone production of *Trichoderma asperellum* Q1 on promoting cucumber growth under salt stress. *J. Integr. Agric.* **2015**, *14*, 1588–1597. [CrossRef]

73. Jain, R.; Garg, V.; Saxena, J. Effect of an organophosphate pesticide, monocrotophos, on phosphate-solubilizing efficiency of soil fungal isolates. *Appl. Biochem. Biotechnol.* **2015**, *175*, 813–824. [CrossRef]
74. Della Mónica, I.F.; Godoy, M.S.; Godeas, A.M.; Scervino, J.M. Fungal extracellular phosphatases: Their role in P cycling under different pH and P sources availability. *J. Appl. Microbiol.* **2017**, *124*, 155–165. [CrossRef]
75. Qiao, H.; Sun, X.R.; Wu, X.Q.; Li, G.E.; Wang, Z.; Li, D.W. The phosphate-solubilizing ability of *Penicillium guanacastense* and its effects on the growth of *Pinus massoniana* in phosphate-limiting conditions. *Biol. Open* **2019**, *8*, bio046797. [CrossRef]
76. Akintokun, A.K.; Akande, G.A.; Akintokun, P.O.; Popoola, T.O.S.; Babalola, A.O. Solubilization of insoluble phosphate by organic acid-producing fungi isolated from Nigerian soil. *Int. J. Soil Sci.* **2007**, *2*, 301–307. [CrossRef]
77. Kanse, O.S.; Whitelaw-Weckert, M.; Kadam, T.A.; Bhosale, H. Phosphate solubilization by stress-tolerant soil fungus *Talaromyces funiculosus* SLS8 isolated from the Neem rhizosphere. *Ann. Microbiol.* **2015**, *65*, 85–93. [CrossRef]
78. Bakri, M.M. Tri-calcium and zinc phosphates solubilization by *Aspergillus niger* and its relation to organic acids production. *BioNanoScience* **2019**, *9*, 238–244. [CrossRef]
79. Li, X.; Sun, X.; Wang, G.; Amombo, E.; Zhou, X.; Du, Z.; Zhang, Y.; Xie, Y.; Fu, J. Inoculation with *Aspergillus aculeatus* alters the performance of perennial ryegrass under phosphorus deficiency. *J. Am. Soc. Hortic. Sci.* **2019**, *144*, 182–192. [CrossRef]
80. Acevedo, E.; Galindo-Castañeda, T.; Prada, F.; Navia, M.; Romero, H.M. Phosphate solubilizing microorganisms associated with the rhizosphere of oil palm (*Elaeis guineensis* Jacq.) in Colombia. *Appl. Soil Ecol.* **2014**, *80*, 26–33. [CrossRef]
81. Xiao, C.; Fang, Y.; Chi, R. Phosphate solubilization in vitro by isolated *Aspergillus niger* and *Aspergillus carbonarius*. *Res. Chem. Intermed.* **2015**, *41*, 2867–2878. [CrossRef]
82. Li, Z.; Bai, T.; Dai, L.; Wang, F.; Tao, J.; Meng, S.; Hu, Y.; Wang, S.; Hu, S. A study of organic acid production in contrasts between two phosphate solubilizing fungi: *Penicillium oxalicum* and *Aspergillus niger*. *Sci. Rep.* **2016**, *6*, 25313. [CrossRef]
83. Qin, W.; Liu, C.; Jiang, W.; Xue, Y.; Wang, G.; Liu, S. A coumarin analogue NFA from endophytic *Aspergillus fumigatus* improves drought resistance in rice as an antioxidant. *BMC Microbiol.* **2019**, *19*, 50. [CrossRef]
84. Li, X.; Zhao, C.; Zhang, T.; Wang, G.; Amombo, E.; Xie, Y.; Fu, J. Exogenous *Aspergillus aculeatus* enhances drought and heat tolerance of perennial ryegrass. *Front. Microbiol.* **2021**, *12*, 593722. [CrossRef]
85. Kaur, R.; Saxena, S. *Penicillium citrinum*, a drought-tolerant endophytic fungus isolated from wheat (*Triticum aestivum* L.) leaves with plant growth-promoting abilities. *Curr. Microbiol.* **2023**, *80*, 184. [CrossRef]
86. Muthuraja, R.; Muthukumar, T.; Natthapol, C. Drought tolerance of *Aspergillus violaceofuscus* and *Bacillus licheniformis* and their influence on tomato growth and potassium uptake in mica amended tropical soils under water-limiting conditions. *Front. Plant Sci.* **2023**, *14*, 1114288. [CrossRef] [PubMed]
87. Ou, T.; Zhang, M.; Gao, H.; Wang, F.; Xu, W.; Liu, X.; Wang, L.; Wang, R.; Xie, J. Study on the potential for stimulating mulberry growth and drought tolerance of plant growth-promoting fungi. *Int. J. Mol. Sci.* **2023**, *24*, 4090. [CrossRef] [PubMed]
88. Rinu, K.; Pandey, A. Temperature-dependent phosphate solubilization by cold- and pH-tolerant species of *Aspergillus* isolated from Himalayan soil. *Mycoscience* **2010**, *51*, 263–271. [CrossRef]
89. Xiao, C.; Chi, R.; Li, X.; Xia, M.; Xia, Z. Biosolubilization of rock phosphate by three stress-tolerant fungal strains. *Appl. Biochem. Biotechnol.* **2011**, *165*, 719–727. [CrossRef]
90. Passamani, F.R.F.; Hernandes, T.; Lopes, N.A.; Bastos, S.C.; Santiago, W.D.; Cardoso, M.G.; Batista, L.R. Effect of temperature, water activity, and pH on growth and production of ochratoxin A by *Aspergillus niger* and *Aspergillus carbonarius* from Brazilian grapes. *J. Food Prot.* **2014**, *77*, 1947–1952. [CrossRef] [PubMed]
91. Kurniati, E.; Arfarita, N.; Imai, T. Potential use of *Aspergillus flavus* strain KRP1 in utilization of mercury contaminant. *Procedia Environ. Sci.* **2014**, *20*, 254–260. [CrossRef]
92. Al Tamie, M.S.S. Effect of salinity on the fungal occurance in Al-Shega area at Al-Qassim, Saudi Arabia. *Res. J. Microbiol.* **2014**, *9*, 287–295. [CrossRef]
93. Vassilev, N.; Eichler-Löbermann, B.; Vassileva, M. Stress-tolerant P-solubilizing microorganisms. *Appl. Microbiol Biotechnol.* **2012**, *95*, 851–859. [CrossRef]
94. Viscardi, S.; Ventorino, V.; Duran, P.; Maggio, A.; De Pascale, S.; Mora, M.L.; Pepe, O. Assessment of plant growth promoting activities and abiotic stress tolerance of *Azotobacter chroococcum* strains for a potential use in sustainable agriculture. *J. Soil Sci. Plant Nutr.* **2016**, *16*, 848–863. [CrossRef]
95. Chaiya, L.; Kumla, J.; Suwannarach, N.; Kiatsiriroat, T.; Lumyong, S. Isolation, characterization, and efficacy of actinobacteria associated with arbuscular mycorrhizal spores in promoting plant growth of chili (*Capsicum flutescens* L.). *Microorganisms* **2021**, *9*, 1274. [CrossRef]
96. Fan, D.; Smith, D.L. Characterization of selected plant growth-promoting rhizobacteria and their non-host growth promotion effects. *Microbiol. Spectr.* **2021**, *9*, e00279-21. [CrossRef] [PubMed]
97. Huang, Y.; Xiao, L.; Li, F.; Xiao, M.; Lin, D.; Long, X.; Wu, Z. Microbial degradation of pesticide residues and an emphasis on the degradation of cypermethrin and 3-phenoxy benzoic acid: A review. *Molecules* **2018**, *23*, 2313. [CrossRef]
98. Spagnoletti, F.N.; Chiocchio, V.M. Tolerance of dark septate endophytic fungi (DSE) to agrochemicals in vitro Tolerancia de hongos endofíticos septados oscuros a agroquímicos in vitro. *Rev. Argent. Microbiol.* **2020**, *52*, 43–49. [CrossRef] [PubMed]
99. Suwannarach, N.; Kumla, J.; Matsui, K.; Lumyong, S. Characterization and efficacy of *Muscodor cinnamomi* in promoting plant growth and controlling Rhizoctonia root rot in tomatoes. *Biol. Control* **2015**, *90*, 25–33. [CrossRef]

100. Karaoglu, S.A.; Bozdeveci, A.; Pehlivan, N. Characterization of local *Trichoderma* spp. as potential bio-control agents, screening of in vitro antagonistic activities and fungicide tolerance. *J. Biol. Chem.* **2018**, *46*, 247–261. [CrossRef]
101. Nuangmek, W.; Aiduang, W.; Kumla, J.; Lumyong, S.; Suwannarach, N. Evaluation of a newly identified endophytic fungus, *Trichoderma phayaoense* for plant growth promotion and biological control of gummy stem blight and wilt of muskmelon. *Front. Microbiol.* **2021**, *12*, e634772. [CrossRef]
102. Eman, A.; Abdel-Megeed, A.; Suliman, A.M.A.; Sadik, M.W.; Sholkamy, E.N. Biodegradation of glyphosate by fungal strains isolated from herbicides polluted-soils in Riyadh area. *Int. J. Curr. Microbiol. Appl. Sci.* **2013**, *2*, 359–381. [CrossRef]
103. Mohiddin, F.A.; Khan, M.R. Tolerance of fungal and bacterial biocontrol agents to six pesticides commonly used in the control of soil borne plant pathogens. *Afr. J. Agric. Res.* **2013**, *8*, 5272–5275.
104. Andy, I.E.; Edu, G.S.; Bassey, I.U.; Markson, A.A.; Umana, E.I.; Udo, S.E. Biodegradation of paraquat. *J. Biopestic. Environ.* **2015**, *1*, 80–85.
105. Wongputtisin, P.; Supo, C.; Suwannarach, N.; Honda, Y.; Nakazawa, T.; Kumla, J.; Lumyong, S.; Khanongnuch, C. Filamentous fungi with high paraquat-degrading activity isolated from contaminated agricultural soils in northern Thailand. *Lett. Appl. Microbiol.* **2020**, *72*, 467–475. [CrossRef]
106. Spinelli, V.; Ceci, A.; Dal Bosco, C.; Gentili, A.; Persiani, A.M. Glyphosate-eating fungi: Study on fungal saprotrophic strains' ability to tolerate and utilise glyphosate as a nutritional source and on the ability of *Purpureocillium lilacinum* to degrade it. *Microorganisms* **2021**, *9*, 2179. [CrossRef] [PubMed]
107. Benito, N.; Magnoli, K.; Carranza, C.S.; Aluffi, M.E.; Magnoli, C.E.; Barberis, C.L. Influence of a glyphosate-based herbicide on growth parameters and aflatoxin B1 production by *Aspergillus* section *Flavi* on maize grains. *Rev. Argent. Microbiol.* **2021**, *53*, 162–170. [CrossRef] [PubMed]
108. Carranza, C.S.; Barberis, C.L.; Aluffi, M.E.; Benito, N.; Magnoli, C.E. Native mycota in agricultural soils exposed to pesticides and *Aspergillus oryzae* tolerance to chlorpyrifos in microcosms assays. *Curr. Res. Environ. Appl. Mycol.* **2017**, *7*, 236–248. [CrossRef]
109. Kumar, S.; Thakur, M.; Rani, A. *Trichoderma*: Mass production, formulation, quality control, delivery and its scope in commercialization in India for the management of plant diseases. *Afr. J. Agric. Res.* **2014**, *9*, 3838–3852. [CrossRef]
110. Raymond, N.S.; Stöver, D.M.; Jensen, L.S.; Håkansson, S. Survival and phosphate solubilisation activity of desiccated formulations of *Penicillium bilaiae* and *Aspergillus niger* influenced by water activity. *J. Microbiol. Methods* **2018**, *150*, 39–46. [CrossRef]
111. Friesen, T.; Hill, G.; Pugsley, T.; Holloway, G.; Zimmerman, D. Experimental determination of viability loss of *Penicillium bilaiae* conidia during convective air drying. *Appl. Microbiol. Biotechnol.* **2005**, *68*, 397–404. [CrossRef]
112. Friesen, T.J.; Holloway, G.; Hill, G.A.; Pugsley, T.S. Effect of conditions and protectants on the survival of *Penicillium bilaiae* during storage. *Biocontrol Sci. Technol.* **2006**, *16*, 89–98. [CrossRef]
113. Pindi, P.K.; Satyanarayana, S.D.V. Liquid microbial consortium—A potential tool for sustainable soil health. *J. Biofertil. Biopestic.* **2012**, *3*, 124. [CrossRef]
114. Phiromtan, M.; Mala, T.; Srinives, P. Effect of various carriers and storage temperatures on survival of *Azotobacter vinelandii* NDD-CK-1 in powder inoculant. *Mod. Appl. Sci.* **2013**, *7*, 81–89. [CrossRef]
115. Raimi, A.; Roopnarain, A.; Adeleke, R. Biofertilizer production in Africa: Current status, factors impeding adoption and strategies for success. *Sci. Afr.* **2021**, *11*, e00694. [CrossRef]
116. Daigle, D.J.; Cotty, P.J. Formulating atoxigenic *Aspergillus flavus* for field release. *Biocontrol Sci. Technol.* **1995**, *5*, 175–184. [CrossRef]
117. Locatelli, G.O.; dos Santos, G.F.; Botelho, P.S.; Finkler, C.L.L.; Bueno, L.A. Development of *Trichoderma* sp. formulations in encapsulated granules (CG) and evaluation of conidia shelf-life. *Biol. Control* **2018**, *117*, 21–29. [CrossRef]
118. Mulatu, A.; Alemu, T.; Megersa, N.; Vetukuri, R.R. Optimization of culture conditions and production of bio-fungicides from *Trichoderma* species under solid-state fermentation using mathematical modeling. *Microorganisms* **2021**, *9*, 1675. [CrossRef] [PubMed]
119. Callaghan, M.O. Microbial inoculation of seed for improved crop performance: Issues and opportunities. *Appl. Microbiol. Biotechnol.* **2016**, *100*, 5729–5746. [CrossRef]
120. Alori, E.T.; Babalola, O.O. Microbial inoculants for improving crop quality and human health in Africa. *Front. Microbiol.* **2018**, *9*, 2213. [CrossRef]
121. Brar, S.K.; Sarma, S.J.; Chaabouni, E. Shelf-life of biofertilizers: An accord between formulations and genetics. *J. Biofertil. Biopestici.* **2012**, *3*, e109. [CrossRef]
122. Melin, P.; Schnürer, J.; Håkansson, S. Formulation and stabilisation of the biocontrol yeast *Pichia anomala*. *Antonie Van Leeuwenhoek* **2011**, *99*, 107–112. [CrossRef]
123. Parnell, J.J.; Berka, R.; Young, H.A.; Sturino, J.M.; Kang, Y.; Barnhart, D.; DiLeo, M.V. From the lab to the farm: An industrial perspective of plant beneficial microorganisms. *Front. Plant Sci.* **2016**, *7*, 1110. [CrossRef]
124. Accinelli, C.; Saccà, M.L.; Abbas, H.K.; Zablotowicz, R.M.; Wilkinson, J.R. Use of a granular bioplastic formulation for carrying conidia of a non-aflatoxigenic strain of *Aspergillus flavus*. *Bioresour. Technol.* **2009**, *100*, 3997–4004. [CrossRef]
125. Jain, R.; Gupta, A.; Sharma, V.; Naik, S.; Saxena, J.; Kumar, V.; Prasad, R. Immobilization-based bio-formulation of *Aspergillus awamori* S29 and evaluation of its shelf life and re-usability in the soil–plant experiment. *Curr. Microbiol.* **2022**, *79*, 163. [CrossRef]
126. Mendes, G.O.; Galvez, A.; Vassileva, M.; Vassilev, N. Fermentation liquid containing microbially solubilized P significantly improved plant growth and P uptake in both soil and soilless experiments. *Appl. Soil Ecol.* **2017**, *117–118*, 208–211. [CrossRef]

127. Tariq, M.R.; Shaheen, F.; Mustafa, S.; Ali, S.; Fatima, A.; Shafiq, M.; Safdar, W.; Sheas, M.N.; Hameed, A.; Nasir, M.A. Phosphate solubilizing microorganisms isolated from medicinal plants improve growth of mint. *PeerJ* **2022**, *10*, e13782. [CrossRef] [PubMed]
128. Lubna; Asaf, S.; Hamayun, M.; Gul, H.; Lee, I.J.; Hussain, A. *Aspergillus niger* CSR3 regulates plant endogenous hormones and secondary metabolites by producing gibberellins and indoleacetic acid. *J. Plant Interact.* **2018**, *13*, 100–111. [CrossRef]
129. Abdel-Motaal, F.; Kamel, N.; El-Zayat, S.; Abou-Ellail, M. Early blight suppression and plant growth promotion potential of the endophyte *Aspergillus flavus* in tomato plant. *Ann. Agric. Sci.* **2020**, *65*, 117–123. [CrossRef]
130. Peng, Q.; Xiao, Y.; Zhang, S.; Zhou, C.; Xie, A.; Li, Z.; Tan, A.; Zhou, L.; Xie, Y.; Zhao, J.; et al. Mutation breeding of *Aspergillus niger* by atmospheric room temperature plasma to enhance phosphorus solubilization ability. *PeerJ* **2022**, *10*, e13076. [CrossRef] [PubMed]
131. Chuang, C.C.; Kuo, Y.L.; Chao, C.C.; Chao, W.L. Solubilization of inorganic phosphates and plant growth promotion by *Aspergillus niger*. *Biol. Fertil. Soils* **2007**, *43*, 575–584. [CrossRef]
132. Shahrajabian, M.H.; Chaski, C.; Polyzos, N.; Petropoulos, S.A. Biostimulants application: A low input cropping management tool for sustainable farming of vegetables. *Biomolecules* **2021**, *11*, 698. [CrossRef]
133. Shahrajabian, M.H.; Chaski, C.; Polyzos, N.; Tzortzakis, N.; Petropoulos, S.A. Sustainable agriculture systems in vegetable production using chitin and chitosan as plant biostimulants. *Biomolecules* **2021**, *11*, 819. [CrossRef]
134. Tsavkelova, E.A.; Cherdyntseva, T.A.; Botina, S.G.; Netrusov, A.I. Bacteria associated with orchid roots and microbial production of auxin. *Microbiol. Res.* **2007**, *162*, 69–76. [CrossRef]
135. Kumla, J.; Suwannarach, N.; Bussaban, B.; Matsui, K.; Lumyong, S. Indole-3-acetic acid production, solubilization of insoluble metal minerals and metal tolerance of some sclerodermatoid fungi collected from northern Thailand. *Ann. Microbiol.* **2014**, *64*, 707–720. [CrossRef]
136. Schwyn, B.; Neiland, J. Universal chemical assay for the detection and determination of siderophores. *Anal. Biochem.* **1987**, *160*, 47–56. [CrossRef] [PubMed]
137. Shin, S.H.; Lim, Y.; Lee, S.E.; Yang, N.W.; Rhee, J.H. CAS agar diffusion assay for the measurement of siderophores in biological fluids. *J. Microbiol. Methods* **2001**, *44*, 89–95. [CrossRef] [PubMed]
138. Fomina, M.A.; Alexander, I.J.; Colpaert, J.V.; Gadd, G.M. Solubilization of toxic metal minerals and metal tolerance of mycorrhizal fungi. *Soil Biol. Biochem.* **2005**, *37*, 851–866. [CrossRef]
139. Elias, F.; Muleta, D.; Woyessa, D. Effects of phosphate solubilizing fungi on growth and yield of haricot bean (*Phaseolus vulgaris* L.) plants. *J. Agric. Sci.* **2016**, *8*, 204–218. [CrossRef]
140. Fiske, C.; Subbarow, Y. The colorimetric determination of phosphorus. *J. Biol. Chem.* **1925**, *66*, 375–400. [CrossRef]
141. Holman, W.I.M. A new technique for the determination of phosphorus by the molybdenum blue method. *Biochem. J.* **1943**, *37*, 256–259. [CrossRef]
142. Gaind, S.; Singh, S. Production, purification and characterization of neutral phytase from thermotolerant *Aspergillus flavus* ITCC 6720. *Int. Biodeterior. Biodegrad.* **2015**, *99*, 15–22. [CrossRef]
143. Adhikari, P.; Pandey, A. Phosphate solubilization potential of endophytic fungi isolated from *Taxus wallichiana* Zucc. roots. *Rhizosphere* **2019**, *9*, 2–9. [CrossRef]
144. Hallsworth, J.E.; Nomura, Y.; Iwahara, M. Ethanol-induced water stress and fungal growth. *J. Ferment. Bioeng.* **1998**, *86*, 451–456. [CrossRef]
145. Tresner, H.D.; Hayes, J.A. Sodium chloride tolerance of terrestrial fungi. *Appl. Microbiol.* **1971**, *22*, 210–213. [CrossRef]
146. Sriram, S.; Roopa, K.P.; Savitha, M.J. Extended shelf-life of liquid fermentation derived talc formulations of *Trichoderma harzianum* with the addition of glycerol in the production medium. *Crop. Prot.* **2011**, *30*, 1334–1339. [CrossRef]
147. Lichtenthaler, H.K.; Wellburn, A.R. Determinations of total carotenoids and chlorophylls *a* and *b* of leaf extracts in different solvents. *Biochem. Soc. Trans.* **1983**, *11*, 591–592. [CrossRef]
148. Liang, Y.; Urano, D.; Liao, K.L.; Hedrick, T.L.; Gao, Y.; Jones, A.M. A nondestructive method to estimate the chlorophyll content of *Arabidopsis* seedlings. *Plant Methods* **2017**, *13*, 26. [CrossRef] [PubMed]
149. Ames, B.N. Assay of inorganic phosphate, total phosphate and phosphatases. *Methods Enzymol.* **1966**, *8*, 115–118. [CrossRef]
150. Wang, L.; Dong, J.; Gao, Z.; Liu, D. The Arabidopsis gene *HYPERSENSITIVE TO PHOSPHATE STARVATION 3* encodes ETHYLENE OVERPRODUCTION 1. *Plant Cell Physiol.* **2012**, *53*, 1093–1105. [CrossRef]


**Disclaimer/Publisher's Note:** The statements, opinions and data contained in all publications are solely those of the individual author(s) and contributor(s) and not of MDPI and/or the editor(s). MDPI and/or the editor(s) disclaim responsibility for any injury to people or property resulting from any ideas, methods, instructions or products referred to in the content.

*plants*

MDPI

*Review*

# The Application of Arbuscular Mycorrhizal Fungi as Microbial Biostimulant, Sustainable Approaches in Modern Agriculture

**Wenli Sun *,† and Mohamad Hesam Shahrajabian †**

Biotechnology Research Institute, Chinese Academy of Agricultural Sciences, Beijing 100081, China; hesamshahrajabian@gmail.com

* Correspondence: sunwenli@caas.cn; Tel.: +86-13-4260-83836

† These authors contributed equally to this work.

**Abstract:** Biostimulant application can be considered an effective, practical, and sustainable nutritional crop supplementation and may lessen the environmental problems related to excessive fertilization. Biostimulants provide beneficial properties to plants by increasing plant metabolism, which promotes crop yield and improves the quality of crops; protecting plants against environmental stresses such as water shortage, soil salinization, and exposure to sub-optimal growth temperatures; and promoting plant growth via higher nutrient uptake. Other important benefits include promoting soil enzymatic and microbial activities, changing the architecture of roots, increasing the solubility and mobility of micronutrients, and enhancing the fertility of the soil, predominantly by nurturing the development of complementary soil microbes. Biostimulants are classified as microbial, such as arbuscular mycorrhizae fungi (AMF), plant-growth-promoting rhizobacteria (PGPR), non-pathogenic fungi, protozoa, and nematodes, or non-microbial, such as seaweed extract, phosphite, humic acid, other inorganic salts, chitin and chitosan derivatives, protein hydrolysates and free amino acids, and complex organic materials. Arbuscular mycorrhizal fungi are among the most prominent microbial biostimulants and have an important role in cultivating better, healthier, and more functional foods in sustainable agriculture. AMF assist plant nutrient and water acquisition; enhance plant stress tolerance against salinity, drought, and heavy metals; and reduce soil erosion. AMF are proven to be a sustainable and environmentally friendly source of crop supplements. The current manuscript gives many examples of the potential of biostimulants for the production of different crops. However, further studies are needed to better understand the effectiveness of different biostimulants in sustainable agriculture. The review focuses on how AMF application can overcome nutrient limitations typical of organic systems by improving nutrient availability, uptake, and assimilation, consequently reducing the gap between organic and conventional yields. The aim of this literature review is to survey the impacts of AMF by presenting case studies and successful paradigms in different crops as well as introducing the main mechanisms of action of the different biostimulant products.

**Keywords:** arbuscular mycorrhizal fungi; biostimulant; microbial biostimulants; mycorrhizal fungi; sustainable agriculture

**Citation:** Sun, W.; Shahrajabian, M.H. The Application of Arbuscular Mycorrhizal Fungi as Microbial Biostimulant, Sustainable Approaches in Modern Agriculture. *Plants* **2023**, *12*, 3101. https://doi.org/10.3390/plants12173101

Academic Editor: Daniela Businelli

Received: 28 July 2023
Revised: 16 August 2023
Accepted: 24 August 2023
Published: 29 August 2023

**Copyright:** © 2023 by the authors. Licensee MDPI, Basel, Switzerland. This article is an open access article distributed under the terms and conditions of the Creative Commons Attribution (CC BY) license (https://creativecommons.org/licenses/by/4.0/).

## 1. Introduction

Biostimulant application is known as an eco-friendly and novel farming practice and is relevant to two otherwise contrasting concepts, namely crop sustainability and intensification [1]. Biostimulant products already form a significant part of the global farming industry, indicating increasing trends over the years and in the future [2]. There are various reports regarding their positive impacts on crops, especially under biotic and abiotic stress conditions, and significant research is continuously conducted to find and/or produce new biostimulatory products, as well as to show the mechanisms of action behind the observed impacts. However, the variance in the composition of biostimulant products, as well as the lack of ordinary application protocols for the different products, may create inconsistencies between the observed results and complicate attempts to reveal the actual mechanisms behind the biostimulatory impacts, which may include physiological procedures, hormonal regulation, and morphological alterations. Biostimulants' beneficial activities include the improvement of nutrient uptake, the induction of root growth, and the production of phytohormones; osmotic adjustment through the synthesis of organic osmolytes has also been confirmed. Biostimulants can also be applied to decrease the application of mineral inorganic fertilizer and are considered environmentally friendly tools with no significant negative impacts on fruit quality or total yield. Humic acids, fulvic acid, protein seaweed extracts, hydrolysates, *N*-containing compounds, botanicals, seaweed extracts, chitosan and other related biopolymers, beneficial bacteria and fungi, and inorganic compounds are the major categories of plant biostimulants. Modern crop production has to cope with abiotic and biotic stressors such as soil and irrigation water salinity, extreme and untimely weather phenomena, water limitations, infections from pathogens, and pests, which severely influence crop performance and the quality of the final products [3–5]. The most important advantages of biostimulants include improved profits, stimulated plant reactions, decreased operating costs, reduced application of fertilizers, improved root protection from soil pathogens, and enhanced drought tolerance; moreover, they repel pests, accelerate root establishment, boost fertilization, enhance stress tolerance, ameliorate fertilization, alleviate leaching, detoxify heavy metals and chemicals, and improve stomata opening and plant transpiration [3–6]. Biostimulatory compounds may also have positive effects on soil biology and are recognized as a good technique for recovering semi-arid areas and degraded ecosystems [6–8]. However, the variable composition of raw materials applied for the production of biostimulant products makes the task of revealing the mechanisms of action more difficult, and long-term research and standardization processes are needed [9]. Different sources of chitin and chitosan in nature are crustaceans (lobster, shrimp, king crab), fungi (*Mucor rouxii*, *Penicillium chrysogenum*, *Aspergillus niger*, *Lactarius vellereus*), insects (ladybug, wax worm, silk worm, butterfly), and mollusks (shell oysters, squid pen). Crustacean shells are the most notable chitin source, and chitin recovery involves three steps consisting of demineralization, deproteination, and the elimination of pigments and lipids [10–12]. Microbial proteases such as *Lactobacillus* sp., *Bacillus* sp., *Pseudomonas* sp., *Serrati marcescens*, etc., are the most significant strains applied in chitin and chitosan production [10].

Biostimulants containing organic substances, humic acids, amino acids, algae extracts, and carbon and boron increased plant growth, yield, and shelf life of onion bulbs [13], and the application of diluted honey extract (DHE) improved photosynthetic parameters, antioxidant activity, biomass production, and yield [14]. The use of seaweed extracts, vermicompost, and a mixture of animal waste increased yield and bulb traits [15]. Foliar application of vermicompost leachate, smoke-water, *Ecklonia maxima* extracts, and indole-3-butyric acid on seedlings of mustard greens grown in soils from goldmines boosted phytoremediation activities through the accumulation of heavy metals [16]. Foliar application of Kelpak SL and Asahi SL increased the nutritional value and improved the storage life of carrots [17], while root and foliar application of protein hydrolysates in lettuce plants grown under salinity conditions mitigated oxidative stress and increased glucosinolate and osmolyte content [18,19]. In intensive cropping sectors such as horticulture and floriculture, biostimulants can also boost nutrient use efficiency, partly substitute chemical fertilizer inputs, and ameliorate the quality and yield of crops [20,21]. Biostimulants based on microorganisms are a subgroup of the heterogeneous family of biostimulants, related to a microorganism (or mix of microorganisms) that can stimulate biochemical and physiological processes that benefit the nutrient efficiency, nutrient uptake, abiotic stress tolerance, crop quality, and/or yield of plants [22], which can moderately mitigate the damaging effects of intensive agriculture [23–25]. The most common microorganisms included in this group of biostimulatory products are the non-pathogenic and non-toxigenic bacteria of *Azotobacter* spp., *Rhizobium* spp., and *Azospirillum* spp., as well as different mycorrhizal fungi [24]. Mycorrhizas are a symbiotic association between fungi and plant roots and are present in several forms according to the fungal taxonomy and the host plant. Two important parameters that influence the distribution of these forms are the climatic and soil conditions and the host plant distribution [26,27] (Hart and Reader, 2002; Yang et al., 2012), and mycorrhiza can significantly boost the efficiency of mineral absorption, falling into two major categories: endotrophic and ectotrophic [28]. The main types of arbuscular mycorrhizal fungi (AMF) are related to the sub-phylum Glomeromycotina of the phylum Mucoromycota [29], and four orders of AMF, namely *Glomerales*, *Paraglomerales*, *Archaeosporales*, and *Diversisporales*, have been recognized in this sub-phylum, which contains 25 genera [30,31]. The protective mechanisms are credited to arbuscular mycorrhizal fungi-assisted alleviation of oxidative stress, rapid water uptake and nutrient absorption, and changes in the transcript levels of genes involved in signaling pathways or stress response [32–34], and the effectiveness of AMF is usually influenced by environmental variables and soil conditions [35]. Considering the numerous literature reports in the past decade related to microbial biostimulants and their impacts on different crops, this review aims to present the most up-to-date key results of microbial biostimulant practical applications on crops and the new tools available to unravel the mechanisms behind the observed impacts. In the present review, all relevant reports in the English language were collected. The literature search was performed by using the keywords of plant biostimulants, microbial biostimulants, arbuscular mycorrhizal fungi, mycorrhizal fungi, and sustainable agriculture in the main indexing systems, including PubMed/MEDLINE, Scopus, the Google Scholar search engine, as well as the Institute for Scientific Information Web of Science, from July 2000 to July 2023.

## 2. Biostimulant Categories

Biostimulants are classified into two distinct groups based on their origin; one category includes products that have biological origins in pathogens or plants, and the second group consists of products that do not have biological origins [36–38]. Another classification approach divides biostimulant products into microbial biostimulants, which are obtained from arbuscular mycorrhizal fungi and plant-growth-promoting bacteria, and non-microbial biostimulants, which include plant micro-algae extracts, humic substances, and biopolymers such as chitosan [39–43]. Different compounds with bioactive properties can be used as biostimulants to boost plant growth and development under normal and stress conditions [44–52]. Salicylic acid is economical and quick in action, environmentally sound, and it also links with other elicitors to boost the biosynthesis of secondary metabolites [53–55]. Humic acid can increase plant growth, retain water, enrich nutrients, and suppress disease [56,57]. Fulvic acids are used in sustainable horticulture and can change plant primary and secondary metabolism and increase nutrient uptake, root growth, and crop tolerance to environmental stresses [58,59]. Protein hydrolysate biostimulants, mostly produced by chemical and enzymatic hydrolysis of plant- and animal-derived proteins, are based on a mixture of soluble amino acids and peptides and can increase the yield and quality of products as well as improve the nutrient uptake and abiotic stress tolerance of plants [60–63]. They are largely prepared from brown seaweeds, such as *Ecklonia maxima*, *Ascophyllum nodosum*, and *Macrocystis pyrifera*, and they include promoting hormones or trace elements such as Zn, Fe, Mn, and Cu [64,65]. Humic-like substances such as fulvic and humic acids may also show biostimulatory activity, since several reports have suggested improved crop performance attributed mainly to auxin- and cytokinin-like impacts; they are obtained from organic matter decomposition and metabolic products of soil microbes, and they have roles in plant growth via the improvement of soil physical–chemical properties and the boosted availability of nutrients in the rhizosphere [66–68]. The actual mechanisms of action seem to be the result of synergy between the several bioactive components in raw materials, although the impacts may change depending on the crop, soil type, and soil microbes present in the rhizosphere [69–72]. The most important impacts of chitin and its derivatives' applications are that they stimulate and protect seed germination, stimulate stress resistance, mitigate negative impacts of abiotic stress, induce plant growth and development, improve soil properties and prevent nutrient leaching, improve the shelf-life of crops, chelate heavy metals, increase crop yield and quality, and protect against pests and pathogens, e.g., bacteria, viruses, fungi, insects, and nematodes [73–75]. Amino acids are the best candidates to boost stress tolerance through osmo-protection, ROS scavenging, metal chelation, and nutrient availability [76], which can notably impact the synthesis and stimulation of some enzymes and gene expression [77–79]. They can also be applied as signal molecules, like for inducing stomatal closure, as sensors of the nutrient contents of cells, or as regulators for inducing their own catabolism. Amino acids can manage the procedure of protein synthesis, strengthening plant growth, photosynthesis, and yield formation. They can increase nutrient assimilation, use, and translocation, as well as increase the quality of constituents [79]. Amino acids are well-known biostimulants due to their positive effects on yield and plant growth, and can mitigate injuries from abiotic stresses [80,81]. Amino acids also have a significant role in ammonium fixation and $C_4$ metabolism and in the biosynthesis of different components, including isoflavonoids, flavonoids, cutin, aurones, sporopollenin, stilbenes, proanthocyanidins, suberin, lignins, catechins, phenylpropenes, lignans, acylated polyamines, and other different alkaloid derivatives. The largest and most diverse group of secondary metabolites in plants is phenols, which have good antioxidant effects and are involved in the regulation of photosynthesis, physiological activities, oxidation reduction procedures, and plant breathing [82]. Phenolic acids and their derivatives are coumarins, stilbenes, quinones, lignans, flavonoids, curcuminoids, and tannins, which have meaningful roles in plant development, especially in pigment and lignin biosynthesis, and of course, they have a significant role in protecting plants from stress [83].

Protein hydrolysate biostimulants, mostly produced by enzymatic and chemical hydrolysis of plant-derived and animal proteins, are based on a mixture of peptides and soluble amino acids, and can increase the quality and yield of products as well as the uptake and abiotic stress tolerance of plants [84]. *Glomus*, the largest and most common genus in the phylum *Glomeromycota*, forms symbiotic relationships with plant roots [85,86], which can boost the drought tolerance of the host plant, mediated by proteins with chaperone-like activity [87]. *Trichoderma* fungi have important functions in nature as plant growth promoters and antagonists of phytopathogenic fungi [88], and as rhizosphere inhabitants, they contribute to interactions with microorganisms, soil, arthropods, and plants at multiple trophic levels [89], and can be used as biocontrol and biopesticide agents [90]. Members of the genus *Trichoderma* are also used in different industry branches, like in the production of biofuel, antibiotics, and enzymes [91]. The main *Trichoderma*–plant interactions include their impacts on plant morphology, plant physiology, nutrient absorption and solubilization, disease resistance, yield improvement, and abiotic stress tolerance [92]. *Trichoderma reesei* is a genus of filamentous fungi and a superior cellulose source for industrial uses, and it can produce proteins, including different enzymes, cellulases, hemicellulases, and hydrophobins [93,94]. The endophytic fungus *Heteroconium chaetospira* can also penetrate through the outer epidermal cells of its host, pass into the inner cortex, and grow all over the cortical cells, consisting of those of the root tip region, without causing apparent pathogenic symptoms [95], and it can provide even more nitrogen to the plant than mineralizing plant-available organic nitrogen [96]. *Arthrobacter* species, which are Gram-positive chemoorganotrophs and obligate aerobes, are commonly identified among soil bacteria [97], being dominant aerobic bacteria under the class of families *Micrococcaceae* and *Actinobacteria* [98], and nutritional versatility is the principal feature of arthrobacters [99]. *Acinetobacter* spp. are Gram-negative coccobacilli that are aerobic, non-motile, and oxidative negative, with no glucose fermentation ability; they can be found in different environments [100] and can fix nitrogen, solubilize minerals, produce siderophores, and act as plant endophytes or epiphytes, which can help hosts in detaching pollutants and tolerating environmental stresses [101]. Moreover, the plant-growth-promoting traits of *Actinobacteria* entail phosphate solubilization, IAA, and siderophores [102]. They can also promote higher phosphorus content and plant growth and increase radical scavenging, plant phenolic components, and antioxidant activity [103]. Other important bacteria are *Enterobacter* spp., *Pseudomonas* spp., *Ochrobactrum* spp., *Bacilus* spp., and *Rhodococcus* spp. [104–113]. Figure 1 shows different classifications of plant biostimulants.

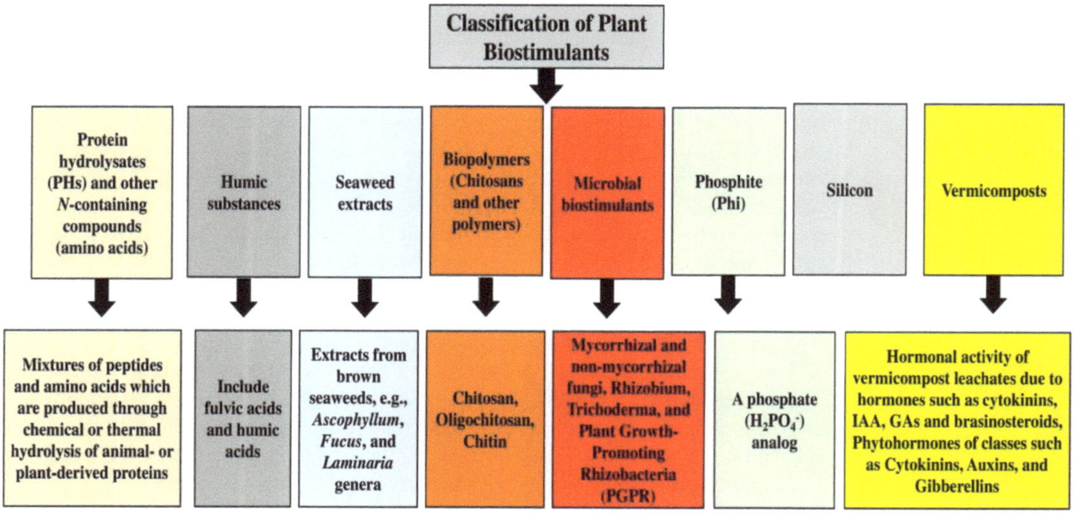


**Figure 1.** Different classifications of plant biostimulants.

## 3. Microbial Biostimulants

There are two kinds of microbial inoculants: biopesticides and biofertilizers. Biofertilizers are categorized under biostimulants [114,115]. They are also known as bioinoculants, which consist of living organisms and promote plant growth via a variety of procedures, such as increasing biomass and root growth, supplying nutrients, and increasing the capacity of nutrient uptake when applied to plants, seeds, or soil [116]. Microbial biostimulants have numerous positive functions associated with the solubilization, uptake, primary and secondary metabolism, and translocation of macro- and micro-nutrients, which induce phytochemical accumulation, the development of robust root systems for foraging surrounding soil, and the improvement of photosynthetic activity to promote growth, increase nutrient use efficiency, and stimulate antioxidant defense systems to decrease the oxidative stress burden [117]. Recently, biostimulants related to living microorganisms have attracted the attention of both academics and industry professionals for the simple reason that the growth and development of a plant can be increased more easily in the field [118,119]. These microbes are effective enzyme producers and are potential alternatives to formulating beneficial microbial consortia that can be used in tandem with gelatin to increase biostimulant activity [120]. Microorganisms acting as biostimulants mostly belong to beneficial fungi groups consisting of arbuscular mycorrhizal fungi and free-living bacteria [116,121,122]. Many parameters are responsible for the development of microbial inoculants as biofertilizers, such as the variety of plants [123,124] and the compatibility with different types of soil, chemical fertilizers, and environmental conditions [123,124]. The activity of microbial inoculants is mainly affected by root exudates (extracellular secretions by plants) and they also act as a substrate for the formation of biologically active substances [125]. AMF play a significant role in stimulating plant growth via different mechanisms: (i) increasing the uptake of water, as AMF boost the surface area of the root so that the plant can easily take up water; (ii) modifications of root architecture; (iii) availability of nutrients, particularly phosphorus, under nutrient-deficient conditions; (iv) alterations in enzymatic and physiological activities, especially for plants that are active in antioxidative responses; and (v) induction of ABA plant hormones, which are mostly involved in stress conditions [126,127]. Polyhydroxyalkanoates (PHAs) are selected as a carbon source for abundant microbial primary degraders [128].

## 4. Mechanisms of Microbial Biostimulant Action

Various notable protective mechanisms in the utilization of microbial biostimulants under different stresses are osmolite production, phytohormone level modulation, increased antioxidant activity, and secretion of extracellular polymeric substances (EPS) under water stress such as flooding and drought [129–131]; phytohormone level modulation, emission of volatile organic components, ice nucleation activity antagonism, delay of senescence, and osmo and thermal protection under thermal stress such as freezing and extreme heat [132–135]; increased soil exploration and mineral nutrient solubilization under nutrient stress [136,137]; and induced system resistance, direct antagonism with pathogens, and phytohormone level modulation under biotic stress [138–142]. Bozhinova [143] reported that the application of the microbial inoculant Europlus® and the protein hydrolysate (PH) Trainer® increased the yield of tobacco by 5.5% and 6.7%, respectively, in comparison with untreated plants. Bozhinova [143] also noted that the application of the microbial inoculant Europlus boosted N, P, and K concentrations in leaves, and the concentrations of Ca, Mn, Zn, and Cu were slightly higher than the control treatment when the microbial inoculant Europlus was applied, and concluded that the use of microbial biostimulants in sustainable tobacco farming was effective in boosting the yield of oriental tobacco and the quality of cured leaves. AMF can increase dry weight, photosynthesis, and seed fresh yield of *Glycine max* under drought stress [144]. Under drought stress, *Gigaspora decipiens* and *Glomus mosseae* can boost the chlorophyll content and growth of *Triticum aestivum*, and *Rhizophagus intraradices* can improve levels of Mg, Zn, Cu, and F in grains of *Triticum durum* [145,146]. Under drought stress, *Rhizophagus intraradices* can increase N, P, and K uptake in *Zea mays* [147]; *Funneliformis geosporus* BEG11 can increase water use efficiency in *Fragaria ananassa* [148]; *Glomus deserticola* can improve the level of proline and the number of leaves in *Antirrhinum majus* [149]; and *Gigaspora gregaria* can increase mineral levels and decrease the level of proline in *Vigna subterranea* [150]. Amid water shortage, *Paraglomus occultum* can increase the rate of water absorption and the length of the hypha in *Poncirus trifoliata* [151]; *Rhizophagus irregularis* can increase the conductivity of stomata and dry matter of shoot in *Digitaria eriantha* [152]; and *Glomus* species can improve the osmotic potential adjustment in *Ipomoea batatas* [153] and increase water uptake and phenolic, metabolite, and glutathione levels in *Saccharum arundinaceum* [154]. Under salinity stress, *Rhizophagus irregularis* can increase the fresh weight of shoots and roots and the number of leaves of *Solanum lycopersicum* [155], *Claroideoglomus etunicatum* can increase the conductivity of stomata and the level of soluble sugars in *Aleurites moluccanus* [156], and *Claroideoglomus etunicatum* can boost the dry mass of shoots and roots as well as the conductivity of stomata in *Aeluropus littoralis* [156]. Under cold stress, *Rhizophagus irregularis* and *Funneliformis mosseae* can increase photochemical reactions, decrease the damage in the membrane, and activate the antioxidant defense system in *Solanum melongena* [157]. *Rhizophagus irregularis* can improve the plant photosynthetic efficiency of *Solanum lycopersicum* under heat stress [158]. AMF containing *Rhizoglomus irregulare* and *Funnerliformis mosseae* can significantly influence the growth, productivity, and nutraceutical and nutritional quality of tomato cultivars, as AMF increase the biosynthesis and nutrient uptake of notable molecules involved in cellular pH and oxidative stress [159]. Six single strains (*Acaulospora laevis*, *Acaulospora scrobiculata*, *Gigaspora gigantea*, *Entrophospora colombiana*, *Glomus manihotis*, and *Scutellospora heterogama*) and a mixture of AMF (*Glomuss mosseae*, *Glomus manihotis*, and *Glomus gigantea*) recultured with *Chloris gayana* increased survival, improved tolerance against stress, and boosted the nutritional status, relative water content, photosynthetic rate, and the contents of P, N, Mg, and Fe in *Vitis vinifera* cv. Pusa Navrang [160]. *Rhizophagus irregularis*, *Pseudomonas fluorescens*, and *Funneliformis mosseae* application lead to an increase in APX and GPX enzyme activities, an enhancement of plant growth parameters, the alleviation of water deficit damage, a decrease in $H_2O_2$ and lipid peroxidation, and an improvement of drought tolerance in *Cupressus arizonica* Green [161]. The utilization of *Glomus mosseae*, *Rhizobium leguminosarum*, and *Arthrobacter protophormiae* induced a decrease in proline content and lipid peroxidation, an improvement of plant weight, an enhancement of nutrient uptake, an increase in pigment content, and the alleviation of salt stress of *Pisum sativum* under salt stress [162]. The application of *Acaulospora* sp., *Claroideoglomus etunicatum*, *Rhizobium* sp., and *Burkholderia* sp. led to an improvement in the absorption of chemical fertilizers and an increase in the wood yield of *Schizolobium parahyba* var. *amazonicum* under salt stress [163]. *Bacillus megaterium*, *Rhizophagus irregularis*, and *Frateuria aurantia* induced an enhancement of nutrient uptake responses, an improvement of low-mobility nutrient uptake such as Zn and Ca, an enhancement of total microbial biomass and microbial metabolism, an increase in plant growth, and an increase in gluten quality of *Triticum aestivum* L. under salt stress [164]. Miceli et al. [165] reported that the application of microbial biostimulants altered seedling growth and responses to different stresses and had a growth-promoting impact on the unstressed seedlings, increasing dry and fresh biomass accumulation, leaf area, and leaf number, being an appropriate option for improving the salinity tolerance of seedlings, particularly under salinity stress.

## 5. Case Studies and Practical Application of Microbial Biostimulants

There are two types of mechanisms of action of microbial biostimulants. For direct mechanisms, it has been reported that microbes are active in the synthesis of components that can boost the adsorption of nutrients, while indirect mechanisms refer to siderophore production, zinc solubilization, phosphorus solubilization, indole acetic acid biosynthesis, antioxidant enzyme production, ammonia and hydrogen cyanide production, biological nitrogen fixation, and phytohormone production [166,167]. Microbial biostimulants can alleviate the adverse effects of environmental stresses by producing hormone-like stimulants with positive impacts on plant growth and final yield [168]. Molecular processes that have roles in the interactions of plants and microorganisms which may lead to the biosynthesis of secondary metabolites are the protective impacts of microbial biostimulants on plants against different stressors [169]. The production of protective molecules is obtained via the shikimate pathway which may consist of the enzyme phenylalanine ammonia lyase (PAL) for the production of phenylpropanoids after microbial eliciting [170], which may have significant effects to adjust the pressure from external factors, called induced systemic resistance (ISR) [171]. The main mechanisms targeted by microorganisms according to various biostimulants can target shoot targets such as stomatal regulation and xylem hydraulic conductance, and root targets such as root zone water availability, root ethylene and auxin levels via ROS scavenging, membrane stability, and osmoprotection [172]. Biostimulants can influence plant phenotype, cellular level, and molecular level. The impact on plant phenotype consists of improved shoot and root growth, improved flowering, higher yield parameters, higher leaf number and vigor, better abiotic stress tolerance, better fruit nutritional quality, increased relative water content, increased stomatal conductance, and higher nutrient acquisition uptake and transportation. The effects at the cellular level include increased antioxidant activity, improved primary and secondary metabolisms, increased chlorophyll content, and higher photosynthetic rates, and those at the molecular level entail increased gene expression of SOD, CAT, APX, nutrient transporters, and stress-related genes [173]. Microbial biostimulants can alleviate abiotic stress, balance plant hormone levels, regulate indole-3-acetic acid (IAA), regulate cytokinins, regulate gibberellins, regulate abscisic acid, produce ACC deaminase, facilitate nutrient availability, modify root biomass and morphology, induce antioxidant enzyme syntheses in crops, accumulate osmolytes in crops, induce gene resistance to drought and salt stress, improve crop organic and inorganic pollutant toxicity tolerance, mediate increased nutrient availability (N, P, K, Fe, and Zn), improve crop cold tolerance, increase crop heat tolerance, and boost crop waterlogging tolerance [174–179]. It has been reported that microbial-based biostimulants boosted plant yield, Cu and Ca, and isochlorogenic acid components, and the influence of biostimulants on the functional and nutritional quality of lettuce was mainly independent of water availability [180]. Microbial biostimulants can significantly improve carbohydrate, protein, sugar, K, Ca, Mn, Zn, and Fe content in chickpea seeds and plants, and the microbiome analysis indicated a positive correlation with soil health and yield [181]. Mrid et al. [182] reported that microbial-based biostimulants can positively influence cropping systems via numerous mechanisms and can increase nutrient uptake and use efficiencies, suppress phytopathogens infection, boost root system development, reduce heavy metal toxicity, and improve crop growth and yield. Natural biostimulants can increase tolerance against abiotic stresses in agricultural and horticultural crops [183,184] and increase crop productivity under environmental stress [185,186]. Alfonzetti et al. [187] showed the potency of microbial biostimulants to positively influence native plant emergence and growth, but the exact impacts are linked to the plant species, type of amendment, and the characteristics of the planting site soil. Microbial biostimulants can increase ornamental plant growth during production and increase crop performance under abiotic stress, and appropriate microbes can also produce secondary metabolites like volatile organic components to improve plant growth. Aside from the direct impacts on molecular procedures, the effects of microbial biostimulants are related to morphological alterations such as changes in root morphology after inoculation with AMF and the increase in root surface, which can both increase the uptake of nutrients and water, thus helping plants cope with the negative impacts of stressors [188]. It has also been reported that the inoculation of water-stressed plants with *Penicillium* sp., *Phoma glomerata, Glomus intraradices, Exophiala* sp., and *Paecilomyces formosus* may lead to greater soil exploration by roots or fungal hyphae with meaningful improved root conductivity [189,190].

## 6. Arbuscular Mycorrhizal Fungi (AMF)

AMF, favorable microorganisms extensively distributed in nature, can establish a symbiotic relationship with most terrestrial herbs. AMF belong to 11 families, 25 genera, and nearly 250 species [191]. Arbuscular mycorrhizal (AM) symbiosis is the widespread mutualistic relationship between fungi and plants and plays an important role in nutrient exchange, ecosystem sustainability, the enhancement of plant stress resistance, and the development of host plants [192–196]. AMF increase the C allocation of the host plant below ground, which can form a microbial community composition [197]. AMF belong to the phylum *Mucoromycota* and subphylum *Glomeromycotina*. AMF are available in our natural environment and are helpful in several ways; they play an important role in increasing plant nutrition acquisition, improving plant tolerance and resistance to stresses, and increasing soil fertility and structure, and have several beneficial applications in agriculture [198–200]. Symbiosis can increase host plant nutrition because plants can absorb nutrients not only through mycorrhizal pathways but also through their own pathways. Besides that, AMF can also influence nutrient availability by altering soil physicochemical characteristics, microbial communities, and nutrient cycling [201–203]. AMF have notable effects on herbaceous plant element stoichiometry, such as plant C:N:P stoichiometry, which depends on plant and fungal functional group identities and soil nutrient availability [204].

AMF promoted the utilization and accumulation of Ca in apple rootstock (*Malus robusta*) by adjusting the expression levels of genes associated with these pathways. GO and KEGG pathway analysis showed that gene expression changes in different critical gene families, such as auxin response (*MdGH3, MdAux/IAAs*, and *MdSAUR*), TCA cycles (*MdCS, MdMDH*, and *MdACO*), phosphate transporters (*MdPHT1;1, MdPHT1;10*, and *MdPHT1;3*), and $Ca^{2+}$ signal transduction pathways (*MdCa*$^{2+}$*/ATPase, MdTPC1, MdCML*, and *MdCDPK*), showing that apple stimulates the expression of genes related to auxin synthesis, organic acid secretion, and calcium transporters and channels, thus promoting the growth of apple root and improving the secretion of organic acids, which may lead to an increase in calcium effectiveness in soil [205]. Uptake of ZnO nanoparticles in barley was increased by AMF, possibly related to the mutualistic connection between the root and the fungi, which promotes water absorption because of the more efficient property of the rhizosphere [206]. The combination of humic substance and AMF remarkably increased plant biomass [207]. The impacts of *Rhizobium* and AMF inoculation were investigated on the growth and yield of *Lablab purpureus*, *Vigna unguiculata*, and *Mucuna pruriens*, and in comparison to single inoculation and uninoculated forages, the dual inoculation with AMF and *Rhizobia* improved the growth and biomass yield of forage legumes, and dual inoculation could be an optimal alternative in attempts to decrease the overreliance of chemical fertilizers on forage production [208]. The effects of mycorrhizal association on plant and root ecosystems are presented in Figure 2.

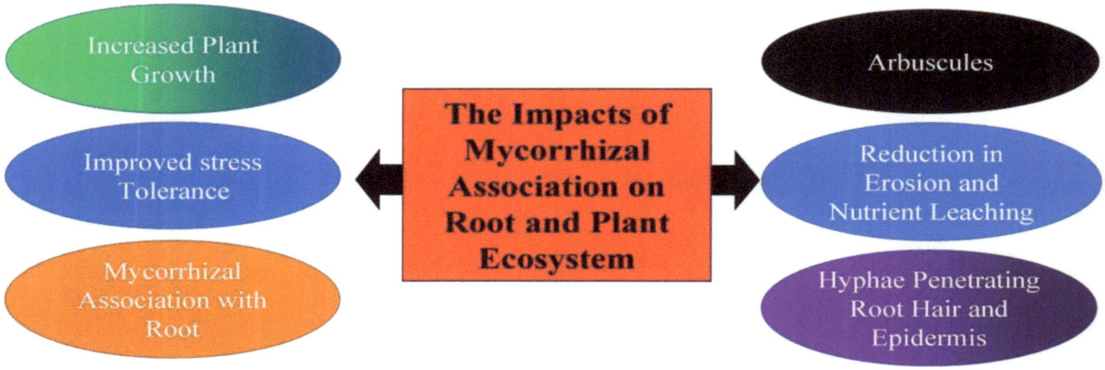


**Figure 2.** The influences of mycorrhizal association on root and plant ecosystem.

When AMF are inoculated in a plant, they increase the plant's tolerance against different stresses such as heavy metals, drought, and high temperatures. AMF form spores and hyphae in the rhizosphere, while inside the root tissues, they shape arbuscular, vesicles, and hyphae to improve the accessibility of plant roots to large soil surface areas by hyphal network knowledge with roots of plants, therefore increasing growth in the plant [209]. AMF synergistically improve plant P nutrition, and feedstock origin determines P accessibility and, as a consequence, mycorrhizal performance [210]. AMF inoculation may lead to the promotion of leaf chlorophyll content, shoot height, leaf photochemical efficiency, RWC, turf quality, and root and shoot biomass, as well as enhanced melatonin (MT) content through the upregulation of the MT biosynthesis genes and the reduction in the levels of malondialdehyde (MDA), electrolyte leakage (EL), chlorophyll catabolic genes (CCGs), hydrogen peroxide ($H_2O_2$), and senescence-associated genes (SAGs) to decrease heat stress in perennial ryegrass [211]. AMF inoculation increased gibberellin (GA), internal MT, and cytokinin (CTK) levels while reducing ABA levels in heat-stressed plants; moreover, the MT biosynthesis genes (*LpASMT1*, *LpTDC1*, *LpCOMT1*, *LpTDC2*, and *LpASMT3*), the abscisic acid (ABA) catabolic gene, the CTK biosynthesis genes, the GA biosynthesis genes, and the associated signal transduction response transcription factors (TFs) (type-B ARRs) showed increased levels, whereas the expression of the GA to ABA exerted decreased levels after MT application and AMF inoculation in heat-stressed plants [211]. AMF can increase earlier flowering of plants, promote fruit and flower production, prolong the total duration of flowering, and increase seed yield, showing potential in promoting seedling growth and seed germination [212].

AMF increased the plant height, root weight, root length, and stem weight of lettuce, and the functional composition profiles showed that several functions were increased, including cell motility and environmental adaptation, and AMF had a significant influence on the lettuce endophytic bacterial network community and structure function [213]. AMF with high Cu components provided adaptive mechanisms for the plant growth and survival of grapevine rootstock; in greenhouse conditions, stomatal conductance and transpiration improved with Cu additions, influencing plant growth [214]. AMF could inhibit the uptake of Hg, particularly methyl-Hg in grains of rice, as it may lead to Hg transfer in the non-edible parts of rice, such as the leaf and stem, and promote the growth indexes and micro-indexes of rice, and it can be a practical remediation technology for soil heavy metal pollution [215]. AMF also responds to Cd stress in kenaf; inoculation with AMF could be applied to kenaf to enhance the removal of Cd from soil in mining areas by phytoremediation, as AMF can adjust the expression levels of key kenaf genes, such as *Hc.AKR*, *Hc.GH3.1*, and *Hc.PHR1*, thus playing an important role in enhancing kenaf's Cd tolerance [216].

Micro-proton-induced X-ray emission was utilized to determine element localization and indicated that AMF improved the nutrient uptake by wetland plants (*Iris pseudacorus*) and suppressed Cr translocation from roots to shoots, which showed that the interaction between plants and AMF could significantly improve the immobilization of high Cr concentrations in semi-aquatic habitats [217]. AMF inoculation and/or legume intercropping uphold the stability of microbial networks by weakening taxonomic interactions and improving modularity under drought events; overall, AMF inoculation causes *Medicago sativa* to need water during a drought event, which worsens soil water deficit and thus may increase interspecific water competition between *Medicago sativa* and *Broussonetia papyrifera*. The results of one experiment revealed that in the roots of tomato plants colonized by AMF, there was a meaningful mutual relationship between the insects feeding on the plant and the fungal species (*Rhizophagus intraradices*, *Funneliformis mosseae*, *R. irregularis*, and *Glomus iranicus*) [218].

AM fungi (*Glomus* spp.) play a role in upregulating plant growth and nutrition status in Fe-ore tailings in technosols of host plants like sorghum, providing the basis for the involvement of AM fungi in the eco-engineered pedogenesis of iron ore tailings [219]. In the ethanolic extracts of leaves of *Anadenanthera colubrina* (Vell.) Brenan, seedlings had more saponins and flavanols and showed significant antioxidant activity when inoculated with AMF (*Gigaspora albida* and *Acaulospora longula*) [220,221]. AMF significantly improved the biomass and nutrient levels of *Iris tectorum* and reduced the constituents of Cr in soils; furthermore, it can boost the abundance of functional genes related to nutrient cycling (N, P) in rhizosphere microbial communities, increase the abundance of functional genes associated with heavy metal resistance and transporters in the rhizosphere microbial community, and increase the complexity and stability of the rhizosphere microbial community [222]. AMF reduce the plant uptake and bioavailability of Cd and As, and the reduction in As and Cd accumulations induced by AMF depends on the plant family [223]. It has also been reported that AMF–*Glomus versiforme* (Gv) symbiosis significantly increased P uptake, plant growth, and photosynthesis in upland rice, and Gv inoculation decreased the expression of *Nramp5*, thus reducing Cd absorption, transfer, and accumulation in upland rice and enhancing the activities of catalase (CAT) and peroxide (POD) in Gv-inoculated upland rice [224]. *Glomeraceae* and *Paraglomeraceae* are the dominant fungi affecting the available P content. Expanding *Glomeraceae* and *Paraglomeraceae* and applying them to *Eucalyptus* plantations may increase soil P availability, and the structure of the AMF community may be a sensitive indicator of the quality of soil-available P in *Eucalyptus* plantations [225]. AMF reduced the toxicity risk of Pb by accumulating Pb in fungal structures, and both flooding and Pb stress reduced the AMF diversity but not the abundance [226].

Inoculation with indigenous AMF such as *Glomus* sp. and *Paraglomus* sp. boosted the biomass of plants and promoted root growth under replanting conditions, and it can also induce the activity of chitinase and β-1,3-glucanase and increase the resistance of apple rootstock by upregulating root reactive oxygen species levels and the antioxidant system under replanting conditions [227]. It is suggested that mycorrhizal fungi can reduce the detrimental impacts of salt stress on *Ligustrum vicaryi* plants, and the mediation of $Ca^{2+}$, $Zn^{2+}$, N, $Mg^{2+}$, and soluble protein components could be the basic mechanism underlying salt tolerance in mycorrhizal plants [228]. AMF and root hairs are functionally redundant in maize, and AMF had a greater effect on microbial structure than root hairs, as well as a greater role in microbial P mineralization than root hairs of maize [229]. Activities of 4-coumaroyl-CoA ligase (4CL), chalcone isomerase (CHI), and phenylalanine ammonia lyase (PAL) and expression levels of PtPAL1 and Pt4CL were induced by *Funneliformis mosseae* inoculation under water stress, and AMF plants showed higher scavenging activity of hydroxyl radical and superoxide ($O_2^-$) by root flavonoid extracts under water stress, together with lower levels of $O_2^-$ and hydrogen peroxide and the degree of membrane lipid peroxidation in comparison with non-AMF plants, as well as increased flavonoid synthesis of trifoliate orange for reducing oxidative damage under water stress [230]. The single inoculation of *Claroideoglomus etunicatum* and *Rhizophagus clarus* promoted the accumulation of nitrogen in the aerial part of cowpea cultivars, and plants inoculated with *C. etunicatum* had similar or superior responses to plants fertilized with NPK or P [231]. Wheat inoculated with *Glomus intraradices* had maximum wheat yield and growth, which mechanically resulted from a higher rhizosphere colonization level, photosynthetic rate, and water use efficiency under drought stress [232].

Fungal inoculations partly improved fruit quality and mineral element components, depending on the fungi species, while the cultured mycorrhiza-like fungus *Piriformospora indica* relatively replaced AMF in citrus plants [233]. Inoculation with AMF provided good dry weight gain in lemon balm (*Melissa officinalis* L.), and notably contributed to high essential oil yield [234]. The mixed AMF inoculation in chamomile cultivation promoted both plant productivity and the quality of flower heads, especially regarding the component of phenolic compounds [235]. AMF inoculation had a positive effect on the yield of raspberry (*Rubus idaeus* L.) [236], while the combined implementation of biochar with AMF improved the colonization potential of AMF and significantly increased the photosynthetic potential of *Tamarindus indica* by increasing the contents of chlorophyll and carotenoids [237]. Different species of filamentous endophytic fungi, such as *Trichoderma*, are capable of controlling the pathogens *Xylella fastidiosa* and *Pseudomonas savastanoi* via the production and release of secondary metabolites; they are also effective against *Oomycetes* sp. and *Colletotrichum* sp. [238]. The combined application of $GA_3$ (Gibberellic acid) and AMF (*Rhizophagus irregularis*) decreased growth impairment under salinity conditions by upregulating the hormonal balance of plants. The utilization of AMF was able to increase the productivity of sweet basil (*Ocimum basilicum*) plants under salinity conditions, and mycorrhizal inoculation significantly promoted water use efficiency and chlorophyll content under salinity stress [239]. AMF formation significantly boosted the high temperature tolerance of lettuce, a finding that could be related to the PSII system's protection from damage under high temperatures [240]. Furthermore, it has been suggested that mycorrhizal symbiosis decreased the $Na^+$ and $Cl^-$ content and improved the relative water content (RWC), the total fresh and dry weight, and the photosynthetic activity of olive plants [241]. Mango (*Mangifera indica* L.) root stocks reacted to AMF inoculation in the nursery and also in the field with improved nutrient uptake, plant growth, and yield [242].

Rouphael et al. [243] concluded that the improvement in the biomass of crops after the application of two beneficial fungi, namely arbuscular mycorrhizal fungi and *Trichoderma koningii* TK7, could be connected to the modulation of the multilayer phytohormone interaction network, as well as a potential improvement in nitrogen use effectiveness via the glutamine oxoglutarate aminotransferase (GS-GOCAT) system. Hashem et al. [244] reported that the negative effects of salt stress in cucumber were ameliorated by AMF inoculation, which improved the activity of antioxidant enzymes that scavenged ROS and protected plant tissues from dehydration stress, including ascorbate peroxidase, catalase, and superoxide dismutase, and enhanced plant biomass and the synthesis of proline, pigments, and glycine betaine. Shekoofeh et al. [245] reported that AMF inoculation protected *Ocimum basilicum* plants from salinity stress by promoting water use efficiency, and increased chlorophyll synthesis and mineral uptake. Other examples of modes of action of AMF include the increased antioxidant activity and the accumulation of osmolytes [246], the adjusting of proline biosynthesis [246,247], and the accumulation of K, Ca, and Mg, which increased chlorophyll production and promoted the activity of enzymes [248,249]. Regarding the mitigating impacts of AMF against salinity stress, Estrada et al. [250] concluded that AMF inoculation restricted both uptake and accumulation of Na by adjusting the expression levels of SOS1, AKT2, and SKOR genes in roots, which allowed them to retain the homeostasis of $Na^+$ and $K^+$. The recent findings in omics science have also helped show that microbial biostimulants' application involves great alterations in secondary and primary metabolites such as amino acids, phenolic acids, lipids, and tricarboxylic acid (TCA) intermediates, as well as alterations in protective mechanisms against stress that involve redox homeostasis, the stabilization of cell membranes, osmoregulation, the production of energy via amino acid degradation, and the increased expression of stress-related genes [150,251]. Qiu et al. [252] reported that the mitigation impacts of AMF increased with increasing AMF root colonization rate, and the mitigation impacts of AMF reduced with increasing soil P and N availability. Hu et al. [253] found that the symbiotic relationship between plant roots and AMF could affect the translocation and accumulation of pharmaceuticals and personal care products (PPCPs) in plants, and that substrate type can affect the function.

The application of humic substances and AMF treatment led to a reduction in carvone content compared to treatment with humic substances alone, and it was suggested that AMF inoculation combined with a high P rate promoted essential oil content, whereas AMF inoculation combined with a high P rate and humic substance addition decreased the carvone content in *Lippia alba* (Mill.) N.E.Br. [254]. AMF can shorten the litter nutrient cycle, which is of great significance for the absorption, storage, and utilization of plant nutrients in the Songnen grassland ecosystem [255]. It has been reported that mycorrhizal colonization increased apple drought tolerance by improving gas exchange capacity, promoting chlorophyll fluorescence parameters, increasing scavenging of reactive oxygen species, creating a greater osmotic adjustment capacity, and using mitogen-activated protein kinase (MAPK) signals for interactions between AMF and their apple plant hosts [256]. Wang et al. [257] showed that AMF increased soil $N_2O$ emissions from legume systems by increasing P acquisition for biological $N_2$ fixation, suggesting that AMF benefit crop production for higher yield and fewer $N_2O$ emissions in legume systems. The AMF *Gigaspora margarita* KKU-BH-01 can effectively increase the growth and aid eucalyptus seedlings to survive leaf blight disease (*Cylindrocladium quinqueseptatum*), as AMF induced plant chitinase and β-1,3-glucanase enzymes, resulting in improved disease resistance in plants [258]. Mycorrhizal oregano and mint plants had higher dry weight and essential oils, and the *Glomus etunicatum* strains were more effective than the *Glomus lamellosum* strain, and mycorrhizal oregano plants had higher dry weight than mycorrhizal mint plants [259]. Lam and Lai [260] reported that AMF treatment increased water spinach (*Ipomoea aquatica* Forsk.) growth and reduced the accumulation of Ni but not Cd. *Glomus mosseae* and *Glomus deserticola* increased shoot length, dry weight, chlorophyll quantity, and total N, P, and K components in *Eucalyptus globulus* shoots [261]. Gheisari Zardak et al. [262] concluded that the application of AMF could be important in the cultivation of medicinal plants under semi-arid and arid conditions. The most important impacts of AMF on different plants are shown in Table 1.

**Table 1.** The most important effects of AMF on plants.

| Type | Plant | Plant Family | Effects | Reference |
|---|---|---|---|---|
| **Arbuscular Mycorrhizal Fungi (AMF)** | | | | |
| | Alfalfa (*Medicago sativa* L.) | Fabaceae | AMF inoculation may increase photosynthetic rates and the content of sugars in leaves. | [263] |
| | Asparagus (*Asparagus officinalis* L.) | Asparagaceae | AMF improved P efficiency via increased P uptake and optimal growth by adding AMF to suitable P fertilization. | [264] |
| | Asparagus (*Asparagus officinalis* L.) | Asparagaceae | AMF inoculation can enhance adaptation to salinity stress. | [265] |
| | Barrel medic (*Medicago truncatula* Gaertn.) | Fabaceae | AMF inoculation may alleviate Pb toxicity by improving the transport of sucrose from shoot to root, increasing the cleaving sucrose in roots, and increasing minor amino acid accumulation. | [266] |
| | Barrel medic (*Medicago truncatula* Gaertn.) | Fabaceae | AMF in combination with K can play an important role in reducing radiocesium uptake and its subsequent translocation to plant shoots. | [267] |
| | Basket willow (*Salix viminalis* L.) | Salicaceae | Willow AMF promoted the dissipation of soil polycyclic aromatic hydrocarbons (PAHs). Willow AMF increased the content of organic acids beneficial to metabolizing PAHs. | [268] |
| | Bishop's flower (*Ammi majus*) | Apiaceae | Its application can induce the accumulation of phyto-molecules, coumarin, which might increase its medicinal and pharmacological applications. | [269] |

**Table 1.** *Cont.*

| Type | Plant | Plant Family | Effects | Reference |
|---|---|---|---|---|
| Arbuscular Mycorrhizal Fungi (AMF) | | | | |
| | Black cumin (*Nigella sativa* Linn.) | Ranunculaceae | Colonization can increase relative water content (RWC), Chl b content, and micronutrient uptake. | [270] |
| | Black locust (*Robinia pseudoacacia* L.) | Fabaceae | AMF associated with black locusts may be useful to be used for improving the phytoremediation of Cd under elevated $CO_2$ ($ECO_2$). | [271] |
| | Cacao (*Theobroma cacao* L.) | Malvaceae | It can improve the overall growth and can positively increase the yield of cacao plants in acidic soils. | [272] |
| | Carob tree (*Ceratonia siliqua* L.) | Fabaceae | The used AMF *inocula* stimulated significantly the height of carob tree as well as the aerial dry weight. | [273] |
| | Castor bean (*Ricinus communis* L.) | Euphorbiaceae | AMF could protect castor bean against drought and salt stress by improving leaf exchanges and photosynthetic capacity and altering concentrations of metabolites. | [274] |
| | Chickpea (*Cicer arietinum* L.) | Fabaceae | AMF inoculation increased the final yield of chickpea. | [275] |
| | Cocoa (*Theobroma cacao* L.) | Malvaceae | AMF application improved the physical (cell wall turgor, root growth) and biochemical (proline, polyamines, enzymatic) characteristics of cocoa seedlings to reduce water stress. | [276] |
| | Durum wheat (*Triticum turgidum* subsp. *durum* (Desf.)) | Gramineae | AMF can decrease water deficiency in cultivars, resulting in the up- and downregulation of many amino acids, phenylpropanoids, lipids, alkaloids, and hormones. | [277] |
| | Fenugreek (*Trigonella foenum-graecum* L.) | Fabaceae | AMF inoculation was effective in improving the tolerance of fenugreek to salinity. | [278] |
| | Fig (*Ficus carica* L.) | Moraceae | Fig was positively responsiveness to mycorrhizal inoculation, and the AMF induced different root architecture models. | [279] |
| | Foxtail millet (*Setaria italica*) | Poaceae | AMF application can decrease heavy metal phytoavailability and post-harvest transfer risks. | [280] |
| | Ginseng (*Panax quinquefolius* L.) | Araliaceae | AMF inoculation can promote plant uptake of N and P by suppressing soil-borne pathogens. | [281] |
| | Italian senna (*Cassia italica* Mill.) | Fabaceae | AMF inoculation increased the chlorophyll, protein, proline, and phenol content and lipid peroxidation. | [282] |
| | Lemongrass (*Cymbopogon citratus*) | Gramineae | AMF inoculation modified lemongrass metabolism with consequences on the essential oil component, composition, and antioxidant properties during growth. | [283] |
| | Maize (*Zea mays* L.) | Gramineae | Inoculation with AMF affects plant P constituent with or without P fertilizers addition.<br>Inoculation may improve P availability in P-unamended and -amended soil. | [284] |
| | Myrrh (*Commiphora myrrha*) | Burseraceae | Mycorrhizal seedlings had higher biomass than non-mycorrhizal seedlings.<br>Mycorrhizal seedlings had higher nutrient concentrations than non-mycorrhizal seedlings. | [285] |
| | Nemesia (*Nemesia × hybridus*) | Scrophulariaceae | AMF inoculation can improve flower yield and growth quality of nemesia.<br>AMF can increase the response of the plant to irrigation with treated wastewater and reduce the cost associated with using other water sources. | [286] |
| | Nitre-bush (*Nitraria tangutorum* Bobr.) | Zygophyllaceae | The combination of AMF and PGPR significantly increased mycorrhizal colonization, promoted biomass accumulation, boosted morphological development, and improved photosynthetic performance, stomatal adjustment ability, and the exchange of water and gas. | [287] |
| | Oregano (*Origanum vulgare*) | Lamiaceae | The synchronous application of AMF and atmospheric $CO_2$ ($eCO_2$) promoted the accumulation of the majority of the detected sugars, amino acids, organic acids, phenolic acids, unsaturated fatty acids, and flavonoids.<br>Both AMF and $eCO_2$ treatments significantly promoted the growth and photosynthesis of oregano plants.<br>Both AMF and $eCO_2$ acted synergistically in improving the antioxidant capacity and anti-lipid peroxidation activity of oregano. | [288] |
| | Rice (*Oryza sativa* L.) | Gramineae | AMF could inhibit the uptake of Hg, particularly methyl-Hg in grains of rice.<br>AMF caused Hg transfer in the non-edible parts of rice, such as leaf and stem.<br>AMF improved the growth index and micro-indexes of rice. | [289] |
| | Rice (*Oryza sativa* L.) | Gramineae | The biochar combined with AMF improved soil nutrient availability and root growth strategy, and then promoted the nutrient absorption capacity of rice. | [290] |
| | Rice (*Oryza sativa* L.) | Gramineae | AMF may have a significant function in wetlands. | [291] |
| | Ryegrass (*Lolium multiflorum*) | Gramineae | AMF had a positive influence on the plant shoot biomass and the contents of P, N, Ca, K, and Mg in plants. | [292] |
| | Prickly pear (*Opuntia ficus-indica*) | Cactaceae | AMF promoted physiological and biochemical factors, and led to a decline in malondialdehyde (MDA) and hydrogen peroxide ($H_2O_2$). | [293] |
| | Seaberry (*Hippophae rhamnoides*) | Elaeagnaceae | AMF had a positive influence on the final yield. | [294] |
| | Sorghum (*Sorghum bicolor* L. Moench) | Gramineae | The total phenolic, carotenoid, flavonoid, and tannin concentrations were significantly higher in AMF–sorghum grain for all cultivars.<br>The total phenolic, carotenoid, flavonoid, and tannin concentrations were significantly higher in AMF–sorghum grain for all cultivars. | [295] |
| | Sunflower (*Helianthus annuus* L.) | Asteraceae | AMF community has greater efficiency in promoting sunflower development and mycorrhizal colonization. | [296] |
| | Sunflower (*Helianthus annuus* L.) | Asteraceae | The combination of AMF and biochar increased antioxidant enzyme activity, nutrient content, osmoprotectants, and relative water content. | [297] |
| | Tomato (*Solanum lycopersicum* L.) | Solanaceae | AMF application has a significant influence on manganese, total nitrogen, and hydrophilic phenol components in the fruit. | [298] |
| | Wheat (*Triticum aestivum* L.) | Geramineae | AMF mitigated earthworm-induced $N_2O$ emissions from upland soil in a rice-rotated wheat farming system. | [299,300] |
| | Zucchini squash (*Cucurbita pepo* L.) | Cucurbitaceae | Inoculation may lead to better nutritional status of P, N, K, Mg, Ca, Zn, B, and Fe and low Al accumulation. | [301] |
| *Glomus mosseae, Glomus etunicatum* | Achnatherum (*Achnatherum sibricium* L.) | Gramineae | Simultaneous infections of both fungi significantly increased total phenolic concentrations. | [302] |
| *Glomus mosseae* | Alfalfa (*Medicago sativa* L.) | Fabaceae | It significantly enhanced Cd uptake by the roots of alfalfa under ET.<br>It has no significant effect on iron-regulated transport 1 (*IRT1*) and natural resistance-associated macrophage protein 1 (*NRAMP1*) gene expression. | [303] |

**Table 1.** *Cont.*

| Type | Plant | Plant Family | Effects | Reference |
|---|---|---|---|---|
| **Arbuscular Mycorrhizal Fungi (AMF)** | | | | |
| *Claroideoglomus etunicatum* | Alfalfa (*Medicago sativa* L.) | Fabaceae | The AMF alone or in combination with Si can alleviate salinity stress in alfalfa. | [304] |
| *Funneliformis mosseae* | Apple (*Malus domestica* Borkh.) | Rosaceae | The synergistic effect of dopamine and AMF improved apple salt resistance, and overexpression of MdTYDC promoted AMF symbiosis. | [305] |
| *Rhizophagus irregularis* | Barley (*Hordeum vulgare* L.) | Gramineae | The inoculation resulted in improved grain and straw Zn concentrations, especially at low soil Zn concentrations. AMF may be more appropriate for improving the quality of barley grain in terms of Zn concentrations, rather than improving yield. | [306] |
| *Rhizophagus intraradices* | Barrel medic (*Medicago truncatula* Gaertn.) | Fabaceae | Mycorrhizal colonization had little effect on root or shoot cesium (Cs) concentrations. | [307] |
| *Glomus caledonium*, *Glomus versiforme* | Bashfulgrass (*Mimosa pudica* L.) | Fabaceae | AMF inoculation significantly increased root mycorrhizal colonization rates and soil acid phosphate activities. | [308] |
| *Glomus mosseae*, *Gigaspora gigantea* | Carrot (*Daucus carota* L.) | Apiaceae | AMF inoculation can lead to successful carrot production under salinity stress. | [309] |
| *Glomus sinuosum*, *Paraglomus occultum* | Cassava (*Manihot esculenta Crantz*) | Euphorbiaceae | It can improve the final yield of cassava. | [310] |
| *Gloms rubiforme*, *Acaulospora scrobiculata*, *Glomus etunicatum*, *Glomus rubiforme*, *Acaulospora tuberculata* | Cassava (*Manihot esculenta Crantz*) | Euphorbiaceae | The inoculation had positive effects on cassava. | [311] |
| *Rhizophagus irregularis*, *Paraglomus* sp. | Castor bean (*Ricinus communis* L.) | Euphorbiaceae | Shoot Cr concentration doubled in non-AMF versus AMF plants; the content was similar. AMF vesicle percentage negatively correlated with Cr root concentration. | [312,313] |
| *Glomus tortuosum* | Chicory (*Cichorium intybus* L.) | Asteraceae | AMF, biochar, and N fertilizer application enhanced biomass. | [314] |
| *Glomus tortuosum* | Chicory (*Cichorium intybus* L.) | Asteraceae | AMF and biochar application increased nutrient absorption and reduced Cd absorption. | [314] |
| *Glomus clarum* | Chili (*Capsicum frutescens*) | Solanaceae | The inoculation increased the growth, flowering, and fruit production, and also increased the P uptake significantly. | [315] |
| *Glomus etunicatum*, *Funneliformis mosseae* | Cinnamomum (*Cinnamomum migao*) | Lauraceae | It markedly upregulated antioxidant enzyme activities and osmotic adjustment substances. | [316] |
| *Rhizophagus clarus* | Coarse Mint (*Mentha arvensis*) | Lamiaceae | Inoculation of coarse mint with AMF *Rhizophagus clarus* and a high dose of P boosted plant growth and the essential oil yield, and it increased carvacrol content. | [317] |
| *Acaulospora* sp., *Glomus* sp. | Common bean (*Phaseolus vulgaris* L.) | Fabaceae | The positive impact of co-infection by AMF and rhizobia on plant growth and the total N content of the plants was reported, along with a synergistic influence on the total P content, the number of nodules, and the mycorrhizal rate of the plants. | [318] |
| *Funneliformis mosseae*, *Rhizophagus irregularis* | Common myrtle (*Myrtus communis*) | Myrtaceae | AMF boosted myrtle drought resistance through enhanced water and nutrient supply and stimulation of antioxidant defense. | [319] |
| *Dominikia disticha*, *Claroideoglomus etunicatum*, *Rhizophagus irregularis* | Cowpea (*Vigna unguiculata* (L.) Walp.) | Fabaceae | Inoculation with all AMF led to high aboveground biomass production and accumulation of N as well as increased P content in plants. | [320] |
| *Glomus* Spp. | Cowpea (*Vigna unguiculata* (L.) Walp.) | Fabaceae | The activity of AMF in alleviating Cd stress in pre-flowering cowpea has been proven. | [321] |
| *Funneliformis mosseae* | Cucumber (*Cucumis sativus* L.) | Cucurbitaceae | The enhanced secondary metabolism and integrated transcriptional regulation might play a crucial role in AMF-mediated alleviation of chilling stress in plants. | [322] |
| *Glomus etunicatum*, *Glomus mosseae*, *Glomus versiforme*, *Glomus margarita* | Cucumber (*Cucumis sativus* L.) | Cucurbitaceae | AMF communities increased plant growth, soluble sugar content, chlorophyll content, and root activity. | [323] |
| *Glomus* spp., *Acaulospora* spp. | Cucumber (*Cucumis sativus* L.) | Cucurbitaceae | The AMF consortium could inhibit *Fusarium* wilt of cucumber, and, consequently, showed promising results as a biological control factor in greenhouse agro-ecosystems. | [324] |
| *Pervetustus simplex*, *Claroideoglomus etunicatum*, *Albahypha drummondii*, *Septoglomus xanthium*, *Funneliformis mosseae*, and *Rhizoglomus irregulare* | Date palm (*Phoenix dactylifera* L.) | Arecaceae | Shoot length and stem diameter were significantly higher in treatments augmented with compost and AMF. | [325] |
| *Glomus monosporus*, *Glomus deserticola*, *Glomus clarum* | Date palm (*Phoenix dactylifera* L.) | Arecaceae | All fungi significantly stimulated shoot height and biomass and increased the number of leaves per plant. | [326] |
| *Glomus iranicum* | Date palm (*Phoenix dactylifera* L.) | Arecaceae | It showed increased biomass production, chlorophyll, and mineral nutrient content. | [327] |
| *Claroideoglomus etunicatum*, *Rhizoglomus irregulare*, *Diversispora versiformis* | Eggplant (*Solanum melongena* L.) | Solanaceae | The inoculation is an effective strategy for alleviating cold stress. | [328] |
| *Rhizoglomus irregulare* | Eggplant (*Solanum melongena* L.) | Solanaceae | AMF improved fruit quality by reducing glycoalkaloid concentration and fruit browning potential. | [329] |
| *Gigaspora* gigantean, *Glomus mosseae* | Eggplant (*Solanum melongena* L.) | Solanaceae | Two mycorrhiza fungi affected plant growth indirectly, and in some situations, they reduced the inputs of chemical pesticides in eggplant. | [330] |
| *Glomus mosseae* | Fennel (*Foeniculum vulgare*) | Apiaceae | The mycorrhiza and growth-promoting bacteria (*Azospirillum*) resulted in the highest yields, total carotenoids, and chlorophyll in fennel plants subjected to water deficit stress. | [331] |
| *Funneliformis mosseae* | Grapevines (*Vitis vinifera*) | Vitaceae | The introduction of *F. mosseae* through donor plants is a suitable field inoculation method for grapevines and can help them to better withstand heat waves. | [332] |
| *Rhizophagus irregularis* | Hemp (*Cannabis sativa* L.) | Cannabaceae | AMF increased the heavy metal tolerance of hemp, and they changed Cd chemical forms by changing the composition of low molecular weight organic acids, which in turn affected soil Cd bioavailability. | [333] |
| *Rhizophagus intraradices* | Holy Basil (*Ocimum tenuiflorum* L.) | Lamiaceae | The inoculation increased the productivity of holy basil and boosted the quality of the final products. | [334] |
| *Rhizophagus clarus*, *Claroideoglomus etunicatum*, *Azospirillum brasilense* | Lemon grass (*Cymbopogon citratus* (DC.) Stapf) | Graminaeae | It is concluded that inoculating lemongrass with AMF enhances plant growth and development and modifies the content and essential oil composition. | [335] |
| *Rhizophagus clarus* | Maize (*Zea mays* L.) | Graminaeae | A combination of AMF (*Rhizophagus clarus*) and PGPR (*Bacillus* sp.) could enhance $^{33}$P uptake in maize plants under soil water stress. | [336] |
| *Funneliformis mosseae* | Maize (*Zea mays* L.) | Graminaeae | AMF massively improved biomass. | [337] |
| *Glomus intraradices* | Maize (*Zea mays* L.) | Graminaeae | After inoculation, there was an increase in leaf and stem ratios but a decrease in ear ratios. | [338] |
| *Funneliformis mosseae*, *Claroideoglomus etunicatum* | Maize (*Zea mays* L.) | Graminaeae | The inoculation increased bacterial diversity, decreased the relative abundances of selenobacteria related to plant Se absorption, and improved bacterial network complexity in selenium (Se)(VI)-stressed soils. | [339] |
| *Glomus clarum*, *Glomus deserticola* | Maize (*Zea mays* L.) | Graminaeae | 30 g of *G. clarum* and *G. deserticola* had biocontrol potential against *Fusarium verticillioides*. | [340] |
| *Glomus intraradices*, *Glomus constrictum*, *Glomus mosseae* | Marigold (*Tagetes erecta* L.) | Asteraceae | It can improve the capability of reactive oxygen species (ROS) scavenging and reduce Cd concentration in plants to alleviate Cd stress in marigolds. | [341] |
| *Glomus constrictum* Trappe | Marigold (*Tagetes erecta* L.) | Asteraceae | AMF affected the host plant positively in growth, pigments, phosphorous content, and flower quality, and thus alleviated the stress imposed by water deficiency. | [342] |
| *Funneliformis mosseae*, *Claroideoglomus etunicatum* | Moldavian balm (Moldavian dragonhead) (*Dracocephalum moldavica* L.) | Lamiaceae | Inoculation may increase growth parameters and salinity tolerance under all salinity levels. | [343] |

**Table 1.** *Cont.*

| Type | Plant | Plant Family | Effects | Reference |
|---|---|---|---|---|
| **Arbuscular Mycorrhizal Fungi (AMF)** | | | | |
| *Glomus mosseae, Glomus intraradices* | Mulberry (*Morus alba* L.) | Moraceae | AMF species colonization increased P and N contents of seedlings. | [344] |
| *Glomus deserticola, Gigaspora margarita* | Olive (*Olea europaea* L.) | Oleaceae | Mycorrhizal symbiosis decreased $Na^+$ and $Cl^-$ content and improved RWC, dry and fresh weight, and photosynthetic activity. | [345] |
| *Rhizophagus irregularis* | Olive (*Olea europaea* L.) | Oleaceae | The inoculation exhibited better performance under drought, especially under partial root-zone drying (PRD) treatment. | [345] |
| *Rhizophagus irregularis* DAOM 197,198 | Olive (*Olea europaea* L.) | Oleaceae | Its colonization with olive roots significantly reduced the deleterious effect of water deficit stress by upregulating the primary and secondary metabolism and preserving a high stem water potential level in olive plants. | [346] |
| *Glomus intraradices* | Olive (*Olea europaea* L.) | Oleaceae | The mycorrhizal inoculation played an important part in the attenuation of the impacts of sulfates contained in gypsum substrate on olive trees. | [347] |
| *Funneliformis mosseae, Funneliformis constrictum, Gigaspora margarita,* and *Rhizophagus irregularis* | Onion (*Allium cepa* L.) | Amryllidaceae | Application of AMF and *Trichoderma viride* to onion plants assisted their growth in nutrient-deficient soils amended with fish waste. | [348] |
| *Rhizophagus intraradices* | Palmarosa (*Cymbopogon maritinii* (Roxb.) Wats. Var. Motia Burk | Gramineae | It may influence palmarosa seedling emergence and growth under salinity conditions, and it is useful for health and significant seedling emergence. | [349] |
| *Glomus intraradices, Glomus mosseae, Glomus etunicatum* | Papaya (*Carica papaya* L.) | Caricaceae | Rhizobacteria and AMF acting together formed a mutualistic relationship that enhanced disease control against *Fusarium oxysporum* and stimulated growth in papaya. | [350] |
| *Gigaspora margarita* | Peanut (*Arachis hypogaea*) | Anacardiaceae | The inoculation significantly enhanced leaf K accumulation, drought resistance, and pod yield under drought stress. | [351] |
| *Glomus intraradices, Gigaspora margarita* | Pepper (*Capsicum annuum* L.) | Solanaceae | Inoculated plants had greater dry weight compared to non-inoculated plants. The inoculation can increase P absorbance and P uptake. | [352,353] |
| *Glomus mosseae, Acaulospora laevis, Glomus manihotis,* and a mixed AMF strain | Pomegranate (*Punica granatum* L.) | Punicaceae | Growth, physiological, and biochemical activities were effectively improved by bio-hardening. | [354] |
| *Funneliformis mossae, Rhizophagus intraradices* | Quince (*Cydonia oblonga* Mill.) | Rosaceae | Inoculation with AMF led to significant enhancements in shoot and root dry weight and leaf chlorophyll content. | [355] |
| Genera *Scutellospora, Acaulospora,* and *Glomus* | Red clover (*Trifolium pratense* L.) | Fabaceae | After inoculation, significant increases in monoterpenes such as myrcene, (-)-β-pinene, and linalool were observed. | [356] |
| *Glomus aggregatum, Funneliformis mosseae, Rhizophagus intraradices* | Red tangerine (*Citrus reticulata* Blanco) | Rutaceae | AMF had a positive effect on red tangerine. | [357] |
| *Claroideoglomus etunicatum* | Redtip (*Photinia fraseri* Dress) | Rosaceae | AMF had an impact on plant height and hyphal length density. | [358] |
| *Rhizophagus intraradices* | Rice (*Oryza sativa* L.) | Gramineae | It significantly reduced total As and inorganic As components in rice grains. | [359,360] |
| *Acaulospora mellea; Glomus formosanum; Rhizoglomus clarum; Glomus* spp. | Rice (*Oryza sativa* L.) | Gramineae | It stimulated plant growth, improved root morphological characteristics, and increased P accumulation in rice plants under salt stress conditions. | [361,362] |
| *Glomus* sp1 | Rose (*Rosa rubiginosa* L.) | Rosaceae | The highest percentage of rose root stock establishment increment was achieved with the application of *Glomus* sp1. | [363] |
| *Glomus etunicatum, Glomus mosseae* | Ryegrass (*Lolium perenne*) | Gramineae | AMF-inoculated plants showed lower Cd toxicity, despite the increase in Cd uptake. | [364] |
| *Funneliformis mosseae, Claroideoglomus etunicatum* | Salt grass (*Puccinellia tenuiflora*) | Gramineae | It is able to alleviate boron (B) toxicity by improving biomass and reducing tissue B concentrations. It can help plants tolerate the combined stresses of salt and drought. | [365] |
| *Rhizophagus intraradices, Funneliformis mosseae* | Saffron (*Crocus sativus* L.) | Iridaceae | The mixture of both species increased the spice yield, quality, antioxidant activity, and bioactive compound contents. | [366] |
| *Glomus mosseae* | Snap bean (*Phaseolus vulgaris* L.) | Fabaceae | AMF increased the concentrations of P, N, Mg, and Ca in roots and shoots. It can be concluded that it may reduce the detrimental impacts of increasing $O_3$ on host plants by improving plant nutrition and growth. | [367] |
| *Glomus mosseae, Glomus intraradices, Glomus hoi* | Sour orange (*Citrus aurantium* L.) | Rutaceae | Under salt stress, mycorrhizal-inoculated plants had higher chlorophyll content, higher growth, lower electrolyte leakage, better water status, greater gas exchange capacity, higher malondialdehyde and hydrogen peroxide content, higher osmolyte accumulation, and better antioxidant defense systems. | [368] |
| *Glomus mosseae, Glomus intraradices* | Sorghum (*Sorghum bicolor* L. Moench) | Gramineae | AMF can change the profile of VOCs emitted by roots as well as root morphology. AMF can positively affect the morphological traits of the host roots, total root length, and specific root length of mycorrhizal plants. | [369] |
| *Glomus* sp. 1, *Glomus* sp. 2, *Glomus* sp. 3, *Glomus aggregatum, Glomus fasciculatum, Acaulospora longula, Glomus occultum, Acaulospora scrobiculata, Acaulospora spinosa, Scutellospora* sp. | Sorghum (*Sorghum bicolor* L. Moench) | Gramineae | AMF application improved P and K uptake in shoots. | [370] |
| *Funneliformis mosseae; Funneliformis geosporum* | Sorghum | Gramineae | Plant height and fresh and dry biomass of AMF-inoculated plants were greater in normal soil, followed by sodic and saline soils. | [371] |
| *Acaulospora saccata, Acaulospora fragilissima, Scutellospora ovalis, Rhizophagus neocaledonicus, Claroideoglomus etunicatum* nc, *Pervetustus simplex* nc | Sorghum (*Sorghum bicolor* L. Moench) | Gramineae | Inoculum of combined AMF isolates is appropriate to obtain higher yields and less contaminated biomass of forage sorghum in ultramafic environments. | [372] |
| *Glomus mosseae, Rhizophagus irregularis* | Soybean (*Glycine max* L.) | Fabaceae | The inoculation can enhance P uptake and soybean productivity. | [373] |
| *Rhizophagus clarus* | Soybean (*Glycine max* L.) | Fabaceae | AMF inoculation positively influenced grain yield, shoot dry weight, and P and N content in leaves. | [374] |
| *Cetraspora pellucida, Claroideoglomus etunicatum* | Strawberry (*Fragaria* × *ananassa* Duch.) | Rosaceae | Plants grown with 9% biochar and inoculated with *C. etunicatum* showed a more profuse root system. | [375] |
| *Rhizophagus clarus* | Strawberry (*Fragaria* × *ananassa* Duch.) | Rosaceae | AMF significantly enhanced plant biomass production by boosting photosynthesis rate, antioxidant enzyme defense, water content and use efficiency, and the nutritional status of Zn, in particular. | [376] |
| *Rhizophagus intraradices* | Sweet flag (*Acorus calamus*) | Acoraceae | Under Cr stress, AMF promoted nutrient uptake by *A. calamus* and increased soil carbon input. AMF significantly increased the synergy between the dominant strains. | [377] |
| *Rhizophagus fasciculatus, Rhizophagus aggregatus, Rhizophagus irregularis* | Tangerine orchard (*Citrus reticulata* L.) | Rutaceae | Inoculation had a positive effect on the final yield. | [378] |
| *Glomus versiforme* | Tobacco (*Nicotiana tabacum* L.) | Solanaceae | AMF can protect tobacco against As uptake, and it can play an important role in food quality and safety. | [379] |

**Table 1.** *Cont.*

| Type | Plant | Plant Family | Effects | Reference |
|---|---|---|---|---|
| **Arbuscular Mycorrhizal Fungi (AMF)** | | | | |
| *Claroideoglomus etunicatum, Claroideoglomus claroideum, Glomus microaggregatum, Rhizophagus intraradices, Funneliformis mosseae, Funneliformis geosporum* | Tomato (*Solanum lycopersicum* L.) | Solanaceae | Mycorrhizal inoculation significantly boosted root colonization levels, height, root dry biomass, total yield, shoot dry biomass, and number of fruits. | [380] |
| *Funneliformis Mosseae* and *Rhizophagus intraradices* | Thyme (*Thymus vulgaris* L.) | Lamiaceae | Their inoculation increased essential oil production in both *Thymus vulgaris* L. and *Thymus daenensis* under water stress conditions. | [381] |
| *Glomus versiforme* | Trifoliate orange (*Poncirus trifoliata*) | Rutaceae | Mycorrhization significantly increased gallic acid, ferulic acid, salicylic acid, and phlorizin acid. | [382] |
| *Rhizophagus intraradices, Funneliformis mosseae* | Valerian (*Valeriana officinalis* L.) | Caprifoliaceae | They considerably improved root proline and total soluble sugars and total phenolics in roots and shoots versus untreated valerian plants. | [383] |
| *Funneliformis mosseae, Glomus versiforme* | Wheat (*Triticum aestivum* L.) | Graminae | Inoculation with AMF could increase Se bioavailability in the rhizosphere. | [384] |
| *Funneliformis mosseae, Glomus versiforme* | Wheat (*Triticum aestivum* L.) | Graminae | AMF combined with 48.76 $mgkg^{-1}$ P applied in soil can not only achieve high grain yield, but also fully exploit the biological potential of Se uptake in wheat. | [385] |
| *Funneliformis mosseae* BGC HEB02, *Rhizophagus intraradices* BGC HEB07D | Wheat (*Triticum aestivum* L.) | Graminae | Zn in wheat grain can be significantly increased by inoculation with AMF, indicating the potential of AMF to cope with Zn deficiency. | [386] |
| *Glomus mosseae, Glomus hoi, Glomus etunicatum, Acaulospora scrobiculata, Acaulospora spinosa* | White yam (*Dioscorea rotundata*) | Dioscoreaceae | AMF can increase yam tuber growth. | [387] |
| *Funneliformis mosseae, Laroideoglomus etunicatum, Rhizophagus intraradices* | Willow (*Salix viminalis*) | Salicaceae | Organic acids including arachidonic acid, octadecanedioic acid, α-linolenic acid, 10,12,14-octadecarachidonic acid, and 5-methoxysalicylic acid were significantly increased under AMF inoculation treatment. AMF inoculation also increased the levels of polyphenol oxidase and dehydrogenase. | [388] |

## 7. Modes of Action of Arbuscular Mycorrhizal Fungi (AMF)

AMF inoculation changed the biochemical procedures of lemongrass with a subsequent effect on growth and increased the suppression of *Fusarium* root rot under greenhouse conditions; additionally, essential oils (EOs) from mycorrhizal lemongrass frequently inhibited *Fusarium Solani* Fs4 development, notably more than the non-inoculated counterpart [389]. AMF can form a symbiosis with most vegetable plants, including the main crops from diverse families, such as Apiaceae (e.g., carrot), Amaryllidaceae (e.g., onion, garlic, leek), Cucurbitaceae (cucumber), Asteraceae (e.g., lettuce), Fabaceae (e.g., pea and bean), and Solanaceae (e.g., bell pepper, tomato) [390]. The most significant beneficial impacts of AMF are the boosted capture of mineral elements such as calcium, phosphorus, copper, zinc, sulfur, and iron, and the increased resistance to various types of abiotic and biotic stress [391,392]. In one experiment, it was reported that the α-diversity of AMF and diazotrophs communities, N-related enzymes, and sediment nutrition constituents were the principle parameters driving the biological nitrogen fixation (BNF) process in mangrove ecosystems [393]. The synchronous utilization of chitosan nanoparticles (CSNPs) and AMF promoted the levels of carotenoids, polyphenols, and tocopherols in the roots, thus increasing antioxidant capacity (33%); therefore, CSNPs can be used as effective biofertilization tools to increase plant growth and fitness, exemplified by the improvement in the health-promoting components in wheat [394]. In the presence of AMF inoculation strains *Rhizophagus irregularis* and *Funneliformis mosseae*, phosphorous components in the leaves of tomato plants were increased under both watering regimes, and even under limited watering conditions, Ca, K, Zn, Mg, and Mn were increased to levels similar to those of non-stressed plants [395]. AMF increased nutrient uptake, particularly phosphorus and other less mobile elements, mostly by increasing the absorptive surface of hyphal exploration in the rhizosphere, and acted as soil reclamation and biocontrol agents for contaminated or polluted soil [396]. Sarkar et al. [397] reported the probability of using arbuscular mycorrhizal fungi on symbiotic *Miscanthus* plants in phytoremediation and growing them in Zn-deficient soils. AMF improved plant growth under toxic trace element (TE) stress. The function of AMF in regulating heavy metal transport is concentration-dependent, and the transport direction of Cd and Zn in plants regulated by AMF changes with TE. AMF increased the absorption of Zn by growth dilution and Cd by biological enrichment, and the interaction between Cd and Zn was affected by AMF and other components [398].

Mycorrhizal fungi respond to metal contamination in different ways. *Funneliformis mosseae* in pepper can promote the dry mass production photosynthesis rate under Cu stress [399], *Glomus versiforme* and *Rhizophagus intraradices* can increase P nutrition and antioxidant activity of *Lotus japonica* under Cd stress [400], *Rhizophagus intraradices* can accumulate Cr in roots and decrease its translocation towards shoots of *Brachiaria mutica* under Cr stress, *Rhizophagus irregularis* may immobilize Cu and Pb in soil for the cultivation of willows [401], *Rhizophagus intraradices* can regulate As transporter in root epidermis cells and decrease its uptake in pea seedlings [402], and *Claroideoglomus* and *Rhizophagus* spp can decrease Cd accumulation in wheat grains [403]. *Funneliformis mosseae* is a famous mycorrhiza among different AMF species for its symbiosis with the roots of various plant species, and it may facilitate the exchange of nutrients via the expansion of roots through the outer membrane of the mycelia and by boosting root development [404]. Some of the most important examples of mycorrhizal fungi mitigating soil nutrient deficiencies are *Rhizophagus irregularis* reducing soil nutrient deficiencies of P and Zn in *Medicago truncatula* by increasing *MtZIP5* and *MtPT4* gene induction [405], *Rhizophagus irregularis* boosting root absorption area and soil P availability in maize, *Rhizophagus irregularis* modifying the ZIP transporter response in barley and boosting grain Zn bioavailability [406], *Funneliformis mosseae* promoting the uptake of nutrients such as Mn, Mg, Zn, Fe, Ca, S, N, P, and K in cucumber [407], and *Rhizophagus irregularis* stimulating MtZIP6 gene expression and boosting the root absorption area in *Medicago truncatula* [406]. Various biochemical and physical mechanisms have been reported to play important roles in the enhancement of common bean plant stress resistance by AMF inoculation against *Rhizoctonia solani* or *Rhizoctonia* root rot disease, namely improved plant growth, increased plant nutrition, cytoplasmic granulation, improved cell wall thickening, and the accumulation of some antimicrobial components such as defense-related enzymes and phenolic substances [408]. Li et al. [409] reported that AMF promoted the accumulation of Cd in the roots of wheat but significantly decreased the amount of Cd in the grains and shoots. Moreover, stomatal conductance, photosynthetic rates, chlorophyll content, transpiration rates, and the accumulation of carbohydrates under Cd stress were elevated by AMF symbiosis. Proteomic analysis showed that AMF led to the expression of two enzymes that have significant functions in the chlorophyll biosynthesis pathway, namely Mg-protoporphyrin IX chelates and coproporphyrinogen oxide; increased the expression of two proteins related to $CO_2$ assimilation, namely malic enzyme and ribulose-1,5-bisphosphate carboxylase; and boosted the expression of S-adenosylmethionine synthase. Under salinity stress, *glomus mosseae* can increase total proline and phenol content, leaf area, and root and shoot dry weights and lengths in *Triticum aestivum* [410]; *Funneliformis mosseae* can increase nutrient uptake, the photosynthetic rate, and the stand establishment rate of *Cucumis sativus* [411]; and *Rhizobium irregularis* and *Funneliformis mosseae* can increase plant biomass and decrease oxidative burst by strengthening the antioxidant enzymatic activities of *Cajanus cajan* [412]. Yu et al. [413] reported that AMF have ecological functions in decreasing Cd loss, with positive effects on maize growth because of leaching from polluted soil, as the AMF changed Cd migration by boosting the contents of glomalin-related soil protein (GRSP), exudates, and root morphology. Selvaraj et al. [414] observed that *Glomus intraradices* can improve both nutrient use efficiency and increase the defense response against herbivorous insects (*Spodoptera litura*) in blackgram. AMF symbiosis reduces temperature stress in host plants by improving the osmotic adjustment [415], increasing the photochemistry of PSII and the photosynthetic rate [416], increasing the uptake of nutrients [417], improving reproductive capacity, and enhancing antioxidant activity [418,419]. Farghaly et al. [420] reported that AMF inoculation can decrease the adverse impacts of alkalinity stress on wheat plants by decreasing Na concentration and pH and improving the availability of P and the productivity of wheat yield.

## 8. Practical Application of Microbial Biostimulants on Crops

Culture-dependent approaches have allowed the isolation of various bacteria taxa from the mycorrhizosphere of *Glomus margarita*, *Glomus versiforme*, *Rhizobium irregular*, *Rhizobium clarus*, and *Funneliformis mosseae* [421–423]. Microbial inoculants in horticulture and agriculture systems have different effects; for example, *Pseudomonas putida* can increase iron uptake in rice, and *Pseudomonas fluorescens* can improve grain yield, plant height, and biomass in rice, sweet potato, and rapeseed [424–426]. The *Streptomyces* strain shows positive effects on the plant growth of tomato and rice [427,428]. Improved dry shoot weight and a significant increase in leaf length were reported after the application of *Azospirillum brasilense* Sp245, and *Aeromonas* spp. increased the root area of rice [429]. *Comamonas acidovorans* and *Bacillus subtilis* have significantly improved plant growth and increased cytokinin constituents in shoots and roots [430,431]. *Bacillus licheniformis* application increased fresh weight, enhanced cell division, and raised the chlorophyll content in cucumber [432], and *Azospirillum lipoferum* improved the root hair density of maize seedlings [433]. *Azospirillum lipoferum* increased tolerance to salinity stress in wheat [434], and *Pseudomonas putida* application led to increased shoot and root biomass and water content of white clover [435]. The utilization of *Alternaria* sp. stimulated drought tolerance in wheat [436,437]. It has also been reported that *Azoarcus* sp. increased plant nitrogen nutrition and root growth and reduced nutrient deficiency in wheat [438], and *Azorhizobium* sp. boosted plant nitrogen nutrition and root growth and alleviated nutrient deficiency [438]. *Rhizobium meliloti* improved plant growth, nitrogen use efficiency, and the quality of pods of peanuts [439,440], and *Rhizobium leguminosarum* boosted growth and yield performance under drought stress in soybean seedlings [441]. *Azotobacter chroococcum* and *Azotobacter vinelandii* increase root and shoot length, chlorophyll content, and root and leaf number [442]. Application of *Bacillus halotolerans*, *Pseudomonas frederiksbergensis RG2*, and *Enterobacter hormaechei* increased germination, growth, and yield; induced better drought resistance; and improved uptake of P, N, and Zn [443]. *Rhizoglomus irregulare* Aoufous consortium enhanced growth traits and physiological characteristics that have functions in the absorption of N and P content, increased protein and sugar content, reduced MDA and $H_2O_2$, alleviated soil pH, and increased the electrical conductivity, organic matter, and total organic carbon of date palm under drought conditions [444]. *Glomus* spp. increased root fresh weigh, root length, shoot height, shoot fresh weight, and germination index and improved the root basal diameter, dry biomass, and seedling length of date palm under heavy metal stress conditions [445]. AMF also can interact with PGPR [446,447]; for example, a positive and significant synergistic interaction has been reported between *Bacillus subtilis* and AMF regarding nitrate, nitrate reductase, and nitrogenase properties and the components of phenols, lipids, fiber, and osmoprotectants such as proline, betaine, and glycine [448]. Zamljen et al. [449] reported that microbial biostimulants added to the nutrient solution showed better results in the volatile content as well as primary metabolism and yield of basil. Anton-Herrero et al. [450] concluded that the origin and composition of microbial biostimulants determine their physiological impacts on pepper plants. Microbial biostimulants increased lipophilic antioxidants and yield and reduced sugars in cherry tomato cultivars [451–463].

## 9. Conclusions and Future Remarks

Biostimulants are sustainable and eco-friendly products that may promote metabolic and enzymatic processes in plants to enhance their yield, productivity, and crop quality and increase resistance to abiotic stresses. Biostimulants contain substances or products containing special formulations, natural components, or microorganisms that can naturally improve plant growth. The most important categories of biostimulants are humic and fulvic acid, fungi, bacteria, amino-acid-containing products, hormone-containing products, etc. Biostimulants are not only considered important alternatives to mineral fertilizers, but are also distinguished in organic farming systems under sustainable crop production management. Increased shoot and root growth, better root growth potential, improved resistance to stressors, and reduced nitrogen fertilizer inputs are some of the most noteworthy effects of biostimulant application in sustainable agriculture systems.

Arbuscular mycorrhizal fungi are widespread root colonization fungi associated with the roots of higher plants. The current multiplication methods of mycorrhizal species under root organ culture (ROC) have become efficient alternatives for the cultivation of specific secondary metabolite compounds. Several kinds of studies have shown that AMF lead to significant changes in the quality and quantity of secondary metabolites that originate from medicinal and aromatic plants of economic interest. AMF can only be grown in the presence of host plants and are extensively used in agriculture and horticulture, especially *Rhizophagus* (formerly known as *Glomus*) *intraradices* and *Funneliformis* (formerly known as *Glomus*) *mosseae*. AMF make a positive contribution to growth, yield, stress tolerance, pathogen protection, and the maintenance of agricultural ecosystem sustainability. The increased activities of secondary metabolite production and antioxidant defense system enzymes related to plant defense have been extensively reported in mycorrhizal plants.

In conclusion, AMF play a key role in plant nutrition and performance due to their capacity to improve plant mineral uptake. AMF, like other microbial biostimulants, can increase the sustainability of agricultural and horticultural production systems as well as improve the quality and quantity of food for the ever-growing world population. Moreover, studying the molecular mechanisms behind the observed activities will help to reveal the physiological and plant metabolism pathways involved in these processes and provide farmers with tailor-made products suitable for various conditions. However, further studies are needed to improve the reproducibility of the positive effects and to standardize the production processes from the lab to an industrial scale. This review article concludes that inoculation with specific arbuscular mycorrhizal fungi can increase the concentration of secondary metabolites in plants. However, future studies are needed to increase the reproducibility of the positive impacts, and more in-depth research is needed to identify the influence of interactions between AMF and biostimulants for promoting plants' physiological characteristics.

**Author Contributions:** W.S., writing—original draft preparation; M.H.S., writing—original draft preparation and editing. All authors have read and agreed to the published version of the manuscript.

**Funding:** This research was supported by the National Key R&D Program of China (Grant No. 2019YFA0904700). This research was funded by the Natural Science Foundation of Beijing, China (Grant No. M21026).

**Institutional Review Board Statement:** Not applicable.

**Informed Consent Statement:** Not applicable.

**Data Availability Statement:** Not applicable.

**Conflicts of Interest:** The authors declare no conflict of interest.

## References

1. Sun, W.; Shahrajabian, M.H. Therapeutic potential of phenolic compounds in medicinal plants-natural health products for human health. *Molecules* **2023**, *28*, 1845. [CrossRef] [PubMed]
2. Sun, W.; Shahrajabian, M.H.; Petropoulos, S.A.; Shahrajabian, N. Developing sustainable agriculture systems in medicinal and aromatic plant production by using chitosan and chitin-based biostimulants. *Plants* **2023**, *12*, 2469. [CrossRef] [PubMed]
3. Shahrajabian, M.H.; Cheng, Q.; Sun, W. The effects of amino acids, phenols and protein hydrolysates as biostimulants on sustainable crop production and alleviated stress. *Recent Pat. Biotechnol.* **2022**, *16*, 319–328. [CrossRef]
4. Shahrajabian, M.H.; Petropoulos, S.A.; Sun, W. Survey of the influences of microbial biostimulants on horticultural crops: Case studies and successful paradigms. *Horticulturae* **2023**, *9*, 193. [CrossRef]
5. Shahrajabian, M.H.; Sun, W. Five important seeds in traditional medicine, and pharmacological benefits. *Seeds* **2023**, *2*, 290–308. [CrossRef]
6. Calvo, P.; Nelson, L.; Kloepper, J.W. Agricultural uses of plant biostimulants. *Plant Soil* **2014**, *383*, 3–41. [CrossRef]
7. Karapouloutidou, S.; Gasparatos, D. Effects of biostimulant and organic amendment on soil properties and nutrient status of *Lactuca sativa* in a Calcareous saline-sodic soil. *Agriculture* **2019**, *9*, 164. [CrossRef]
8. Tejada, M.; Benitez, C.; Gomez, I.; Parrado, J. Use of biostimulants on soil restoration: Effects on soil biochemical properties and microbial community. *Appl. Soil Ecol.* **2011**, *49*, 11–17. [CrossRef]
9. Yakhin, O.I.; Lubyanov, A.A.; Yakhin, I.A.; Brown, P.H. Biostimulants in plants science: A global perspective. *Front. Plant Sci.* **2017**, *7*, 2049. [CrossRef]
10. Philibert, T.; Lee, B.H.; Fabien, N. Current status and new perspectives on chitin and chitosan as functional biopolymers. *Appl. Biochem. Biotechnol.* **2017**, *181*, 1314–1337. [CrossRef]

11. Meramo-Hurtado, S.I.; Gonzalez-Delgado, A.D. Application of techno-economic and sensitivity analyses as decision-making tools for assessing emerging large-scale technologies for production of chitosan-based adsorbents. *ACS Omega* **2020**, *5*, 17601–17610. [CrossRef] [PubMed]
12. Koleska, I.; Hasanagic, D.; Todorovic, V.; Murtic, S.; Klokic, I.; Paradikovic, N.; Kukavica, B. Biostimulants prevents yield loss and reduces oxidative damage in tomato plants grown on reduced NPK nutrition. *J. Plant Interact.* **2017**, *12*, 209–218. [CrossRef]
13. Shehata, S.A.; AbdelGawad, K.F.; Elmogy, M. Quality and shelf-life of onion bulbs influenced by biostimulants. *Int. J. Veg. Sci.* **2017**, *23*, 362–371. [CrossRef]
14. Semida, W.M.; El-Mageed, T.A.A.; Hemida, K.; Rady, M.M. Natural bee-honey based biostimulants confer salt tolerance in onion via modulation of the antioxidant defense system. *J. Hortic. Sci. Biotechnol.* **2019**, *94*, 632–642. [CrossRef]
15. Anbarasi, D.; Haripriya, K. Response of aggregatum onion (*Allium cepa* L. var. *aggregatum Don.*) to organic inputs, biofertilizer and biostimulants. *Plant Arch.* **2020**, *20*, 759–762.
16. Arthur, G.D.; Aremua, A.O.; Kulkarni, M.G.; Okem, A.; Stirk, W.A.; Davies, T.C.; Van Staden, J. Can the use of natural biostimulants be a potential means of phytoremediating contaminated soils from goldmines in South Africa? *Int. J. Phytoremediation* **2015**, *18*, 427–434. [CrossRef] [PubMed]
17. Wszelaczynska, E.; Szczepanek, M.; Poberezny, J.; Kazula, M.J. Effect of biostimulant application and long-term storage on the nutritional value of carrot. *Hortic. Bras.* **2019**, *37*, 451–457. [CrossRef]
18. Halpern, M.; Bar-Tal, A.; Ofek, M.; Minz, D.; Muller, T.; Yermiyahu, U. The use of biostimulants for enhancing nutrient uptake. *Adv. Agron.* **2015**, *130*, 141–174. [CrossRef]
19. Lucini, L.; Rouphael, Y.; Cardarelli, M.; Canaguier, R.; Kumar, P.; Colla, G. The effect of a plant-derived biostimulant on metabolic profiling and crop performance of lettuce grown under saline condition. *Sci. Hortic.* **2015**, *182*, 124–133. [CrossRef]
20. Petropoulos, S.A.; Fernandes, A.; Plexida, S.; Chrysargyris, A.; Tzortzakis, N.; Barreira, J.C.M.; Barros, L.; Ferreira, I.C.F.R. Biostimulants application alleviates water stress effects on yield and chemical composition of greenhouse green bean (*Phaseolus vulgaris* L.). *Agronomy* **2020**, *10*, 181. [CrossRef]
21. Shahrajabian, M.H.; Chaski, C.; Polyzos, N.; Tzortzakis, N.; Petropoulos, S.A. Sustainable agriculture systems in vegetable production using chitin and chitosan as plant biostimulants. *Biomolecules* **2021**, *11*, 819. [CrossRef] [PubMed]
22. Joly, P.; Calteau, A.; Wauquier, A.; Dumas, R.; Beuvin, M.; Vallenet, D.; Crovadore, J.; Cochard, B.; Lefort, F.; Berthon, J.-Y. From strain characterization to field authorization: Highlights on *Bacillus velezensis* strain B2 beneficial properties for plants and its activities on phytopathogenic fungi. *Microorganisms* **2021**, *9*, 1924. [CrossRef] [PubMed]
23. Tomas, M.S.J.; Carrasco, M.G.; Lobo, C.B.; Alessandrello, M.J.; Sanchez, L.; Ferrero, M.A. PAH removal simultaneous and sequential inoculation of *Pseudomonas monteilii* P26 and *Gordonia* sp. H19 in the presence of biostimulants. *Int. Biodeterior. Biodegrad.* **2019**, *144*, 104752. [CrossRef]
24. Barros-Rodriguez, A.; Rangseekawe, P.; Lasudee, K.; Pathom-aree, W.; Manzanera, M. Regulatory risks associated with bacteria as biostimulants and biofertilizers in the frame of the European Regulation (EU) 2019/1009. *Sci. Total Environ.* **2020**, *740*, 140239. [CrossRef] [PubMed]
25. Mickan, B.S.; Alsharmani, A.R.; Solaiman, Z.M.; Leopold, M.; Abbott, L.K. Plant-dependent soil bacterial responses following amendment with a multispecies microbial biostimulant compared to rock mineral and chemical fertilizers. *Front. Plant Sci.* **2021**, *11*, 550169. [CrossRef]
26. Hart, M.M.; Reader, R.J. Taxonomic basis for varition in the colonizatio strategy of arbuscular mycorrhizal fungi. *New Phytol.* **2002**, *153*, 335–344. [CrossRef]
27. Yang, H.; Zang, Y.; Yuan, Y.; Tang, J.; Chen, X. Selectivity by host plants affects the distribution of arbuscular mycorrhizal funi: Evidence from ITS rDNA sequence metadata. *BMC Evol. Biol.* **2012**, *12*, 50. [CrossRef]
28. Cruz, C.; Vishwakarma, K.; Kumar, D.; Varma, A. (Eds.) *Soil Nitrogen Ecology*; Springer Nature Switzerland AG: Cham, Switzerland, 2021. [CrossRef]
29. Spatafora, J.W.; Chang, Y.; Benny, G.L.; Lazarus, K.; Smith, M.E.; Berbee, M.L.; Bonito, G.; Corradi, N.; Grigoriev, I.; Gryganskyi, A.; et al. A phylum-level phylogenetic classification of zygomycete fungi based on genome-scale data. *Mycologia* **2016**, *108*, 1028–1046. [CrossRef]
30. Redecker, D.; Schussler, A.; Stockinger, H.; Sturmer, S.L.; Morton, J.B.; Walker, C. An evidence-based consensus for the classification of arbuscular mycorrhizal fungi (*Glomeromycota*). *Mycorrhiza* **2013**, *23*, 515–531. [CrossRef]
31. Zhu, B.; Gao, T.; Zhang, D.; Fing, K.; Li, C.; Ma, F. Functions of arbuscular mycorrhizal fungi in horticultural crops. *Sci. Hortic.* **2022**, *303*, 111219. [CrossRef]
32. Cui, X.; Jia, B.; Diao, F.; Li, X.; Xu, J.; Zhang, Z.; Li, F.Y.; Guo, W. Transcriptomic analysis reveals the molecular mechanisms of arbuscular mycorrhizal fungi and nitrilotriacetic acid on *Suaeda salsa* tolerance to combined stress of cadmium and salt. *Process Saf. Environ. Prot.* **2022**, *160*, 210–220. [CrossRef]
33. Sales, L.R.; Silva, A.O.; Sales, F.R.; Rodrigues, T.L.; Barbosa, M.V.; Santos, J.V.D.; Kemmelmeier, K.; Siqueira, J.O.; Carneiro, M.A.C. *On farm* inoculation of native arbuscular mycorrhizal fungi improves efficiency in increasing sugarcane productivity in the field. *Rhizosphere* **2022**, *22*, 100539. [CrossRef]
34. Liang, M.; Wu, Y.; Zhao, Q.; Jiang, Y.; Sun, W.; Liu, G.; Ma, L.; Xue, S. Secondary vegetation succession on the Loess Plateau altered the interaction between arbuscular mycorrhizal fungi and nitrogen-fixing bacteria. *For. Ecol. Manag.* **2023**, *530*, 120744. [CrossRef]

35. Paymaneh, Z.; Sarcheshmehpour, M.; Mohammadi, H.; Hesni, M.A. Vermicompost and/or compost and arbuscular mycorrhizal fungi are conducive to improving the growth of pistachio seedlings to drought stress. *Appl. Soil Ecol.* **2023**, *182*, 104717. [CrossRef]
36. Rafiee, H.; Badi, H.N.; Mehrafarin, A.; Qaderi, A.; Zarinpanjeh, N.; Sekara, A.; Zand, E. Application of plant biostimulants as new approach to improve the biological responses of medicinal plants—A critical review. *J. Med. Plants.* **2016**, *15*, 6–39.
37. Bosi, S.; Negri, L.; Accorsi, M.; Baffoni, L.; Gaggia, F.; Gioia, D.D.; Dinelli, G.; Marotti, I. Biostimulants for sustainable management of sport turfgrass. *Plants* **2023**, *12*, 539. [CrossRef]
38. Vasconsuelo, A.; Boland, R. Molecular aspects of the early stages of elicitation of secondary metabolites in plants. *Plant Sci.* **2007**, *172*, 861–875. [CrossRef]
39. Ahemad, M.; Kibret, M. Mechanisms and applications of plant growth promoting rhizobacteria: Current perspective. *J. King Saud Univ. Sci.* **2014**, *26*, 1–20. [CrossRef]
40. Perez-Montano, F.; Alias-Villegas, C.; Bellogin, R.; del Cerro, P.; Espuny, M.; Jimenez-Guerrero, I.; Lopez-Baena, F.; Ollero, F.; Cubo, T. Plant growth promotion in cereal and leguminous agricultural important plants: From microorganism capacities to crop production. *Microbiol. Res.* **2014**, *169*, 325–336. [CrossRef]
41. Lugtenberg, B.; Kamilova, F. Plant-growth-promoting rhizobacteria. *Annu. Rev. Microbiol.* **2009**, *63*, 541–556. [CrossRef]
42. De Vries, F.T.; Griffiths, R.I.; Knight, C.G.; Nicolitch, O.; Williams, A. Harnessing rhizosphere microbiomes for drought-resilient crop production. *Science* **2020**, *368*, 270–274. [CrossRef] [PubMed]
43. Rouphael, Y.; Colla, G. Toward a sustainable agriculture through plant biostimulants: From experimental data to practical applications. *Agronomy* **2020**, *10*, 1461. [CrossRef]
44. Askari-Khorasgani, O.; Hatterman-Valenti, H.; Pardo, F.B.F.; Pessarakli, M. Plant and symbiont metabolic regulation and biostimulants application improve symbiotic performance and cold acclimation. *J. Plant Nutr.* **2019**, *42*, 2151–2163. [CrossRef]
45. Massa, D.; Lenzi, A.; Montoneri, E.; Ginepro, M.; Prisa, D.; Burchi, G. Plant response to biowaste soluble hydrolysates in hibiscus grown under limiting nutrient availability. *J. Plant Nutr.* **2017**, *41*, 396–409. [CrossRef]
46. Kloepper, J.W.; Ryu, C.-M.; Zhang, S. Induced systemic resistance and promotion of plant growth by *Bacillus* spp. *Phytopathology* **2004**, *94*, 1259–1266. [CrossRef]
47. Glick, R.B. Bacteria with ACC deaminase can promote plant growth and help to feed the world. *Microbiol. Res.* **2014**, *169*, 30–39. [CrossRef]
48. Ruzzi, M.; Aroca, R. Plant growth-promoting rhizobacteria act as biostimulants in horticulture. *Sci. Hortic.* **2015**, *196*, 124–134. [CrossRef]
49. Rouphael, Y.; Colla, G. Editorial: Biostimulants in Agriculture. *Front. Plant Sci.* **2020**, *11*, 40. [CrossRef]
50. Shahrajabian, M.H.; Sun, W.; Cheng, C. The importance of flavonoids and phytochemicals of medicinal plants with antiviral activities. *Mini Rev. Org. Chem.* **2022**, *19*, 293–318. [CrossRef]
51. Nazir, A.; Shafiq, M.; Bareen, F. Fungal biostimulant-driven phytoextraction of heavy metals from tannery solid waste contaminated soil. *Int. J. Phytoremed.* **2021**, *24*, 47–58. [CrossRef]
52. Shahrajabian, M.H.; Sun, W. The golden spice for life: Turmeric with the pharmacological benefits of curcuminoids components, including curcumin, bisdemethoxycurcumin, and demethoxycurcumin. *Curr. Org. Synth.* **2023**, *20*, 1–12. [CrossRef]
53. Ali, B. Salicylic acid: An efficient elicitor of secondary metabolite production in plants. *Biocatal. Agric. Biotechnol.* **2021**, *31*, 101884. [CrossRef]
54. Eraslan, F.; Inal, A.; Gunes, A.; Alpaslan, M. Impact of exogenous salicylic acid on the growth, antioxidant activity and physiology of carrot plants subjected to combined salinity and boron toxicity. *Sci. Hortic.* **2007**, *113*, 120–128. [CrossRef]
55. Lefevere, H.; Bauters, L.; Gheysen, G. Salicylic acid biosynthesis in plants. *Front. Plant Sci.* **2020**, *11*, 338. [CrossRef] [PubMed]
56. Guo, X.; Liu, H.; Wu, S. Humic substances developed during organic waste composting: Formation mechanisms, structural properties, and agronomic functions. *Sci. Total Environ.* **2019**, *662*, 501–510. [CrossRef]
57. Xiang, Y.; Kang, F.; Xiang, Y.; Jiao, Y. Effects of humic acid-modified magnetic $Fe_3O_4$/MgAl-layered double hydroxide on the plant growth, soil enzyme activity, and metal availability. *Ecotoxicol. Environ. Saf.* **2019**, *182*, 109424. [CrossRef]
58. Canellas, L.P.; Olivares, F.L.; Aguiar, N.O.; Jones, D.L.; Nebbioso, A.; Mazzeri, P.; Piccolo, A. Humic and fulvic acids as biostimulants in horticulture. *Sci. Hortic.* **2015**, *196*, 15–27. [CrossRef]
59. Ouni, Y.; Ghnaya, T.; Montemurro, F.; Abdelly, C.; Lakhdar, A. The role of humic substances in mitigating the harmful effects of soil salinity and improve plant productivity. *Int. J. Plant Prod.* **2014**, *8*, 353–374.
60. Bah, C.S.F.; Bekhit, A.E.D.A.; Carne, A.; McConnell, M.A. Production of bioactive peptide hydrolysates from deer, sheep, and pig plasma using plant and fungal protease preparations. *Food Chem.* **2015**, *176*, 54–63. [CrossRef]
61. Colla, G.; Nardi, S.; Cardarelli, M.; Ertani, A.; Lucini, L.; Canaguier, R.; Rouphael, Y. Protein hydrolysates as biostimulants in horticulture. *Sci. Hortic.* **2015**, *196*, 28–38. [CrossRef]
62. Sun, W.; Shahrajabian, M.H.; Lin, M. Research progress of fermented functional foods and protein factory-microbial fermentation technology. *Fermentation* **2022**, *8*, 688. [CrossRef]
63. Battacharyya, D.; Babgohari, M.Z.; Rathor, P.; Prithiviraj, B. Seaweed extracts as biostimulants in horticulture. *Sci. Hortic.* **2015**, *196*, 39–48. [CrossRef]
64. Gupta, V.; Kumar, M.; Brahmbhatt, H.; Reddy, C.; Seth, A.; Jha, B. Simultaneous determination of different endogenetic plant growth regulators in common green seaweeds using dispersive liquid-liquid microextraction method. *Plant Physiol. Biochem.* **2011**, *49*, 1259–1263. [CrossRef]

65. Sivasankari, S.; Venkatesalu, V.; Anantharaj, M.; Chandrasekaran, M. Effect of seaweed extract on the growth and biochemical constituents of *Vigna sinensis*. *Bioresour. Technol.* **2006**, *97*, 1745–1751. [CrossRef] [PubMed]
66. Pizzeghello, D.; Francioso, O.; Ertani, A.; Muscolo, A.; Nardi, S. Isopentenyladenosine and cytokinin-like activity of different humic substances. *J. Geochem. Explor.* **2013**, *129*, 70–75. [CrossRef]
67. Puglisi, I.; Barone, V.; Sidella, S.; Coppa, M.; Broccanello, C.; Gennari, M.; Baglieri, A Biostimulant activity of humic-like substances from agro-industrial waste on *Chlorella vulgaris* and *Scenedesmus quadricauda*. *Eur. J. Phycol.* **2018**, *53*, 433–442. [CrossRef]
68. Vujinovic, T.; Zanin, L.; Venuti, S.; Contin, M.; Ceccon, P.; Tomasi, N.; Pinton, R.; Cesco, S.; De Nobili, M. Biostimulant action of dissolved humic substances from a conventionally and an organically managed soil on nitrate acquisition in maize plants. *Front. Plant Sci.* **2020**, *10*, 1652. [CrossRef]
69. Olaetxea, M.; De Hita, D.; Garcia, C.A.; Fuentes, M.; Baigorri, R.; Mora, V.; Garnica, M.; Urrutia, O.; Erro, J.; Zamarreno, A.M.; et al. Hypothetical framework integrating the main mechanisms involved in the promoting action of rhizospheric humic substances on plant root- and shoot- growth. *Appl. Soil Ecol.* **2018**, *123*, 521–537. [CrossRef]
70. Conselvan, G.B.; Fuentes, D.; Merchant, A.; Peggion, C.; Francioso, O.; Carletti, P. Effects of humic substances and indole-3-acetic acid on *Arabidopsis* sugar and amino acid metabolic profile. *Plant Soil* **2018**, *426*, 17–32. [CrossRef]
71. Garcia, A.C.; Olaetxea, M.; Santos, L.A.; Mora, V.; Baigorri, R.; Fuentes, M.; Zamarreno, A.M.; Berbara, R.L.; Garcia-Mina, J.M. Involvement of hormone- and ROS- signaling pathways in the beneficial action of humic substances on plants growing under normal and stressing conditions. *BioMed Res. Int.* **2016**, *2016*, 3747501. [CrossRef]
72. Jindo, K.; Martim, S.A.; Navarro, E.C.; Perez-Alfocea, F.; Hernandez, T.; Garcia, C.; Aguiar, N.O.; Canellas, L P. Root growth promotion by humic acids from composted and non-composted urban organic wastes. *Plant Soil* **2012**, *353*, 209–220. [CrossRef]
73. Goni, O.; Quille, P.; O'Connell, S. Production of chitosan oligosaccharides for inclusion in a plant biostimulant. *Pure Appl. Chem.* **2016**, *88*, 881–889. [CrossRef]
74. Xu, C.; Mou, B. Chitosan as soil amendment affects lettuce growth, photochemical efficiency, and gas exchange. *HortTechnology* **2018**, *28*, 476–480. [CrossRef]
75. Shahrajabian, M.H.; Sun, W. Survey on medicinal plants and herbs in traditional Iranian medicine with anti-oxidant, anti-viral, anti-microbial, and anti-inflammation properties. *Lett. Drud Des. Discov.* **2023**, *20*, 1707–1743. [CrossRef]
76. Oosten, M.J.V.; Pepe, O.; Pascale, S.D.; Silletti, S.; Maggio, A. The role of biostimulants and bioeffectors as alleviators of abiotic stress in crop plants. *Chem. Biol. Technol. Agric.* **2017**, *4*, 5. [CrossRef]
77. Rai, V.K. Role of amino acids in plant responses to stresses. *Biol. Plant.* **2002**, *45*, 481–487. [CrossRef]
78. Shahrajabian, M.H.; Sun, W. Sustainable approaches to boost yield and chemical constituents of aromatic and medicinal plants by application of biostimulants. *Recent Adv. Food Nutr. Agric.* **2022**, *13*, 72–92. [CrossRef]
79. Shahrajabian, M.H.; Sun, W. Mechanism of action of collagen and epidermal growth factor: A review on theory and research methods. *Mini Rev. Med. Chem.* **2023**, *23*. [CrossRef]
80. Popko, M.; Michalak, I.; Wilk, R.; Gramza, M.; Chojnacka, K.; Gorecki, H. Effect of the new plant growth biostimulants based on amino acids on yield and grain quality of winter wheat. *Molecules* **2018**, *23*, 470. [CrossRef]
81. Sadak, S.H.; Abdelhamid, M.T.; Schmidhalter, U. Effect of foliar application of amino acids on plant yield and physiological parameters in bean plants irrigated with seawater. *Acta. Biol. Colomb.* **2015**, *20*, 141–152. [CrossRef]
82. Ozyigit, I.I.; Kahraman, M.V.; Ercan, O. Relation between explants age, total phenol and regeneration response in tissue cultured cotton (*Gossypium hirsutum* L.). *Afr. J. Biotechnol.* **2007**, *6*, 3–8.
83. Sharma, A.; Shahzad, B.; Rehman, A.; Bhardwaj, R.; Landi, M.; Zheng, B. Response of phenylpropanoid pathway and the role of polyphenols in plants under abiotic stress. *Molecules* **2019**, *24*, 2452. [CrossRef]
84. Schaafsma, G. Safety of protein hydrolysates, fractions thereof and bioactive peptides in human nutrition. *Eur. J. Clin. Nutr.* **2009**, *63*, 1161–1168. [CrossRef]
85. Albertsen, A.; Ravnskov, S.; Green, H.; Jensen, D.F.; Larsen, J. Interactions between the external mycelium of the *Mycorrhizal* fungus, *Glomus intraradices* and other soil microorganisms as affected by organic matter. *Soil Biol. Biochem.* **2006**, *38*, 1008–1014. [CrossRef]
86. Scervino, J.M.; Ponce, M.A.; Erra-Bassells, R.; Vierheilig, H.; Ocampo, J.A.; Godeas, A. *Arbuscular mycorrhizal* colonization of tomato by *Gigaspora* and *Glomus* species in the presence of root flavonoids. *J. Plant Physiol.* **2005**, *162*, 625–633. [CrossRef]
87. Porcel, R.; Aroca, R.; Cano, C.; Bago, A.; Ruiz-Lozano, J.M. A gene from the *Arbuscular mycorrhical* fungus *Glomus intraradices* encoding a binding protein is up-regulated by drought stress in some mycorrhizal plants. *Environ. Exp. Bot.* **2007**, *60*, 251–256. [CrossRef]
88. Schubert, M.; Mourad, S.; Fink, S.; Schwarze, F.W.M.R. Ecophysiological responses of the biocontrol agent *Trichoderma atroviride* (T-15603.1) to combined environmental parameters. *Biol. Control* **2005**, *49*, 84–90. [CrossRef]
89. Reiter, B.; Sessitsch, A. Bacterial endophytes of the wildflower *Crocus albiflorus* analyzed by characterization of isolates and by a cultivation-independent approach. *Can. J. Microbiol.* **2006**, *52*, 140–149. [CrossRef]
90. Niznansky, L.; Varecka, L.; Kyrstofova, S. Disruption of GABA shunt affects *Trichoderma atroviride* response to nutritional and environmental stimuli. *Acta Chim. Solv.* **2016**, *9*, 109–113. [CrossRef]
91. Blaszczyk, L.; Siwulski, M.; Sobieralski, K.; Lisiecka, J.; Jedryczka, M. *Trichoderma* spp.—Application and prospects for use in organic farming and industry. *J. Plant. Prot. Res.* **2014**, *54*, 309–317. [CrossRef]

92. Wuczkowski, M.; Druzhinina, I.; Gherbawy, Y.; Klug, B.; Prillinger, H.; Kubicek, C.P. Species pattern and genetic diversity of *trichoderma* in a mid-European, primeval floodplain-forest. *Microbiol. Res.* **2003**, *158*, 125–133. [CrossRef]
93. Katenkamp, U.; Jacob, H.-E.; Kerns, G.; Dalchow, E. Hybridization of *Trichoderma reesei* protoplasts by electrofusion. *Bioelectrochem. Bioenerg.* **1989**, *22*, 57–67. [CrossRef]
94. Keshavarz, B.; Khalesi, M. *Trichoderma reesi*, a superior cellulose source for industrial applications. *Biofuels* **2016**, *7*, 713–721. [CrossRef]
95. Hashiba, T.; Narisawa, K. The development and endophytic nature of the fungus *Heteroconium chaetrospira*. *FEMS Microbiol. Lett.* **2005**, *252*, 191–196. [CrossRef]
96. Ohki, T.; Yonezawa, M.; Hashiba, T.; Masuya, H.; Usuki, F.; Narisawa, K.; Narisawa, K. Colonization process of the root endophytic fungus *Heteroconium chaetospira* in roots of Chinese cabbage. *Mycoscience* **2002**, *43*, 191–194. [CrossRef]
97. Orlandini, V.; Maida, I.; Fondi, M.; Perrin, E.; Papaleo, M.C.; Bosi, E.; Pascale, D.D.; Tutino, M.L.; Michaud, L.; Giudice, A.L.; et al. Genomic analysis of three sponge-associated *Arthrobacter antarcitic* strains, inhibiting the growth of *Burkholderia cepacia* complex bacteria by synthesizing volatile organic compounds. *Microbiol. Res.* **2014**, *169*, 593–601. [CrossRef]
98. Mukhia, S.; Khatri, A.; Acharya, V.; Kumar, R. Comparative genomics and molecular adaptational analysis of *Arthrobacter* from Sikkim Himalaya provided insights into its survivability under multiple high-altitude stress. *Genomics* **2021**, *113*, 151–158. [CrossRef]
99. Cacciari, I.; Lippi, D. *Arthrobacters*: Successful arid soil bacteria: A review. *Arid. Soil. Res. Rehabil.* **1987**, *1*, 1–30. [CrossRef]
100. Wu, H.-G.; Liu, W.-S.; Zhu, M.; Li, X.-X. Research and analysis of 74 bloodstream infection cases of *Acinetobacter baumannii* and drug resistance. *Eur. Rev. Med. Pharmacol. Sci.* **2018**, *22*, 1782–1786.
101. Khaksar, G.; Treesubsuntorn, C.; Thiravetyan, P. Impact of endophytic colonization patterns on *Zamioculcas amiifolia* stress response and in regulating ROS, tryptophan and IAA levels under airbone formaldehyde and formaldehyde-contaminated soil conditions. *Plant Physiol. Biochem.* **2017**, *114*, 1–9. [CrossRef]
102. Lin, H.-R.; Shu, H.-Y.; Lin, G.-H. Biological roles of indole-3-acetic acid in *Acinetobacter baumannii*. *Microbiol. Res.* **2018**, *216*, 30–39. [CrossRef] [PubMed]
103. Joe, M.M.; Devaraj, S.; Benson, A.; Sa, T. Isolation of phosphate solubilizing endophytic bacteria from *Phyllanthus amarus* schum & thonn: Evaluation of plant growth promotion an antioxidant activity under salt stress. *J. Appl. Res. Med. Aromat. Plants* **2016**, *3*, 71–77. [CrossRef]
104. Abraham, J.; Silambarasan, S. Biodegradation of carbendazim by *Rhodococcus erythropolis* and its plant growth-promoting traits. *Biol. Environ.* **2018**, *118*, 69–80. [CrossRef]
105. Hashem, A.; Tabassum, B.; Fathi Abd Allah, E. *Bacillus subtilis*: A plant-growth promoting rhizobacterium that also impacts biotic stress. *Saudi J. Biol. Sci.* **2019**, *26*, 1291–1297. [CrossRef] [PubMed]
106. Sun, W.; Shahrajabian, M.H.; Cheng, Q. Anise (*Pimpinella anisum* L.), a dominant spice and traditional medicinal herb for both food and medicinal purposes. *Cogent Biol.* **2019**, *5*, 1673688. [CrossRef]
107. Sun, W.; Shahrajabian, M.H.; Cheng, Q. Natural dietary and medicinal plants with anti-obesity therapeutics activities for treatment and prevention of obesity during lock down and in post-COVID-19 era. *Appl. Sci.* **2021**, *11*, 7889. [CrossRef]
108. Sun, W.; Shahrajabian, M.H.; Cheng, Q. Fenugreek cultivation with emphasis on historical aspects and its uses in traditional medicine and modern pharmaceutical science. *Mini Rev. Med. Chem.* **2021**, *21*, 724–730. [CrossRef]
109. Sun, W.; Shahrajabian, M.H.; Cheng, Q. Nitrogen fixation and diazotrophs—A review. *Rom. Biotechnol. Lett.* **2021**, *26*, 2834–2845. [CrossRef]
110. Shahrajabian, M.H.; Sun, W.; Cheng, Q. Traditional herbal medicine for the prevention and treatment of Cold and Flu in the autumn of 2020, overlapped with COVID-19. *Nat. Prod. Commun.* **2020**, *15*, 1934578X20951431. [CrossRef]
111. Shahrajabian, M.H.; Sun, W.; Soleymani, A.; Cheng, Q. Traditional herbal medicines to overcome stress, anxiety and improve mental health in outbreaks of human coronaviruses. *Phytother. Res.* **2020**, *35*, 1237–1247. [CrossRef]
112. Francis, P.B.; Earnest, L.D.; Bryant, K. Maize growth and yield response to a biostimulant amendment. *J. Crop Improv.* **2016**, *6*, 632–640. [CrossRef]
113. Marmitt, D.; Shahrajabian, M.H. Plant species used in Brazil and Asia regions with toxic properties. *Phytother. Res.* **2021**, *35*, 4703–4726. [CrossRef] [PubMed]
114. Bashan, Y.; Holguin, G. Proposal for division of plant growth promoting rhizobacteria into two classification: Biocontrol-PGPB (Plant growth-promoting bacteria) and PGPB. *Soil Biol. Biochem.* **1998**, *30*, 1225–1228. [CrossRef]
115. Tejada, M.; Rodriguez-Morgado, B.; Gomez, I.; Parrado, J. Degradation of chlorpyrifos using different biostimulants/biofertilizers: Effects on soil biochemical properties and microbial community. *Appl. Soil Ecol.* **2014**, *84*, 158–165. [CrossRef]
116. Vessey, J.K. Plant growth promoting rhizobacteria as biofertilizers. *Plant Soil* **2003**, *255*, 1225–1228. [CrossRef]
117. Younes, N.A.; Anik, T.R.; Rahman, M.M.; Wardany, A.A.; Dawood, M.F.A.; Tran, L.-S.P.; Latef, A.A.H.A.; Mostofa, M.G. Effects of microbial biostimulants (*Trichoderma album* and *Bacillus megaterium*) on growth, quality attributes, and yield of onion under field conditions. *Heliyon* **2023**, *9*, e14203. [CrossRef]
118. Jacobs-Hoffman, I.; Hills, P.N. Effects of the commercial biostimulant BC204 on the rhizosphere microbial community of *Solanum lycopersicum* L. *S. Afr. J. Bot.* **2021**, *143*, 52–60. [CrossRef]

119. Mandal, S.; Anand, U.; Lopez-Bucio, J.; Radha; Kumar, M.; Lal, M.K.; Tiwari, R.K.; Dey, A. Biostimulants and environmental stress mitigation in crops: A novel and emerging approach for agricultural sustainability under climate change. *Environ. Res.* **2023**, *233*, 116357. [CrossRef]
120. Costa, O.Y.A.; Pijl, A.; Houbraken, J.; va Lith, W.; Kuramae, E.E. Soil substrate source drives the microbes involved in the degradation of gelatin used as a biostimulant. *Appl. Soil Ecol.* **2023**, *189*, 104906. [CrossRef]
121. Berg, G. Plant-microbe interactions promoting plant growth and health: Perspectives for controlled use of microorganisms in agriculture. *Appl. Microbiol. Biotechnol.* **2009**, *84*, 11–18. [CrossRef]
122. Doss, I.C.; Ruiz-Lozano, J.M. Microbial enhancement of crop resource use efficiency. *Curr. Opin. Biotechnol.* **2012**, *23*, 236–242. [CrossRef]
123. Dalmastri, C.; Chiarini, L.; Cantale, C.; Bevivino, A.; Tabacchioni, S. Soil type and maize cultivar affect the genetic diversity of maize root-associated *Burkholderia cepacia* populations. *Microb. Ecol.* **1999**, *38*, 273–284. [CrossRef] [PubMed]
124. Remans, R.; Beebe, S.; Blair, M.; Manrique, G.; Tovar, E.; Rao, I.; Vanderleyden, J. Physiological and genetic analysis of root responsiveness to auxin-producing plant growth-promoting bacteria in common bean (*Phaseolus vulgaris* L.). *Plant Soil* **2008**, *302*, 149–161. [CrossRef]
125. Khalid, A.; Arshad, M.; Zahir, Z.A. Screening plant growth-promoting rhizobacteria for improving growth and yield of wheat. *J. Appl. Microbiol.* **2004**, *96*, 473–480. [CrossRef]
126. Begum, N.; Qin, C.; Ahanger, M.A.; Raza, S.; Khan, M.I.; Ashraf, M.; Ahmed, N.; Zhang, L. Role of arbuscular mycorrhizal fungi in plant growth regulation: Implications in abiotic stress tolerance. *Front. Plant Sci.* **2019**, *10*, 1068. [CrossRef]
127. Querejeta, J.I.; Barea, J.M.; Allen, M.F.; Caravaca, F.; Roldan, A. Differential response of δ13C and water use efficiency to arbuscular mycorrhizal infection in two arid land woody plant species. *Oecologia* **2003**, *135*, 510–515. [CrossRef]
128. Botti, A.; Musmeci, E.; Negroni, A.; Capuozzo, R.; Fava, F.; Biagi, E.; Zanaroli, G. Site-specific response of sediment microbial community to supplementation of polyhydroxyalkanoates as biostimulants for PCB reductive dechlorination. *Sci. Total Environ.* **2023**, *898*, 165485. [CrossRef]
129. Lindow, S.E.; Brandl, M.T. Microbiology of the phyllosphere. *Appl. Env. Microbiol.* **2003**, *69*, 1875–1883. [CrossRef]
130. Lim, J.H.; Kim, S.D. Induction of drought stress resistance by multi-functional PGPR *Bacillus licheniformis* K11 in pepper. *Plant Pathol. J.* **2013**, *29*, 201–208. [CrossRef]
131. Saia, S.; Colla, G.; Raimondi, G.; Di Stasio, E.; Cardarelli, M.; Bonini, P.; Vitaglione, P.; De Pascale, S.; Rouphael, Y. An endophytic fungi-based biostimulant modulated lettuce yield, physiological and functional quality responses to both moderate and severe water limitation. *Sci. Hortic.* **2019**, *256*, 108595. [CrossRef]
132. Mishra, P.K.; Mishra, S.; Selvakumar, G.; Bisht, S.C.; Bisht, J.K.; Kundu, S.; Gupta, H.S. Characterization of a psychrotolerant plant growth promoting *Pseudomonas* sp. strain PGERs17 (MTCC 9000) isolated from North Western Indian Himalayas. *Ann. Microbiol.* **2008**, *58*, 561–568. [CrossRef]
133. Ali, S.Z.; Sandhya, V.; Grover, M.; Kishore, N.; Rao, L.V.; Venkateswarlu, B. *Pseudomonas* sp. strain AKM-P6 enhances tolerance of sorghum seedlings to elevated temperatures. *Biol. Fertil. Soils* **2009**, *46*, 45–55. [CrossRef]
134. Theocharis, A.; Bordiex, S.; Fernandez, O.; Paquis, S.; Dhondt-Cordelier, S.; Baillieul, F.; Clement, C.; Barka, E.A. *Burkholderia phytofirmans* PsJN primes *Vitis vinifera* L. and confers a better tolerance to low nonfreezing temperatures. *Mol. Plant Microbe Interact.* **2012**, *25*, 241–249. [CrossRef] [PubMed]
135. Tiryaki, D.; Aydin, I.; Atici, P. Psychrotolerant bacteria isolated from the lead apoplast of cold-adapted wild plants improve the cold resistance of bean (*Phaseolus vulgaris* L.) under low temperature. *Cyrobiology* **2019**, *86*, 111–119. [CrossRef]
136. Bhalerao, R.P.; Eklof, J.; Ljung, K.; Marchant, A.; Bennett, M.; Sandberg, G. Shoot-derived auxin is essential for early lateral root emergence in Arabidopsis seedlings. *Plant J.* **2002**, *29*, 32–33. [CrossRef]
137. Romano, I.; Ventorino, V.; Pepe, O. Effectiveness of plant beneficial microbes: Overview of the methodological approaches for the assessment of root colonization and persistence. *Front. Plant Sci.* **2020**, *11*, 6. [CrossRef]
138. Mangman, J.S.; Deaker, R.; Rogers, G. Optimal plant growth-promoting concentration of *Azospirillum brasilense* inoculated to cucumber, lettuce, and tomato seeds varies between bacterial strains. *Isr. J. Plant Sci.* **2015**, *62*, 145–152. [CrossRef]
139. He, Y.; Pantigoso, H.A.; Wu, Z.; Vivanco, J.M. Co-inoculation of *Bacillus* sp. and *Pseudomonas putida* at different development stages acts as a biostimulant to promote growth, yield, and nutrient uptake of tomato. *J. Appl. Microbiol.* **2019**, *127*, 196–207. [CrossRef]
140. Pellegrino, M.; Spera, D.; Ercole, C.; del Gallo, M. *Allium cepa* L. seed inoculation with a consortium of plant growth-promoting bacteria: Effects on plant growth and development and soil fertility status and microbial community. *Proceedings* **2020**, *6*, 20. [CrossRef]
141. Kumar, P.; Erturk, V.S.; Almusawa, H. Mathematical structure of mosaic disease using microbial biostimulants via Caputo and Atangana-Baleanu derivatives. *Results Phys.* **2021**, *24*, 104186. [CrossRef]
142. Tanveer, Y.; Jahangir, S.; Shah, Z.A.; Yasmin, H.; Nosheen, A.; Hassan, M.N.; Illyas, N.; Bajguz, A.; El-Sheikh, M.A.; Ahmad, P. Zinc oxide nanoparticles mediated biostimulant impact on cadmium detoxification and *in silico* analysis of zinc oxide-cadmium networks in *Zea mays* L. regulome. *Environ. Pollut.* **2023**, *316*, 120641. [CrossRef] [PubMed]
143. Bozhinova, R. Yield and chemical composition of oriental tobacco as affected by biostimulant application. *Bulg. J. Agric. Sci.* **2023**, *29*, 89–96.

144. Pavithra, D.; Yapa, N. Arbuscular mycorrhizal fungi inoculation enhances drought stress tolerance of plants. *Groundw. Sustain Dev.* **2018**, *7*, 49–494. [CrossRef]
145. Goicoechea, N.; Antolin, M.C. Increased nutritional value in food crops. *Microb. Biotechnol.* **2017**, *10*, 1004–1007. [CrossRef] [PubMed]
146. Goicoechea, N.; Bettoni, M.M.; Fuertes-Mendizabal, T.; Gonzalez-Murua, C.; Aranjuelo, I.; Goicoechea, N.; Bettoni, M.M.; Fuertes-Mendizabal, T.; Gonzalez-Murua, C.; Aranjuelo, I. Durum wheat quality traits affected by mycorrhizal inoculation, water availability and atmospheric $CO_2$ concentration. *Crop Pasture Sci.* **2016**, *67*, 147–155. [CrossRef]
147. Zhao, R.; Guo, W.; Bi, M.; Guo, J.; Wang, L.; Zhao, J.; Zhang, J. Arbuscular mycorrhizal fungi affect the growth, nutrient uptake and water status of maize (*Zea mays* L.) grown in two types of coal mine spoils under drought stress. *Appl. Soil Ecol.* **2015**, *88*, 41–49. [CrossRef]
148. Boyer, L.R.; Brain, P.; Xu, X.M.; Jeffries, P. Inoculation of drought-stressed strawberry with a mixed inoculum of two arbuscular mycorrhizal fungi: Effects on population dynamics of fungal species in roots and consequential plant tolerance to water deficiency *Mycorrhiza* **2015**, *25*, 215–227. [CrossRef]
149. Asrar, A.-W.A.; Elhindi, K.M. Alleviation of drought stress of marigold (*Tagetes erecta*) plants by using arbuscular mycorrhizal fungi. *Saudi J. Bio. Sci.* **2011**, *18*, 93–98. [CrossRef]
150. Tsoata, E.; Njock, S.R.; Youmbi, E.; Nwaga, D. Early effects of water stress on some biochemical and mineral parameters of mycorrhizal *Vigna subterranea* (L.) Verdc. (Fabaceae) cultivated in cameroon. *Int. J. Agron. Agric. Res.* **2015**, *7*, 21–35.
151. Zhang, F.; He, J.D.; Ni, Q.D.; Wu, Q.S.; Zou, Y.N. Enhancement of drought tolerance in trifoliate orange by mycorrhiza: Changes in root sucrose and proline metabolism. *Not. Bot. Horti Agrobot.* **2018**, *46*, 270–276. [CrossRef]
152. Pedranzani, H.; Rodriguez-Rivera, M.; Gutierrez, M.; Porcel, R.; Hause, B.; Ruiz-Lozano, J.M. Arbuscular mycorrhizal symbiosis regulates physiology and performance of *Digitaria eriantha* plants subjected to abiotic stresses by modulating antioxidant and jasmonate levels. *Mycorrhiza* **2016**, *26*, 141–152. [CrossRef] [PubMed]
153. Yooyongwech, S.; Samphumphuang, T.; Tisarum, R.; Theerawitaya, C.; Cha-Um, S. Arbuscular mycorrhizal fungi (AMF) improved water deficit tolerance in two different sweet potato genotypes involves osmotic adjustments via soluble sugar and free proline. *Sci. Hortic.* **2016**, *198*, 107–117. [CrossRef]
154. Mirshad, P.P.; Puthur, J.T. Arbuscular mycorrhizal association enhances drought tolerance potential of promising bioenergy grass (*Saccharum arundinaceum* Retz.). *Environ. Monit. Assess.* **2016**, *188*, 425. [CrossRef] [PubMed]
155. Khalloufi, M.; Martinez-Andujar, C.; Lachaal, M.; Karray-Bouraoui, N.; Perez-Alfocea, F.; Albacete, A. The interaction between foliar $GA_3$ application and arbuscular mycorrhizal fungi inoculation improves growth in salinized tomato (*Solanum lycopersicum* L.) plants by modifying the hormonal balance. *J. Plant Physiol.* **2017**, *214*, 134–144. [CrossRef]
156. Hajiboland, R.; Dashtebani, F.; Aliasgharzad, N. Physiological responses of halophytic $C_4$ grass *Aeluropus littoralis* to salinity and arbuscular mycorrhizal fungi colonization. *Photosynthetica* **2015**, *53*, 572–584. [CrossRef]
157. Pasbani, B.; Salimi, A.; Aliasgharzad, N.; Hajiboland, R. Colonization with arbuscular mycorrhizal fungi mitigates cold stress through improvement of antioxidant defense and accumulation of protecting molecules in eggplants. *Sci. Hortic.* **2020**, *272*, 109575. [CrossRef]
158. Calvo-Polanco, M.; Sanchez-Romera, B.; Aroca, R.; Asins, M.J.; Declerck, S.; Dodd, I.C.; Martinez-Andujar, C.; Albacete, A.; Ruiz-Lozano, J.M. Exploring the use of recombinant inbreed lines in combination with beneficial microbial inoculants (AM Fungus and PGPR) to improve drought stress tolerance in tomato. *Environ. Exp. Bot.* **2016**, *131*, 47–57. [CrossRef]
159. Carillo, P.; Kyratzis, A.; Kyriacou, M.C.; Dell'Aversana, E.; Fusco, G.M.; Corrado, G.; Rouphael, Y. Biostimulatory action of arbuscular mycorrhizal fungi enhances productivity, functional and sensory quality in Piennolo del Vesuvio cherry tomato landraces. *Agronomy* **2020**, *10*, 911. [CrossRef]
160. Krishna, H.; Singh, S.K.; Sharma, R.R.; Khawale, R.N.; Grover, M.; Patel, V.B. Biochemical changes in microporpagated grape (*Vitis vinifera* L.) plantlets due to arbuscular-mycorrhizal fungi (AMF) inoculation during ex vitro acclimatization. *Sci. Hortic.* **2005**, *106*, 554–567. [CrossRef]
161. Aalipour, H.; Nikbakht, A.; Etemadi, N.; Rejali, F.; Soleimani, M. Biochemical response and interactions between arbuscular mycorrhizal fungi and plant growth promoting rhizobacteria during establishment and stimulating growth of Arizona cypress (*Cupressus arizonica* G.) under drought stress. *Sci. Hortic.* **2020**, *261*, 108923. [CrossRef]
162. Barnawal, D.; Bharti, M.; Maji, D.; Chanotiya, C.S.; Kalra, A. ACC deaminase-containing Arthrobacter protophormiae induces NaCl stress tolerance thtough reduced ACC oxidase activity and ethylene production resulting in improved nodulation and mycorrhization in *Pisum sativum*. *J. Plant Physiol.* **2014**, *171*, 884–894. [CrossRef] [PubMed]
163. Cely, M.V.T.; Siviero, M.A.; Emiliano, J.; Spago, F.R.; Freiras, V.F.; Barazetti, A.R.; Goya, E.T.; Lamberti, G.D.S.; Dos Santos, I.M.O.; De Oliveira, A.G.; et al. Inoculation of *Schizolobium parahyba* with mycorrhizal fungi and plant growth-promoting rhizobacteria increases wood yield under field conditions. *Front. Plant Sci.* **2016**, *7*, 1708. [CrossRef] [PubMed]
164. Dal Cortivo, C.; Barion, G.; Visioli, G.; Mattarozzi, M.; Mosca, G.; Vamerali, T. Increased root growth and nitrogen accumulation in common wheat following PGPR inoculation: Assessment of plant-microbe interactions by ESEM. *Agric. Ecosyst. Environ.* **2017**, *247*, 396–408. [CrossRef]
165. Miceli, A.; Moncada, A.; Vetrano, F. Use of microbial biostimulants to increase the salinity tolerance of vegetable transplants. *Agronomy* **2021**, *11*, 1143. [CrossRef]

166. Kumar, M.; Poonam; Ahmad, S.; Singh, R. Plant growth promoting microbes: Diverse roles for sustainable and ecofriendly agriculture. *Energy Nexus* **2022**, *7*, 100133. [CrossRef]
167. Kumari, M.; Swarupa, P.; Kesari, K.K.; Kumar, A. Microbial inoculants as plant biostimulants: A review on risk status. *Life* **2023**, *13*, 12. [CrossRef]
168. Joshi, M.; Parewa, H.P.; Joshi, S.; Sharma, J.K.; Shukla, U.N.; Paliwal, A.; Gupta, V. Chapter 5—Use of Microbial Biostimulants in Organic Farming. In *Advances in Organic Farming*; Meena, V.S., Meena, S.K., Srinivasarao, C., Eds.; Woodhead Publishing: Sawston, UK, 2021; pp. 59–73.
169. Ganugi, P.; Martinelli, E.; Lucini, L. Microbial biostimulants as a sustainable approach to improve the functional quality in plant-based foods: A review. *Curr. Opin. Food Sci.* **2021**, *41*, 217–223. [CrossRef]
170. Mansoor, S.; Sharma, V.; Mir, M.A.; Mir, J.I.; Nabi, S.U.; Ahmed, N.; Alkahtani, J.; Alwahibi, M.S.; Masoodi, K.Z. Quantification of polyphenolic compounds and relative gene expression studies of phenylpropanoid pathway in apple (*Malus domestica* Borkh) in response to Venturia inaequalis infection. *Saudi J. Biol. Sci.* **2020**, *27*, 3397–3404. [CrossRef]
171. Heil, M.; Bostock, R. Induced systemic resistance (ISR) against pathogens in the content of induced plant defences. *Ann. Bot.* **2002**, *89*, 503–512. [CrossRef]
172. Shahrajabian, M.H.; Sun, W.; Cheng, Q. The importance of *rhizobium*, agrobacterium, bradyrhizobium, herbaspirillum, sinorhizobium in sustainable agricultural production. *Not. Bot. Horti Agrobot.* **2021**, *49*, 12183. [CrossRef]
173. Shahrajabian, M.H.; Cheng, Q.; Sun, W. Using bacteria and fungi as plant biostimulants for sustainable agricultural systems. *Recent Pat. Biotechnol.* **2020**, *17*, 206–244. [CrossRef] [PubMed]
174. Behera, B.; Supraja, K.V.; Paramasivan, B. Integrated microalgal biorefinery for the production and application of biostimulants in circular bioeconomy. *Bioresour. Technol.* **2021**, *339*, 125588. [CrossRef] [PubMed]
175. Mian, G.; Belfiore, N.; Musetti, R.; Tomasi, D.; Cantone, P.; Lovat, L.; Lupinelli, S.; Iacumin, L.; Celotti, E.; Golinelli, F. Effect of a triacontanol-rich biostimulant on the ripening dynamic and wine must technological parameters in *Vitis vinifera* cv. Ribolla Gialla. *Plant Physiol. Biochem.* **2022**, *188*, 60–69. [CrossRef] [PubMed]
176. Su, H.; Kuang, X.; Ren, Y.; Luo, L. Biostimulants promote biodegraation of n-hexadecane by *Raoultella planticola*: Generation of lipopeptide biosurfactants. *J. Environ. Chem. Engin.* **2022**, *10*, 108382. [CrossRef]
177. Etesami, H.; Jeong, B.R.; Glick, B.R. Potential use of *Bacillus* spp. as an effective biostimulant against abiotic stresses in crops—A review. *Curr. Res. Biotechnol.* **2023**, *5*, 100128. [CrossRef]
178. Hao, J.; Tan, J.; Zhang, Y.; Wang, S.; Zhang, X.; Wang, Z.; Li, J. Metabolomics reveals the molecular mechanism of sewage sludge-derived nutrients and biostimulants stimulating resistance enhancement and the redistribution of carbon and nitrogen metabolism in *pakchoi* cabbage. *Sci. Total Environ.* **2023**, *891*, 164330. [CrossRef]
179. Tolisano, C.; Del Buono, D. Biobased: Biostimulants and biogenic nanoparticles enter the scene. *Sci. Total Environ.* **2023**, *885*, 163912. [CrossRef]
180. Saia, S.; Corrado, G.; Vitaglione, P.; Colla, G.; Bonini, P.; Giordano, M.; Stasio, E.D.; Raimondi, G.; Sacchi, R.; Rouphael, Y. An endophytic fungi-based biostimulant modulates volatile and non-volatile secondary metabolites and yield of greenhouse basil (*Ocimum basilicum* L.) through variable mechanisms dependent on salinity stress level. *Pathogens* **2021**, *10*, 797 [CrossRef]
181. Mukherjee, A.; Singh, S.; Gaurav, A.K.; Chouhan, G.K.; Jaiswal, D.K.; Pereiro, A.P.D.A.; Passari, A.K.; Abdel-Azeem, A.M.; Verma, J.P. Harnessing of phytomicrobiome for developing potential biostimulant consortium for enhancing the productivity of chickpea and soil health under sustainable agriculture. *Sci. Total Environ.* **2022**, *836*, 155550. [CrossRef]
182. Mrid, R.B.; Benmrid, B.; Hafsa, J.; Boukcim, H.; Sobeh, M.; Yasri, A. Secondary metabolites as biostimulant and bioprotectant agents: A review. *Sci. Total Environ.* **2021**, *777*, 146204. [CrossRef]
183. Kaushal, P.; Ali, N.; Saini, S.; Pati, P.K.; Pati, A.M. Physiological and molecular insight of microbial biostimulants for sustainable agriculture. *Front. Plant Sci.* **2023**, *14*, 1041413. [CrossRef] [PubMed]
184. Zulfiqar, F.; Casadesus, A.; Brockman, H.; Munne-Bosch, S. An overview of plant-based natural biostimulants for sustainable horticulture with a particular focus on moringa leaf extracts. *Plant Sci.* **2020**, *295*, 110194. [CrossRef]
185. Buono, D.D. Can biostimulants be used to mitigate the effect of anthropogenic climate change on agriculture? It is time to respond. *Sci. Total Environ.* **2021**, *751*, 141763. [CrossRef] [PubMed]
186. Mattarozzi, M.; Di Zinno, J.; Montanini, B.; Manfredi, M.; Marengo, E.; Fornasier, F.; Ferrarini, A.; Careri, M.; Visioli, G. Biostimulants applied to maize seeds modulate the enzymatic activity and metaproteome of the rhizosphere. *Appl. Soil Ecol.* **2020**, *148*, 103480. [CrossRef]
187. Alfonzetti, M.; Doleac, S.; Mills, C.H.; Gallagher, R.V.; Tetu, S. Characterizing effects of microbial biostimulants and whole-soil inoculums for native plant revegetation. *Microorganisms* **2023**, *11*, 55. [CrossRef] [PubMed]
188. Lephatsi, M.; Nephali, L.; Meyer, V.; Piater, L.A.; Buthelezi, N.; Dubery, I.A.; Opperman, H.; Brand, M.; Huyser, J.; Tugizimana, F. Molecular mechanisms associated with microbial biostimulant-mediated growth enhancement, priming and drought stress tolerance in maize plants. *Sci. Rep.* **2022**, *12*, 10450. [CrossRef]
189. Aroca, R.; Rosa, P.; Ruiz-Lozano, J.M. How does arbuscular mycorrhizal symbiosis regulate root hydraulic properties and plasma membrane aquaporins in *Phaseolus vulgaris* under drought, cold or salinity stresses? *N. Phytol.* **2007**, *173*, 808–816. [CrossRef]
190. Khan, A.L.; Hussain, J.; Al-Harrasi, A.; Al-Rawahi, A.; Lee, I.J. Endophytic fungi: Resource for gibberellins and crop abiotic stress resistance. *Crit. Rev. Biotechnol.* **2015**, *35*, 62–76. [CrossRef]

191. Shubler, A.; Schwarzott, D.; Walker, C. A new fungal phylum, the *Glomeromycota*: Phylogeny and evolution. *Mycol. Res.* **2001**, *105*, 1413–1421. [CrossRef]
192. Shahrajabian, M.H.; Chaski, C.; Polyzos, N.; Petropoulos, S.A. Biostimulants application: A low input cropping management tool for sustainable farming of vegetables. *Biomolecues* **2021**, *11*, 698. [CrossRef]
193. Sun, W.; Shahrajabian, M.H. The effectiveness of *rhizobium* bacteria on soil fertility and sustainable crop production under cover and catch crops management and green manuring. *Not. Bot. Horti Agrobot.* **2022**, *50*, 12560. [CrossRef]
194. Sun, W.; Shahrajabian, M.H.; Cheng, Q. Archaea, bacteria and termite, nitrogen fixation and sustainable plants production. *Not. Bot. Horti Agrobot.* **2021**, *49*, 12172. [CrossRef]
195. Shahrajabian, M.H.; Sun, W. Potential roles of longan as a natural remedy with tremendous nutraceutical values. *Curr. Nutr. Food Sci.* **2023**, *19*, 888–895. [CrossRef]
196. Lanfranco, L.; Bonfante, P. Lessons from arbuscular mycorrhizal fungal genomes. *Curr. Opin. Microbiol.* **2023**, *75*, 102357. [CrossRef]
197. Huang, H.; Liu, S.; Du, Y.; Tang, J.; Hu, L.; Chen, X. Carbon allocation mediated by arbuscular mycorrhizal fungi alters the soil microbial community under various phosphorus levels. *Fungal Ecol.* **2023**, *62*, 101227. [CrossRef]
198. Chen, E.C.H.; Morin, E.; Beaudet, D.; Noel, J.; Yildirir, G.; Ndikumana, S.; Charron, P.; St-Onge, C.; Giorgi, J.; Kruger, M.; et al. High intraspecific genome diversity in the model arbuscular mycorrhizal symbiont *Rhizophagus irregularis*. *N. Phytol.* **2018**, *220*, 1161–1171. [CrossRef]
199. Tester, M.; Smith, S.E.; Smith, F.A. The phenomenon of nonmycorrhizal plants. *Can. J. Bot.* **1987**, *65*, 419–431. [CrossRef]
200. Wang, B.; Qiu, Y.L. Phylogenetic distribution and evolution of mycorrhizas in land plants. *Mycorrhiza* **2006**, *16*, 299–363. [CrossRef]
201. Kan, Z.-R.; Wei, Q.; Yang, R.; Li, Y.; Zhou, J.; Qi, J.; Li, F.-M.; Yang, H. Arbuscular mycorrhizal fungi mitigate earthworm-induced $N_2O$ emissions from upland soil in a rice-rotated wheat farming system. *Appl. Soil Ecol.* **2023**, *189*, 104981. [CrossRef]
202. Ng, A.; Wilson, B.A.L.; Frew, A. Belowground crop responses to root herbivory are associated with the community structure of native arbuscular mycorrhizal fungi. *Appl. Soil Ecol.* **2023**, *185*, 104797. [CrossRef]
203. Zheng, X.; Liu, Q.; Chen, X.; Cao, M.; Wu, F.; Li, W.; Zhang, L.; Liu, S.; Jiang, J. Arbuscular mycorrhizal fungi decrease soil ammonium availability and nitrous oxide emissions under nitrogen input. *Agric. For. Meteorol.* **2023**, *333*, 109385. [CrossRef]
204. Yang, X.; Ma, Y.; Zhang, J.; Bai, H.; Shen, Y. How arbuscular mycorrhizal fungi drives herbaceous plants' C: N: P stoichiometry? A meta-analysis. *Sci. Total Environ.* **2023**, *862*, 160807. [CrossRef] [PubMed]
205. Fu, W.; Yan, M.; Zhao, L.; Zeng, X.; Cai, B.; Qu, S.; Wang, S. Inoculation with arbuscular mycorrhizal fungi increase calcium uptake in *Malus robusta*. *Sci. Hortic.* **2023**, *321*, 112295. [CrossRef]
206. Kimura, T.; Ueno, M.; Petchsang, N.; Nishiyama, K. Barley uptake of ZnO nanoparticles enhanced by asrbuscular mycorrhizal fungi: Implications for biomarker systems. *J. Mol. Liq.* **2023**, *386*, 122496. [CrossRef]
207. Zhao, Y.; Naeth, M.A. Soil amendment with a humic substance and arbuscular mycorrhizal fungi enhance coal mine reclamation. *Sci. Total Environ.* **2022**, *823*, 153696. [CrossRef]
208. Mpongwana, S.; Manyevere, A.; Mupangwa, J.; Mpendulo, C.T.; Mashamaite, C.V. Optimizing biomass yield of three herbaceous forage legumes through dual inoculation of Arbuscular Mycorrhizal Fungi and *Rhizobia*. *S. Afr. J. Bot.* **2023**, *159*, 61–71. [CrossRef]
209. Khaliq, A.; Perveen, S.; Alamer, K.H.; Zia Ul Haq, M.; Rafique, Z.; Alsudays, I.M.; Althobaiti, A.T.; Saleh, M.A.; Hussain, S.; Attia, H. Arbuscular mycorrhizal fungi symbiosis to enhance plant-soil interaction. *Sustainability* **2022**, *14*, 7840. [CrossRef]
210. Figueira-Galan, D.; Heupel, S.; Duelli, G.; Morgano, M.T.; Stapf, D.; Requena, N. Exploring the synergistic effects of biochar and arbuscular mycorrhizal fungi on phosphorus acquisition in tomato plants by using gene expression analyses. *Sci. Total Environ.* **2023**, *884*, 163506. [CrossRef]
211. Wei, H.; He, W.; Kuang, Y.; Wang, Z.; Wang, Y.; Hu, W.; Tang, M.; Chen, H. Arbuscular mycorrhizal symbiosis and melatonin synergistically suppress heat-induced leaf senescence involves in abscisic acidm gibberellin, and cytokinin-mediated pathways in perennial ryegrass. *Enviorn. Experim. Bot.* **2023**, *213*, 105436. [CrossRef]
212. Wang, L.; Chen, X.; Wang, S.; Du, Y.; Zhang, D.; Tang, Z. Effects of arbuscular mycorrhizal symbiosis on the growth and reproduction of cherry tomato can be persistent to the next generation. *Eur. J. Soil Biol.* **2022**, *112*, 103429. [CrossRef]
213. Han, Z.; Zhang, Z.; Li, Y.; Wang, B.; Xiao, Q.; Li, Z.; Geng, X.; Lin, K.; Huang, T.; Li, X.; et al. Effect of arbuscular mycorrhizal fungi (AMF) inoculation on endophytic bacteria of lettuce. *Physiol. Mol. Plant Pathol.* **2023**, *126*, 102036. [CrossRef]
214. Betancur-Agudelo, M.; Meyer, E.; Lovato, P.E. Increased copper concentrations in soil affect indigenous arbuscular mycorrhizal fungi and physiology of grapevine plantlets. *Rhizosphere* **2023**, *27*, 100711. [CrossRef]
215. Li, X.; Zhou, M.; Shi, F.; Meng, B.; Liu, J.; Mi, Y.; Dong, C.; Su, H.; Liu, X.; Wang, F.; et al. Influence of arbuscular mycorrhizal fungi on mercury accumulation in rice (*Oryza sativa* L.): From enriches isotope tracing perspective. *Ecotoxicol. Environ. Saf.* **2023**, *255*, 114776. [CrossRef] [PubMed]
216. Pan, J.; Cao, S.; Xu, G.; Rehman, M.; Li, X.; Luo, D.; Wang, C.; Fang, W.; Xiao, H.; Liao, C.; et al. Comprehensive analysis reveals the underlying mechanism of arbuscular mycorrhizal fungi in kenaf cadmium stress alleviation. *Chemosphere* **2023**, *314*, 137566. [CrossRef] [PubMed]
217. Chen, Z.; Hu, B.; Hu, S.; Vogel-Mikus, K.; Pongrac, P.; Vymazal, J. Immobilization of chromium enhanced by arbuscular mycorrhizal fungi in semi-aquatic habitats with biochar addition. *J. Hazard. Mat.* **2022**, *439*, 129562. [CrossRef] [PubMed]
218. Shafiei, F.; Shahidi-Noghabi, S.; Sedaghati, E. The impact of arbuscular mycorrhizal fungi on tomato plant resistance against *Tuta absoluta* (Meyrick) in greenhouse conditions. *J. Asia Pac. Entomol.* **2022**, *25*, 101971. [CrossRef]

219. Li, Z.; Wu, S.; Liu, Y.; Yi, Q.; Hall, M.; Saha, N.; Wang, J.; Huang, Y.; Huang, L. Arbuscular mycorrhizal fungi regulate plant mineral nutrient uptake and partitioning in iron ore tailings undergoing eco-engingeered pedogenesis. *Pedosphere* **2023**. [CrossRef]
220. Falcao, E.L.; Muniz, B.C.; Filh, C.J.A.B.; Kapoor, R.; da Silva, F.S.B. Soil microbial respiration and pH modulated by arbuscular mycorrhizal fungi influence the biosynthesis of health-promoting compounds in *Anadenanthera colubrina* (Vell.) Brenan. *Rhizosphere* **2023**, *26*, 100685. [CrossRef]
221. Qin, Z.; Peng, Y.; Yang, G.; Feng, G.; Christie, P.; Zhou, J.; Zhang, J.; Li, X.; Gai, J. Relationship between phosphorus uptake via indigenous arbuscular mycorrhizal fungi and crop response: A $^{32}$P-labeling study. *Appl. Soil Ecol.* **2022**, *180*, 104624. [CrossRef]
222. Zhao, W.; Chen, Z.; Yang, X.; Sheng, L.; Mao, H.; Zhu, S. Metagenomics reveal arbuscular mycorrhizal fungi altering functional gene expression of rhizosphere microbial community to enhance *Iris tectorum*'s resistance to Cr stress. *Sci. Total Environ.* **2023**, *895*, 164970. [CrossRef]
223. Tan, Q.; Guo, Q.; Wei, R.; Zhu, G.; Du, C.; Hu, H. Influence of arbuscular mycorrhizal fungi on bioaccumulation and bioavailability of As and Cd: A meta-analysis. *Environ. Pollut.* **2023**, *316*, 120619. [CrossRef] [PubMed]
224. Zhu, Q.; Xu, P.; Lei, L.; Jing, Y. Transcriptome analysis reveals decreased accumulation and toxicity of Cd in upland rice inoculated with arbuscular mycorrhizal fungi. *Appl. Soil Ecol.* **2022**, *177*, 104501. [CrossRef]
225. Liu, S.; Lu, X.; Yang, G.; He, C.; Shi, Y.; Li, C.; Liu, S.; Wang, Y.; Wang, Z.; Chen, L.; et al. Variation of arbuscular mycorrhizal fungi communities in the rhizosphere soil of *Eucalyptus* plantations based on different stand ages and its effect on phosphorus fractionation. *Appl. Soil Ecol.* **2023**, *189*, 104908. [CrossRef]
226. Lv, Y.; Liu, J.; Fan, Z.; Fang, M.; Xu, Z.; Ban, Y. The function and community structure of arbuscular mycorrhizal fungi in ecological floating beds used for remediation of Pb contaminated wastewater. *Sci. Total Environ.* **2023**, *872*, 162233. [CrossRef]
227. Wang, M.; Zhang, R.; Zhao, L.; Wang, H.; Chen, X.; Mao, Z.; Yin, C. Indigenous arbuscular mycorrhizal fungi enhance resistance of apple rootstock M9T337 to apple replant disease. *Physiol. Mol. Plant Pathol.* **2021**, *116*, 101717. [CrossRef]
228. Qiu, Y.-J.; Zhang, N.-L.; Zhang, L.-L.; Zhang, X.-L.; Wu, A.-P.; Huang, J.-Y.; Yu, S.-Q.; Wang, Y.-H. Mediation of arbuscular mycorrhizal fungi on growth and biochemical parameters of *Ligustrum vicaryi* in response to salinity. *Physiol. Mol. Plant Pathol.* **2020**, *112*, 101522. [CrossRef]
229. Zhou, J.; Zhang, L.; Feng, G.; George, T.S. Arbuscular mycorrhizal fungi have a greater role than root hairs of maize for priming the rhizosphere microbial community and enhancing rhizosphere organic P mineralization. *Soil Biol. Biochem.* **2022**, *171*, 108713. [CrossRef]
230. Liu, X.-Q.; Cheng, S.; Aroca, R.; Zou, Y.-N.; Wu, Q.-S. Arbuscular mycorrhizal fungi induce flavonoid synthesis for mitigating oxidative damage of trifoliate orange under water stress. *Environ. Exp. Bot.* **2022**, *204*, 105089. [CrossRef]
231. Oliveira, E.P.D.; Soares, P.P.D.S.; Santos, G.L.; Coutrim, R.L.; Assis, F.G.D.V.D.; Miguesl, D.L.; Leal, P.L. Single inoculation with arbuscular mycorrhizal fungi promotes superior or similar effects on cowpea growth compared to co-inoculation with *Bradyrhizobium*. *S. Afr. J. Bot.* **2022**, *151*, 941–948. [CrossRef]
232. Naseer, M.; Zhu, Y.; Li, F.-M.; Yang, Y.-M.; Wang, S.; Xiong, Y.-C. Nano-enabled improvements of growth and colonization rate in wheat inoculated with arbuscular mycorrhizal fungi. *Environ. Pollut.* **2022**, *295*, 118724. [CrossRef]
233. Cheng, X.-F.; Xie, M.-M.; Li, Y.; Liu, B.-Y.; Liu, C.-Y.; Wu, Q.-S.; Kuca, K. Effects of field inoculation with arbuscular mycorrhizal fungi and endophytic fungi on fruit quality and soil properties of Newhall navel orange. *Appl. Soil Ecol.* **2022**, *170*, 104308. [CrossRef]
234. Alves de Assis, R.M.; Carneiro, J.J.; Medeiros, A.P.R.; Carvalho, A.A.D.; Honorato, A.D.C.; Carneiro, M.A.C.; Bertolucci, S.K.V.; Pinto, J.E.B.P. Arbuscular mycorrhizal fungi and organic manure enhance growth and accumulation of citral, total phenols, and flavonoids in *Melissa officinalis* L. *Ind. Crops Prod.* **2020**, *158*, 112981. [CrossRef]
235. Baczek, K.B.; Wisniewska, M.; Przybyl, J.L.; Kosakowska, O.; Weglarz, Z. Arbuscular mycorrhizal fungi in chamomile (*Matricaria recutita* L.) organic cultivation. *Ind. Crops Prod.* **2019**, *140*, 111562. [CrossRef]
236. Chen, K.; Kleijn, D.; Scheper, J.; Fijen, T.P.M. Additive and synergistic effects of arbuscular mycorrhizal fungi, insect pollination and nutrient availability in a perennial fruit crop. *Agric. Ecosyst. Environ.* **2022**, *325*, 107742. [CrossRef]
237. Ndiate, N.I.; Qun, C.L.; Nkoh, J.N. Importance of soil amendments with biochar and/or *Arbuscular Mycorrhizal* fungi to mitigate aluminum toxicity in tamarind (*Tamarindus indica* L.) on an acidic soil: A greenhouse study. *Heliyon* **2022**, *8*, e09009. [CrossRef] [PubMed]
238. Poveda, J.; Baptista, P. Filamentous fungi as biocontrol agents in olive (*Olea europaea* L.) diseases: Mycorrhizal and endophytic fungi. *Crop Prot.* **2021**, *146*, 105672. [CrossRef]
239. Elhindi, K.M.; El-Din, A.S.; Elgorban, A.M. The impact of arbuscular mycorrhizal fungi in mitigating salt-induced adverse effects in sweet basil (*Ocimum basilicum* L.). *Saudi J. Biol. Sci.* **2017**, *24*, 170–179. [CrossRef]
240. Yan, Z.; Ma, T.; Guo, S.; Liu, R.; Li, M. Leaf anatomy, photosynthesis and chlorophyll fluorescence of lettuce as influenced by arbuscular mycorrhizal fungi under high temperature stress. *Sci. Hortic.* **2021**, *280*, 109933. [CrossRef]
241. Hassena, A.B.; Zouari, M.; Trabelsi, L.; Decou, R.; Amar, F.B.; Chaari, A.; Sousa, N.; Labrousse, P.; Khabou, W; Zouari, N. Potential effects of arbuscular mycorrhizal fungi in mitigating the salinity of treated wastewater in young olive plants (*Olea europaea* L. cv. Chetoui). *Agric. Water Manag.* **2021**, *245*, 106635. [CrossRef]
242. Mohandas, S. Arbuscular mycorrhizal fungi benefit mango (*Mangifera indica* L.) plant growth in the field. *Sci. Hortic.* **2012**, *143*, 43–48. [CrossRef]

243. Rouphael, Y.; Lucini, L.; Miras-Moreno, B.; Colla, G.; Bonini, P.; Cardarelli, M. Metabolomic responses of maize shoots and roots elicited by combinatorial seed treatments with microbial and non-microbial biostimulants. *Front. Microbiol.* **2020**, *11*, 664. [CrossRef] [PubMed]
244. Hashem, A.; Alqarawi, A.A.; Radhakrishnan, R.; Al-Arjani, A.-B.F.; Aldehaish, H.A.; Egamberdieva, D.; Allah, E.F.A. Arbuscular mycorrhizal fungi regulate the oxidative system, hormones and ionic equilibrium to trigger salt stress tolerance in *Cucumis sativus* L. *Saudi J. Biol. Sci.* **2018**, *25*, 1102–1114. [CrossRef] [PubMed]
245. Shekoofeh, E.; Sepideh, H.; Roya, R. Role of mycorrhizal fungi and salicylic acid in salinity tolerance of *Ocimum basilicum* resistance to salinity. *J. Biotech.* **2012**, *11*, 2223–2235. [CrossRef]
246. Qun, H.Z.; Zing, H.C.; Bin, Z.Z.; Rong, Z.Z.; Song, W.H. Changes of antioxidative enzymes and cell membrane osmosis in tomato colonized by arbuscular mycorrhizae under NaCL stress. *Coll. Surf. B Bioint.* **2007**, *59*, 128–133. [CrossRef]
247. Iqbal, N.; Umar, S.; Khan, N.A. Nitrogen availability regulates proline and ethylene production and alleviates salinity stress in mustard (*Brassica juncea*). *J. Plant Physiol.* **2015**, *178*, 84–91. [CrossRef]
248. Mo, Y.; Wang, Y.; Yang, R.; Zheng, J.; Liu, C.; Li, H.; Ma, J.; Zhang, Y.; Wei, C.; Zhang, X. Regulation of plant growth, photosynthesis, antioxidation and osmosis by an arbuscular mycorrhizal fungus in watermelon seedlings under well-watered and drought conditions. *Front. Plant Sci.* **2016**, *7*, 644. [CrossRef]
249. Parre, E.; Ghars, M.A.; Leprince, A.S.; Thiery, L.; Lefebvre, D.; Bordenave, M.; Richard, L.; Mazars, C.; Abdelly, C.; Savoure, A. Calcium signaling via phospholipase C is essential for proline accumulation upon ionic but not nonionic hyperosmotic stresses in Arabidopsis. *Plant Physiol.* **2007**, *144*, 503–512. [CrossRef]
250. Estrada, B.; Aroca, R.; Maathuis, F.J.M.; Barea, J.M.; Ruiz-Lozano, J.M. Arbuscular mycorrhizal fungi native from a Mediterranean saline area enhance maize tolerance to salinity through improved ion homeostasis. *Plant Cell Environ.* **2013**, *36*, 1771–1782. [CrossRef]
251. Nephali, L.; Moodley, V.; Piater, L.; Steenkamp, P.; Buthelezi, N.; Dubery, I.; Burgess, K.; Huyser, J.; Tugizimana, F. A metabolomic landscape of mize plants treated with a microbial biostimulant under well-watered and drought conditions. *Front. Plant Sci.* **2021**, *12*, 676632. [CrossRef]
252. Qiu, Q.; Bender, S.F.; Mgelwa, A.S.; Hu, Y. Arbuscular mycorrhizal fungi mitigate soil nitrogen and phsophorus losses: A meta-analysis. *Sci. Total Environ.* **2022**, *807*, 150857. [CrossRef]
253. Hu, B.; Hu, S.; Vymazal, J.; Chen, Z. Application of arbuscular mycorrhizal fungi for pharmaceuticals and personal care productions removal in constructed wetlands with different substrate. *J. Clean. Prod.* **2022**, *339*, 130760. [CrossRef]
254. Lermen, C.; Cruz, R.M.S.D.; Pinc, M.M.; Zardeto, G.; Lorencete, M.D.S.; Schwengber, R.P.; Dias-Arieira, C.R.; Alberton, O. Essential oil of bushy lippia inoculated with arbuscular mycorrhizal fungi under different levels of humic substances and phosphorus. *Rhizosphere* **2023**, *25*, 100660. [CrossRef]
255. Mei, L.; Zhang, P.; Cui, G.; Yang, X.; Zhang, T.; Guo, J. Arbuscular mycorrhizal fungi promote litter decomposition and alleviate nutrient limitations of soil microbes under warming and nitrogen application. *Appl. Soil Ecol.* **2022**, *171*, 104318. [CrossRef]
256. Huang, D.; Ma, M.; Wang, Q.; Zhang, M.; Jing, G.; Li, C.; Ma, F. Arbuscular mycorrhizal fungi enhanced drought resistance in apple by regulating genes in the MAPK pathway. *Plant Physiol. Biochem.* **2020**, *149*, 245–255. [CrossRef]
257. Wang, L.; Liu, Y.; Zhu, X.; Zhang, Y.; Yang, H.; Dobbie, S.; Zhang, X.; Deng, A.; Qian, H.; Zhang, W. Effects of arbuscular mycorrhizal fungi on crop growth and soil $N_2O$ emissions in the legume system. *Agric. Ecosyst. Environ.* **2021**, *322*, 107641. [CrossRef]
258. Karagiannidis, N.; Thomidis, T.; Lazari, D.; Panou-Filotheou, E.; Karagiannidou, C. Effect of three Greek arbuscular mycorrhizal fungi in improving the growth, nutrient concentration, and production of essential oils of oregano and mint plants. *Sci. Hortic.* **2011**, *129*, 329–334. [CrossRef]
259. Klinsukon, C.; Ekprasert, J.; Boonlue, S. Using arbuscular mycorrhizal fungi (*Gigaspora margarita*) as a growth promoter and biocontrol of leaf blight disease in eucalyptus seedlings caused by *Cylindrocladium quinqueseptatum*. *Rhizosphere* **2021**, *20*, 100450. [CrossRef]
260. Lam, C.-M.; Lai, H.-Y. Effect of inoculation with arbuscular mycorrhizal fungi and blanching on the bioaccessibility of heavy metals in water spinach (*Ipomoea aquatica* Forsk.). *Ecotoxicol. Environ. Saf.* **2018**, *162*, 563–570. [CrossRef]
261. Arriagada, C.A.; Herrera, M.A.; Ocampo, J.A. Beneficial effect of saprobe and arbuscular mycorhizal fungi on growth of *Eucalyptus globulus* co-cultured with *Glycine max* in soil contaminated with heavy metals. *J. Environ. Manag.* **2007**, *84*, 93–99. [CrossRef]
262. Gheisari Zardak, S.; Dehnavi, M.M.; Salehi, A.; Ghlamhoseini, M. Effects of using arbuscular mycorrhizal fungi to alleviate drought stress on the physiological traits and essential oil yield of fennel. *Rhizosphere* **2018**, *6*, 31–38. [CrossRef]
263. Goicoechea, N.; Baslam, M.; Erice, G.; Irigoyen, J.J. Increased photosynthetic acclimation in alfalfa associated with arbuscular mycorrhizal fungi (AMF) and cultivated in greenhouse under elevated $CO_2$. *J. Plant Physiol.* **2014**, *171*, 1774–1781. [CrossRef] [PubMed]
264. Xu, P.; Liang, L.-Z.; Dong, X.-Y.; Xu, J.; Jiang, P.-K.; Shen, R.-F. Response of soil phosphorus required for maximum growth of *Asparagus officinalis* L. to inoculation of arbuscular mycorrhizal fungi. *Pedosphere* **2014**, *24*, 776–782. [CrossRef]
265. Zhang, X.; Han, C.; Gao, H.; Cao, Y. Comparative transcriptome analysis of the garden asparagus (*Asparagus officinalis* L.) reveals the molecular mechanism for growth with arbuscular mycorrhizal fungi under salinity stress. *Plant Physiol. Biochem.* **2019**, *141*, 20–29. [CrossRef] [PubMed]

266. Zhang, X.; Zhang, H.; Zhang, Y.; Liu, Y.; Zhang, H.; Tang, M. Arbuscular mycorrhizal fungi alter carbohydrate distribution and amino acid accumulation in *Medicago truncatula* uncer lead stress. *Environ. Exp. Bot.* **2020**, *171*, 103950. [CrossRef]
267. Gyuricza, V.; Declerck, S.; Boulois, H.D.D. Arbuscular mycorrhizal fungi decrease radiocesium accumulation in *Medicago truncatula. J. Environ. Radioact.* **2010**, *101*, 591–596. [CrossRef]
268. Li, X.; Kang, X.; Zou, J.; Yin, J.; Wang, Y.; Li, A.; Ma, X. Allochthonous arbuscular mycorrhizal fungi promote *Salix viminalis* L.—Mediated phytoremediation of polycyclic aromatic hydrocarbons characterized by increasing the release of organic acids and enzymes in soils. *Ecotoxicol. Environ. Saf.* **2023**, *249*, 114461. [CrossRef]
269. Mohammed, A.E.; Alotaibi, M.O.; Elobeid, M. Interactive influence of elevated $CO_2$ and arbuscular mycorrhizal fungi on sucrose and coumarin metabolism in *Ammi majus. Plant Physiol. Biochem.* **2022**, *185*, 45–54. [CrossRef]
270. Darakeh, S.A.S.S.; Weisany, W.; Tahir, N.A.-R.; Schenk, P.M. Physiological and biochemical responses of black cumin to vermicompost and plant biostimulants: Arbuscular mycorrhizal and plant growth-promoting rhizobacteria. *Ind. Crops Prod.* **2022**, *188*, 115557. [CrossRef]
271. Wang, L.; Jia, X.; Zhao, Y.; Zhang, C.-Y.; Gao, Y.; Li, X.; Cao, K.; Zhang, N. Effects of elevated $CO_2$ on arbusular mycorrhizal fungi associated with *Robinia pseudoacacia* L. grown in cadmium-contaminated soils. *Sci. Total Environ.* **2021**, *768*, 144453. [CrossRef]
272. Aggangan, N.S.; Cortes, A.D.; Reaño, C.E. Growth response of cacao (*Theobroma cacao* L.) plant as affected by bamboo biochar and arbuscular mycorrhizal fungi in sterilized and unsterilized soil. *Biocatal. Agric. Biotechnol.* **2019**, *22*, 101347. [CrossRef]
273. Outamamat, E.; Dounas, H.; Aziz, F.; Barguaz, A.; Duponnois, R.; Ouahmane, L. The first use of morphologically isolated arbuscular mycorrhizal fungi single-species from Moroccan ecosystems to improve growth, nutrients uptake and photosynthesis in *Ceratonia siliqua* seedlings under nursery conditions. *Saudi J. Biol. Sci.* **2022**, *29*, 2121–2130. [CrossRef] [PubMed]
274. Zhang, T.; Hu, Y.; Zhang, K.; Tian, C.; Guo, J. Arbuscular mycorrhizal fungi improve plant growth of *Ricinus communis* by altering photosynthetic properties and increasing pigments under drought and salt stress. *Ind. Crops Prod.* **2018**, *117*, 13–19. [CrossRef]
275. Pellegrino, E.; Bedini, S. Corrigendum to "Enhancing ecosystem services in sustainable agriculture: Biofertilization and biofortification of chickpea (*Cicer arietinum* L.) by arbuscular mycorrhizal fungi". *Soil Biol. Biochem.* **2014**, *75*, 314–315. [CrossRef]
276. Kaba, J.S.; Abunyewa, A.A.; Kugbe, J.; Kwashie, G.K.S.; Ansah, E.O.; Andoh, H. Arbuscular mycorrhizal fungi and potassium fertilizer as plant biostimulants and alternative research for enhancing plants adaptation to drought stress: Opportunities for enhancing drought tolerance in cocoa (*Theobroma cacao* L.). *Sustain. Environ.* **2021**, *7*, 1963927. [CrossRef]
277. Ganugi, P.; Pathan, S.I.; Zhang, L.; Arfaioli, P.; Benedettelli, S.; Masoni, A.; Pietramellara, G.; Lucini, L. The pivotal role of cultivar affinity to arbuscular mycorrhizal fungi in determining mycorrhizal responsiveness to water deficit. *Phytochemistry* **2022**, *203*, 113381. [CrossRef]
278. Metwally, R.A.; Abdelhameed, R.E. Synergistic effect of arbuscular mycorrhizal fungi on growth and physiology salt-stressed *Trigonella foenum-graecum* plants. *Biocatal. Agric. Biotechnol.* **2018**, *16*, 538–544. [CrossRef]
279. Caruso, T.; Mafrica, R.; Bruno, M.; Vescio, R.; Sorgona, A. Root architectural traits of rooted cuttings of two fig cultivars: Treatments with arbuscular mycorrhizal fungi formulation. *Sci. Hortic.* **2021**, *283*, 110083. [CrossRef]
280. Shen, W.; Feng, Z.; Song, H.; Jin, D.; Fu, Y.; Cheng, F. Effects of solid waste-based soil conditioner and arbuscular mycorrhizal fungi on crop productivity and heavy metal distribution in foxtail millet (*Setaria italica*). *J. Environ. Manag.* **2022**, *313*, 114974. [CrossRef]
281. Liu, N.; Shao, C.; Sun, H.; Liu, Z.; Guan, Y.; Wu, L.; Zhang, L.; Pan, X.; Zhang, Z.; Zhang, Y.; et al. Arbuscular mycorrhizal fungi biofertilizer improves American ginseng (*Panax quinquefolius* L.) growth under the continuous cropping regime. *Geoderma* **2020**, *363*, 114155. [CrossRef]
282. Hashem, A.; Abd-Allah, E.F.; Alqarawi, A.A.; Egamberdieva, D. Bioremediation of adverse impact of cadmium toxicity on *Cassia italica* Mill by arbuscular mycorrhizal fungi. *Saudi J. Biol. Sci.* **2016**, *23*, 39–47. [CrossRef]
283. Fokom, R.; Adamou, S.; Essono, D.; Ngwasiri, D.P.; Eke, P.; Mofor, C.T.; Tchoumbougnang, F.; Fekam, B.F.; Zollo, P.H.A.; Nwaga, D.; et al. Growth, essential oil content, chemical composition and antioxidant properties of lemongrass as affected by harvest period and arbuscular mycorrhizal fungi in field conditions. *Ind. Crops Prod.* **2019**, *138*, 111477. [CrossRef]
284. Cozzolino, V.; Meo, V.D.; Piccolo, A. Impact of arbuscular mycorrhizal fungi applications on maize production and soil phosphorus availability. *J. Geochem. Explor.* **2013**, *129*, 40–44. [CrossRef]
285. Birhane, E.; Bongers, F.; Damtew, A.; Tesfay, A.; Norgrove, L.; Kuyper, T.W. Arbuscular mycorrhizal fungi improve nutrient status of *Commiphora myrrha* seedlings under drought. *J. Arid Environ.* **2023**, *209*, 104877. [CrossRef]
286. El-Nashar, Y.I.; Hassan, B.A.; Aboelsaadat, E.M. Response of *Nemesia* (*Nemesia* $\times$ *hybridus*) plants to different irrigation water sources and arbuscular mycorrhizal fungi inoculation. *Agric. Water Manag.* **2021**, *243*, 106416. [CrossRef]
287. Ostadi, A.; Javanmard, A.; Amani Machiani, M.; Sadeghpour, A.; Maggi, F.; Nouraein, M.; Morshedloo, M.R.; Hano, C.; Lorenzo, J.M. Co-application of $TiO_2$ nanoparticles and arbuscular mycorrhizal fungi improves essential oil quality and quality of Sage (*Salvia officinalis* L.) in drought stress conditions. *Plants* **2022**, *11*, 1659. [CrossRef]
288. Saleh, A.M.; Abdel-Mawgoud, M.; Hassan, A.R.; Habeeb, T.H.; Yehia, R.S.; El-gawad, H. Global metabolic changes induced by arbuscular mycorrhizal fungi in oregano plants grown under ambient and elevated levels of atmospheric $CO_2$. *Plant Physiol. Biochem.* **2020**, *151*, 255–263. [CrossRef]
289. Chen, D.; Saeed, M.; Ali, M.N.H.A.; Raheel, M.; Ashraf, W.; Hassan, Z.; Hassan, M.Z.; Farooq, U.; Hakim, M.F.; Rao, M.J.; et al. Plant growth promoting rhizobacteria (PGPR) and arbuscular mycorrhizal fungi combined application reveals enhanced soil fertility and rice production. *Agronomy* **2023**, *13*, 550. [CrossRef]

290. Wen, Z.; Chen, Y.; Liu, Z.; Meng, J. Biochar and arbuscular mycorrhizal fungi stimulate rice root growth strategy and soil nutrient availability. *Eur. J. Soil Biol.* **2022**, *113*, 103448. [CrossRef]
291. Bao, X.; Wang, Y.; Olsson, P.A. Arbuscular mycorrhiza under water-carbon-phosphorus exchange between rice and arbuscular mycorrhizal fungi under different flooding regimes. *Soil Biol. Biochem.* **2019**, *129*, 169–177. [CrossRef]
292. Liu, M.; Che, Y.; Wang, L.; Zhao, Z.; Zhang, Y.; Wei, L.; Xiao, Y. Rice straw bioachr and phosphorus inputs have more positive effects on the yield and nutrient uptake of *Lolium multiflorum* than arbuscular mycorrhizal fungi in acidic Cd-contaminated soils. *Chemosphere* **2019**, *235*, 32–39. [CrossRef]
293. Lahbouki, S.; Ben-Laouane, R.; Anli, M.; Boutasknit, A.; Ait-Rahou, Y.; Ait-El-Mokhtar, M.; El Gabardi, S.; Douira, A.; Wahbi, S.; Outzourhit, A.; et al. Arbuscular mycorrhizal fungi and/or organic amendment enhance the tolerance of prickly pear (*Opuntia ficus-indica*) under drought stress. *J. Arid Environ.* **2022**, *199*, 104703. [CrossRef]
294. Minggui, G.; Ming, T.; Qiaoming, Z.; Xinxin, F. Effects of climatic and edaphic factors on arbuscular mycorrhizal fungi in the rhizosphere of *Hippophae rhamnoides* in the Loess Plateau, China. *Acta Ecol. Sin.* **2012**, *32*, 62–67. [CrossRef]
295. Abdelhalim, T.S.; Tia, N.A.J.; Elkhatim, K.A.; Othman, M.H.; Joergensen, R.G.; Almaiman, S.A.; Hassan, A.B. Exploring the potential of arbuscular mycorrhizal fungi (AMF) for improving health-promoting phytochemicals in sorghum. *Rhizosphere* **2022**, *24*, 100596. [CrossRef]
296. Dantas, L.V.D.A.; Silva, E.N.; Silva, D.K.A.D.; Beckmann-Cavalcante, M.Z.; Yano-Melo, A.M. Impact of long-term application of paclobutrazol in communities of arbuscular mycorrhiza fungi and their efficiency in the development of *Helianthus annuus* L. *Appl. Soil Ecol.* **2023**, *191*, 105029. [CrossRef]
297. Langeroodi, A.R.S.; Mancinelli, R.; Radicetti, E. Contribution of biochar and arbuscular mycorrhizal fungi to sustainable cultivation of sunflower under semi-arid environment. *Field Crops Res.* **2021**, *273*, 108292. [CrossRef]
298. Paskovic, I.; Soldo, B.; Ban, S.G.; Radic, T.; Lukic, M.; Urlic, B.; Mimica, M.; Bubola, K.B.; Colla, G.; Rouphael, Y.; et al. Fruit quality and volatile compound composition of processing tomato as affected by fertilisation practices and arbuscular mycorrhizal fungi application. *Food Chem.* **2021**, *359*, 129961. [CrossRef]
299. Duan, J.; Tian, H.; Drijber, R.A.; Gao, Y. Systemic and local regulation of phosphate and nitrogen transporter genes by arbuscular mycorrhizal fungi in roots of winter wheat (*Triticum aestivum* L.). *Plant Physiol. Biochem.* **2015**, *96*, 199–208. [CrossRef]
300. Jerbi, M.; Labidi, S.; Bahri, B.A.; Laruelle, F.; Tisserant, B.; Jeddi, F.B.; Sahraoui, A.L.-H. Soil properties and climate affects arbuscular mycorrhizal fungi and soil microbial communities in Mediterranean rainfed cereal cropping systems. *Pedobiologia* **2021**, *87–88*, 150748. [CrossRef]
301. Rouphael, Y.; Cardarelli, M.; Colla, G. Role of arbuscular mycorrhizal fungi in alleviating the adverse effects of acidity and aluminium toxicity in zucchini squash. *Sci. Hortic.* **2015**, *188*, 97–105. [CrossRef]
302. Zhou, Y.; Li, X.; Qin, J.; Liu, H.; Chen, Q.; Niu, Y.; Ren, A.; Gao, Y. Effects of simultaneous infections of endophytic fungi and arbuscular mycorrhizal fungi on the growth of their shared host grass *Achnatherum sibiricum* under varying N and P supply. *Fungal Ecol.* **2016**, *20*, 56–65. [CrossRef]
303. Gao, Y.; Jia, X.; Zhao, Y.; Zhao, J.; Ding, X.; Zhang, C.; Feng, X. Effect of arbuscular mycorrhizal fungi (*Glomus mosseae*) and elevated air temperature on Cd migration in the rhizosphere soil of alfalfa. *Ecotoxicol. Environ. Saf.* **2022**, *248*, 114342. [CrossRef]
304. Borj, M.A.; Etesami, H.; Alikhani, H.A. Silicon improves the effect of phosphate solubilizing bacterium and arbuscular mycorrhizal fungus on phosphorus concentration of salinity-stressed alfalfa (*Medicago sativa* L.). *Rhizosphere* **2022**, *24*, 100619. [CrossRef]
305. Gao, T.; Liu, X.; Shan, L.; Wu, Q.; Liu, Y.; Zhang, Z.; Ma, F.; Li, C. Dopamine and arbuscular mycorrhizal fungi act synergistically to promote apple growth under salt stress. *Environ. Experim. Bot.* **2020**, *178*, 104159. [CrossRef]
306. Watts-Williams, S.J.; Tyerman, S.D.; Cavagnaro, T.R. The dual benefit of arbuscular mycorrhizal fungi under soil zinc deficiency and toxicity: Linking plant physiology and gene expression. *Plant Soil.* **2017**, *420*, 375–388. [CrossRef]
307. Wiesel, L.; Dubchak, S.; Turnau, K.; Broadley, M.R.; White, P.J. Caesium inhibits the colonization of *Medicago truncatula* by arbuscular mycorrhizal fungi. *J. Environ. Radioact.* **2015**, *141*, 57–61. [CrossRef] [PubMed]
308. Hu, J.; Wang, H.; Wu, F.; Wu, S.; Cao, Z.; Lin, X.; Wong, M.H. Arbuscular mycorrhizal fungi influence the accumulation and partitioning of Cd and P in bashfulgrass (*Mimosa pudica* L.) grown on a moderately Cd-contaminated soil. *Appl. Soil Ecol.* **2014**, *73*, 51–57. [CrossRef]
309. Yadav, V.K.; Jha, R.K.; Kaushik, P.; Altalayan, H.; Balawi, T.A.; Alam, P. Traversing arbuscular mycorrhizal fungi and *Pseudomonas fluorescens* for carrot production under salinity. *Saudi J. Biol. Sci.* **2021**, *28*, 4217–4223. [CrossRef]
310. Sarr, P.S.; Sugiyama, A.; Begoude, A.D.B.; Yazaki, K.; Araki, S.; Nawata, E. Diversity and distribution of arbuscular mycorrhizal fungi in cassava *(Manihot esculenta* Crantz) croplands in Cameroon as revealed by Illumina MiSeq. *Rhizosphere* **2019**, *10*, 100147. [CrossRef]
311. Straker, C.J.; Hilditch, A.J.; Rey, M.E.C. Arbuscular mycorrhizal fungi associated with cassava (*Manihot esculenta* Crantz) in South Africa. *S. Afr. J. Bot.* **2010**, *76*, 102–111. [CrossRef]
312. Sehoane, E.; Mogni, V.; Pagani, A.; Gil-Cardeza, M.L. Taxonomical and functional analysis of four arbuscular mycorrhizal fungi populations obtained from a *Ricinus communis* rhizospheric Cr(VI) polluted soil. *Environ. Adv.* **2023**, *11*, 100343. [CrossRef]
313. Gil-Cardeza, M.L.; Muller, D.R.; Amaya-Martin, S.M.; Viassolo, R.; Gomez, E. Differential responses to high soil chromium of two arbuscular mycorrhizal fungi communities isolated from Cr-polluted and non-polluted rhizospheres of *Ricinus communis*. *Sci. Total Environ.* **2018**, *625*, 1113–1121. [CrossRef] [PubMed]

314. Zhao, Z.; Chen, L.; Xiao, Y. The combined use of arbuscular mycorrhizal fungi, biochar and nitrogen fertilizer is most beneficial to cultivate *Cichorium intybus* L. in Cd-contaminated soil. *Ecotoxicol. Environ. Saf.* **2021**, *217*, 112154. [CrossRef] [PubMed]
315. Boonlue, S.; Surapat, W.; Pukahuta, C.; Suwanarit, P.; Suwanarit, A.; Morinaga, T. Diversity and efficiency of arbuscular mycorrhizal fungi in soils from organic chili (*Capsicum frutescens*) farms. *Mycoscience* **2012**, *53*, 10–16. [CrossRef]
316. Xiao, X.; Chen, J.; Liao, X.; Yan, Q.; Liang, G.; Liu, J.; Wang, D.; Guan, R. Different arbuscular mycorrhizal fungi established by two inoculation methods improve growth and droughts resistance of Cinnamomum Migao seedlings differently. *Biology* **2022**, *11*, 220. [CrossRef] [PubMed]
317. Merlin, E.; Melato, E.; Lourenco, E.L.B.; Jacomassi, E.; Gasparotto Junior, A.; Sete da Cruz, R.M.; Otenio, J.K.; Da Silva, C.; Alberton, O. Inoculation of arbuscular mycorrhizal fungi and phosphorus addition increase coarse mint (*Plectranthus amboinicus* Lour.) plant growth and essential oil content. *Rhizosphere* **2020**, *15*, 100217. [CrossRef]
318. Razakatiana, A.T.E.; Trap, J.; Baohanta, R.H.; Raherimandimby, M.; Le Roux, C.; Duponnois, R.; Ramanankierana, H.; Becquer, T. Benefits of dual inoculation with arbuscular mycorrhizal fungi and rhizobia on *Phaseolus vulgaris* planted in a low-fertility tropical soil. *Pedobiologia* **2020**, *83*, 150685. [CrossRef]
319. Azizi, S.; Kouchaksaraei, M.T.; Hadian, J.; Nosrat Abad, A.R.; Sanavi, S.A.M.M.; Ammer, C.; Bader, M.K.-F. Dual inoculation of arbuscular mycorrhizal fungi and plant growth-promoting rhizobacteria boost drought resistance and essential oil yield of common myrtle. *For. Ecol. Manag.* **2021**, *497*, 119478. [CrossRef]
320. Kavadia, A.; Omirou, M.; Fasoula, D.A.; Louka, F.; Ehaliotis, C.; Ioannides, I.M. Co-inoculations with rhizobia and arbuscular mycorrhizal fungi alters mycorrhizal composition and lead to synergistic growth effects in cowpea that are fungal combination-dependent. *Appl. Soil Ecol.* **2021**, *167*, 104013. [CrossRef]
321. Ogunkunle, C.O.; El-Imam, A.M.A.; Bassey, E.; Vishwakarma, V.; Fatoba, P.O. Co-application of indigenous arbuscular mycorrhizal fungi and nano-$TiO_2$ reduced Cd uptake and oxidative stress in pre-flowering cowpea plants. *Environ. Technol. Innov.* **2020**, *20*, 101163. [CrossRef]
322. Chen, S.; Jin, W.; Liu, A.; Zhang, S.; Liu, D.; Wang, F.; Lin, X.; He, C. Arbuscular mycorrhizal fungi (AMF) increase growth and secondary metabolism in cucumber subjected to low temperatures stress. *Sci. Hortic.* **2013**, *160*, 222–229. [CrossRef]
323. Li, Y.; Chen, Y.-L.; Li, M.; Lin, X.-G.; Liu, R.-J. Effects of arbuscular mycorrhizal fungi communities on soil quality and the growth of cucumber seedlings in a greenhouse soil of continuously planting cucumber. *Pedosphere* **2012**, *22*, 79–87. [CrossRef]
324. Hu, J.-L.; Lin, X.-G.; Wang, J.-H.; Shen, W.-S.; Wu, S.; Peng, S.-P.; Mao, T.-T. Arbuscular mycorrhizal fungal inoculation enhances suppression of cucumber *Fusarium* wilt in greenhouse soils. *Pedosphere* **2010**, *20*, 586–593. [CrossRef]
325. Hilali, R.E.; Symanczik, S.; Kinany, S.E.; Oehl, F.; Ouahmane, L.; Bouamri, R. Cultivation, identification, and application of arbuscular mycorrhizal fungi associated with date palm plants in Draa-Tafilalet oasis. *Rhizosphere* **2022**, *22*, 100521. [CrossRef]
326. Jaiti, F.; Meddich, A.; El-Hadrami, I. Effectiveness of arbuscular mycorrhizal fungi in the protection of date palm (*Phoenic dactylifera* L.) against bayound disease. *Physiol. Mol. Plant Pathol.* **2007**, *71*, 166–173. [CrossRef]
327. Kinany, S.E.; Achbani, E.; Faggroud, M.; Ouahmane, L.; El-Hilali, R.; Haggoud, A.; Bouamri, R. Effect of organic fertilizer and commercial arbuscular mycorrhizal fungi on the growth of micropropagated date plam cv. Feggouss. *J. Saudi Soc. Agric. Sci.* **2019**, *18*, 411–417. [CrossRef]
328. Shahrajabian, M.H.; Kuang, Y.; Cui, H.; Fu, L.; Sun, W. Metabolic changes of active components of important medicinal plants on the basis of traditional Chinese medicine under different environmental stress. *Curr. Org. Chem.* **2023**, 23. [CrossRef]
329. Di Miceli, G.; Vultaggio, L.; Sabatino, L.; De Pasquale, C.; La Bella, S.; Consentino, B.B. Synergistic effect of a plant-derived protein hydrolysate and arbuscular mycorrhizal fungi on eggplant grown in open fields: A two-year study. *Horticulturae* **2023**, *9*, 592. [CrossRef]
330. Sharma, M.; Saini, I.; Kaushik, P.; Aldawsari, M.M.; Balawi, T.; Alam, P. Mycorrhizal fungi and *Pseudomonas fluorescens* application reduces root-knot nematode (*Meloidogyne javanica*) infestation in eggplant. *Saudi J. Biol. Sci.* **2021**, *28*, 3685–3691. [CrossRef]
331. Alipour, A.; Rahimi, M.M.; Hosseini, S.M.A.; Bahrani, A. Mycorrhizal fungi and growth-promoting bacteria improves fennel essential oil yield under water stress. *Ind. Crops Prod.* **2021**, *170*, 113792. [CrossRef]
332. Nogales, A.; Rottier, E.; Campos, C.; Victorino, G.; Costa, J.M.; Coito, J.L.; Pereira, H.S.; Viegas, W.; Lopes, C. The effects of field inoculation of arbuscular mycorrhizal fungi through rye donor plants on grapevine performance and soil properties. *Agric. Ecosyst. Environ.* **2021**, *313*, 107369. [CrossRef]
333. Sun, S.; Fan, X.; Feng, Y.; Wang, X.; Gao, H.; Song, F. Arbuscular mycorrhizal fungi influence the uptake of cadmium in industrial hemp (*Cannabis sativa* L.). *Chemosphere* **2023**, *330*, 138728. [CrossRef] [PubMed]
334. Gaveliene, V.; Socik, B.; Jankovska-Bortkevic, E.; Jurkoniene, S. Plant microbial biostimulants as a promising tool to enhance the productivity and quality of carrot root crops. *Microorganisms* **2021**, *9*, 1850. [CrossRef] [PubMed]
335. Souza, B.C.D.; Cruz, R.M.S.D.; Lourenco, E.L.B.; Pinc, M.M.; Dalmagro, M.; Silva, C.D.; Nunes, M.G.L.F.; Souza, S.G.H.D.; Alberton, O. Inoculation of lemongrass with arbuscular mycorrhizal fungi and rhizobacteria alters plants growth and essential oil production. *Rhizosphere* **2022**, *22*, 100514. [CrossRef]
336. Silva, A.M.M.; Jones, D.L.; Chadwick, D.R.; Qi, X.; Cotta, S.R.; Araujo, V.L.V.P.; Matteoli, F.P.; Lacerda-Junior, G.V.; Pereira, A.P.A.; Fernandes-Junior, P.I.; et al. Can arbuscular mycorrhizal fungi and rhizobacteria facilitate $^{33}P$ uptake in maize plants under water stress? *Microbiol. Res.* **2023**, *271*, 127350. [CrossRef] [PubMed]

337. Yang, Y.-M.; Naseer, M.; Zhu, Y.; Zhu, S.-G.; Wang, S.; Wang, B.-Z.; Wang, J.; Zhu, H.; Wang, W.; Tao, H.-Y.; et al. Dual effects of nZVI on maize growth and water use are positively mediated by arbuscular mycorrhizal fungi via rhizosphere interactions. *Environ. Pollut.* **2022**, *308*, 119661. [CrossRef]
338. Celebi, S.Z.; Demir, S.; Celebi, R.; Durak, E.D.; Yilmaz, I.H. The effect of arbuscular mycorrhizal fungi (AMF) applications on the silage maize (*Zea mays* L.) yield in different irrigation regimes. *Eur. J. Soil Biol.* **2010**, *46*, 302–305. [CrossRef]
339. Sun, C.; Guo, Q.; Zeeshan, M.; Milham, P.; Qin, S.; Ma, J.; Yang, Y.; Lai, H.; Huang, J. Dual RNA and 16S ribosomal DNA sequencing reveal arbuscular mycorrhizal fungi-mediated mitigation of selenate stress in *Zea mays* L. and reshaping of soil microbiota. *Ecotoxicol. Environ. Saf.* **2022**, *247*, 114217. [CrossRef]
340. Olowe, O.M.; Olawuyi, O.J.; Sobowale, A.A.; Odebode, A.C. Role of arbuscular mycorrhizal fungi as biocontrol agents against *Fusarium verticillioides* causing ear rot of *Zea mays* L. (Maize). *Curr. Plant Biol.* **2018**, *15*, 30–37. [CrossRef]
341. Liu, L.-Z.; Gong, Z.-Q.; Zhang, Y.-L.; Li, P.-J. Growth, cadmium accumulation and physiology of marigold (*Tagetes erecta* L.) as affected by arbuscular mycorrhizal fungi. *Pedosphere* **2011**, *21*, 319–327. [CrossRef]
342. Asrar, A.A.; Abdel-Fattah, G.M.; Elhindi, K.M. Improving growth, flower yield, and water relations of snapdragon (*Antirhinum majus* L.) plants grown under well-watered and water-stress conditions using arbuscular mycorrhizal fungi. *Photosynthetica* **2021**, *50*, 305–316. [CrossRef]
343. Alizadeh, S.; Gharagoz, S.F.; Pourakbar, L.; Moghaddam, S.S.; Jamalomidi, M. Arbuscular mycorrhizal fungi alleviate salinity stress and alter phenolic compounds of Moldavian balm. *Rhizosphere* **2021**, *19*, 100417. [CrossRef]
344. Lu, N.; Zhou, X.; Cui, M.; Yu, M.; Zhou, J.; Qin, Y.; Li, Y. Colonization with arbuscular mycorrhizal fungi promotes the growth of *Morus alba* L. seedlings under greenhouse conditions. *Forests* **2015**, *6*, 734–747. [CrossRef]
345. Aganchich, B.; Wahbi, S.; Yaakoubi, A.; El-Aououad, H.; Bota, J. Effect of arbuscular mycorrhizal fungi inoculation on growth and physiology performance of olive trees under regulated deficit irrigation and partial rootzone drying. *S. Afr. J. Bot.* **2022**, *148*, 1–10. [CrossRef]
346. Tekaya, M.; Dabbaghi, O.; Guesmi, A.; Attia, F.; Chehab, H.; Khezami, L.; Algathami, F.K.; Hamadi, N.B.; Hammadi, M.; Prinsen, E.; et al. Arbuscular mycorrhizas modulate carbohydrate, phenolic compounds and hormonal metabolism to enhance water deficit tolerance of olive trees (*Olea europaea*). *Agric. Water Manag.* **2022**, *274*, 107947. [CrossRef]
347. Khabou, W.; Hajji, B.; Zouari, M.; Rigane, H.; Abdallah, F.B. Arbuscular mycorrhizal fungi improve growth and mineral uptake of olive tree under gypsum substrate. *Ecol. Eng.* **2014**, *73*, 290–296. [CrossRef]
348. Metwally, R.A.; Soliman, S.A.; Latef, A.A.H.A.; Abdelhameed, R.E. The individual and interactive role of arbuscular mycorrhizal fungi and *Trichoderma viride* on growth, protein content, amino acids fractionation, and phosphatases enzyme activities of onion plants amended with fish waste. *Ecotoxicol. Environ. Saf.* **2021**, *214*, 112072. [CrossRef]
349. Pankaj, U.; Kurmi, A.; Lothe, N.B.; Verma, R.K. Influence of the seedlings emergence and initial growth of palmarosa (*Cymbopogon martinii* (Roxb.) Wats. var. Motia Burk) by arbuscular mycorrhizal fungi in soil salinity conditions. *J. Appl. Res. Med. Aromat. Plants* **2021**, *24*, 100317. [CrossRef]
350. Hernandez-Montiel, L.G.; Rueda-Puente, E.O.; Cordoba-Matson, M.V.; Holguin-Pena, J.R.; Zulueta-Rodriguez, R. Mutualistic interaction of rhizobacteria with arbuscular mycorrhizal fungi and its antagonistic effect on *Fusarium oxysporum* in *Carica papaya* seedlings. *Crop Prot.* **2013**, *47*, 61–66. [CrossRef]
351. Xu, F.-J.; Zhang, A.-Y.; Yu, Y.-Y.; Sun, K.; Tang, M.-J.; Zhang, W.; Xie, X.-G.; Dai, C.C. Soil legacy of arbuscular mycorrhizal fungus *Gigaspora margarita*: The potassium-sequestering glomalin improves peanut (*Arachis hypogaea*) drought resistance and pod yield. *Microbiol. Res.* **2021**, *249*, 126774. [CrossRef]
352. Sensoy, S.; Demir, S.; Turkmen, O.; Erdinc, C.; Savur, O.B. Responses of some different pepper (*Capsicum annuum* L.) genotypes to inoculation with two different arbuscular mycorrhizal fungi. *Sci. Hortic.* **2007**, *113*, 92–95. [CrossRef]
353. Sharif, M.; Claassen, N. Action mechanisms of arbuscular mycorrhizal fungi in phosphorus uptake by *Capsicum annuum* L. *Pedosphere* **2011**, *21*, 502–511. [CrossRef]
354. Singh, N.V.; Singh, S.K.; Singh, A.K.; Meshram, D.T.; Suroshe, S.S.; Mishra, D.C. Arbuscular mycorrhizal fungi (AMF) induced hardening of micropropagated pomegranate (*Punica granatum* L.) plantlets. *Sci. Hortic.* **2012**, *136*, 122–127. [CrossRef]
355. Rahimi, S.; Baninasab, B.; Talebi, M.; Gholami, M.; Zarei, M. Arbuscular mycorrhizal fungi inoculation improves iron deficiency in quince via alterations in host root phenolic compounds and expression of genes. *Sci. Hortic.* **2021**, *285*, 110165. [CrossRef]
356. Medina, C.; Mutis, A.; Bardehle, L.; Hormazabal, E.; Borie, F.; Aguilera, P.; Ortega, F.; Quiroz, A. Arbuscular mycorrhizal fungi enhance monoterpene production in red clover (*Trifolium pratense* L.): A potential tool for pest control. *Nat. Prod. Res.* **2023**, *37*, 981–984. [CrossRef] [PubMed]
357. Wang, P.; Shu, B.; Wang, Y.; Zhang, D.J.; Liu, J.F.; Xia, R.X. Diversity of arbuscular mycorrhizal fungi in red tangerine (*Citrus reticulata* Blanco) rootstock rhizospheric soils from hillside citrus orchards. *Pedobiologia* **2013**, *56*, 161–167. [CrossRef]
358. Li, Y.; Wang, S.; Lu, M.; Zhang, Z.; Chen, M.; Li, S.; Cao, R. Rhizosphere interactions between earthworms and arbuscular mycorrhizal fungi increase nutrient availability and plant growth in the desertification soils. *Soil Tillag Res.* **2019**, *186*, 146–151. [CrossRef]
359. Li, H.; Chen, X.W.; Wong, M.H. Arbuscular mycorrhizal fungi reduced the ratios of inorganic/organic arsenic in rice grains. *Chemosphere* **2016**, *145*, 224–230. [CrossRef]
360. Li, H.; Luo, N.; Zhang, L.J.; Zhao, H.M.; Li, Y.W.; Cai, Q.Y.; Wong, M.H.; Mo, C.H. Do arbuscular mycorrhizal fungi affect cadmium uptake kinetics, subcellular distribution and chemical forms in rice? *Sci. Total Environ.* **2016**, *571*, 1183–1190. [CrossRef]

361. Baki, M.Z.I.; Suzuki, K.; Takahashi, K.; Chowdhury, S.A.; Asiloglu, R.; Harada, N. Molecular genetic characterization of arbuscular mycorrhizal fungi associated with upland rice in Bangladesh. *Rhizosphere* **2021**, *18*, 100357. [CrossRef]
362. Silva, H.F.O.D.; Tavares, O.C.H.; Silva, L.D.S.D.; Zonta, E.; Silva, E.M.R.D.; Saggin Junior, O.J.; Nobre, C.P.; Berbara, R.L.L.; Garcia, A.C. Arbuscular mycorrhizal fungi and humic substances increased the salinity tolerance of rice plants. *Biocatal. Agric. Biotechnol.* **2022**, *44*, 102472. [CrossRef]
363. Beyene, B.B.; Tuji, F.A. Inoculation of rose (*Rosa rubigniosa* L., eglantine or the briar rose) flower root stocks with consortia of endophytic bacteria and arbuscular mycorrhizal fungi improves its establishment and success rate under greenhouse conditions. *Rhizoshphere* **2023**, *26*, 100698. [CrossRef]
364. Han, Y.; Zveushe, O.K.; Dong, F.; Ling, Q.; Chen, Y.; Sajid, S.; Zhou, L.; Dios, V.R.D. Unraveling the effects of arbuscular mycorrhizal fungi on cadmium uptake and detoxification mechanisms in perennial ryegrass (*Lolium perenne*). *Sci. Total Environ.* **2021**, *798*, 149222. [CrossRef] [PubMed]
365. Liu, C.; Dai, Z.; Cui, M.; Lu, W.; Sun, H. Arbuscular mycorrhizal fungi alleviate boron toxicity in *Puccinellia tenuiflora* under the combined stresses of salt and drought. *Environ. Pollut.* **2018**, *240*, 557–565. [CrossRef]
366. Caser, M.; Demasi, S.; Victorino, I.M.M.; Donno, D.; Faccio, A.; Lumini, E.; Bianciotto, V.; Scariot, V. Arbuscular mycorrhizal fungi modulate the crop performance and metabolic profile in saffron in soilless cultivation. *Agronomy* **2019**, *9*, 232. [CrossRef]
367. Wang, S.; Feng, Z.; Wang, X.; Gong, W. Arbuscular mycorrhizal fungi alter the response of growth and nutrient uptake of snap bean (*Phaseolus vulgaris* L.) to $O_3$. *J. Environ. Sci.* **2011**, *23*, 968–974. [CrossRef]
368. Hadian-Deljou, M.; Esna-Ashari, M.; Mirzaie-asl, A. Alleviation of salt stress and expression of stress-responsive gene through the symbiosis of arbuscular mycorrhizal fungi with sour orange seedlings. *Sci. Hortic.* **2020**, *268*, 109373. [CrossRef]
369. Sun, X.-G.; Tang, M. Effect of arbuscular mycorrhizal fungi inoculation on root traits and root volatile organic compound emissions of *Sorghum bicolor*. *S. Afr. J. Bot.* **2013**, *88*, 373–379. [CrossRef]
370. Nakmee, P.S.; Techapinyawat, S.; Ngamprasit, S. Comparative potentials of native arbuscular mycorrhizal fungi to improve nutrient uptake and biomass of *Sorghum bicolor* Linn. *Agric. Nat. Resour.* **2016**, *50*, 173–178. [CrossRef]
371. Chandra, P.; Singh, A.; Prajapat, K.; Rai, A.K.; Yadav, R.K. Native arbuscular mycorrhizal fungi improve growth, biomass yield, and phosphorus nutrition of sorghum in saline and sodic soils of the semi-arid region. *Environ. Experim. Bot.* **2022**, *201*, 104982. [CrossRef]
372. Crossay, T.; Cavaloc, Y.; Majorel, C.; Redecker, D.; Medevielle, V.; Amir, H. Combinations of different arbuscular mycorrhizal fungi improve fitness and metal tolerance of sorghum in ultramafic soil. *Rhizosphere* **2020**, *14*, 100204. [CrossRef]
373. Thiub, M.; Ewusi-Mensah, N.; Sarkodie-Addo, J.; Adjei-Gyapong, T. Arbuscular mycorrhizal fungi inoculation enhances phosphorus use efficiency and soybean productivity on a Haplic Acrisol. *Soil Till. Res.* **2019**, *192*, 174–186. [CrossRef]
374. Barazetti, A.R.; Simionato, A.S.; Navarro, M.O.P.; Santos, I.M.O.D.; Modolon, F.; Andreata, M.F.D.L.; Liuti, G.; Cely, M.V.T.; Chryssafidis, A.L.; Dealis, M.L.; et al. Formulations of arbuscular mycorrhizal fungi inoculum applied to soybean and corn plant under controlled and field conditions. *Appl. Soil Ecol.* **2019**, *142*, 25–33. [CrossRef]
375. Chiomento, J.L.T.; Nardi, F.S.D.; Filippi, D.; Trentin, T.D.S.; Dornelles, A.G.; Fornari, M.; Nienow, A.A.; Calvete, E.O. Morpho-horticultural performance of strawberry cultivated on substrate with arbuscular mycorrhizal fungi and biochar. *Sci. Hortic.* **2021**, *282*, 110053. [CrossRef]
376. Moradtalab, N.; Hajoboland, R.; Aliasgharzad, N.; Hartmann, T.E.; Neumann, G. Silicon and the association with an arbuscular-mycorrhizal fungus (*Rhizophagus clarus*) mitigate the adverse effects of drought stress on strawberry. *Agronomy* **2019**, *9*, 41. [CrossRef]
377. Wei, Z.; Sixi, Z.; Xiuqing, Y.; Guodong, X.; Baichun, W.; Baojing, G. Arbuscular mycorrhizal fungi alter rhizosphere bacterial community characteristics to improve Cr tolerance of *Acorus calamus*. *Ecotoxicol. Environ. Saf.* **2023**, *253*, 114652. [CrossRef] [PubMed]
378. Xiao, L.; Lai, S.; Chen, M.; Long, X.; Fun, X.; Yang, H. Effects of grass cultivation on soil arbuscular mycorrhizal fungi community in a tangerine orchard. *Rhizosphere* **2022**, *24*, 100583. [CrossRef]
379. Hua, J.; Lin, X.; Yin, R.; Jiang, Q.; Shao, Y. Effects of arbuscular mycorrhizal fungi inoculation on arsenic accumulation by tobacco (*Nicotiana tabacum* L.). *J. Environ. Sci.* **2009**, *21*, 1214–1220. [CrossRef]
380. Ziane, H.; Hamza, N.; Meddad-Hamza, A. Arbuscular mycorrhizal fungi and fertilization rates optimize tomato (*Solanum lycopersicum* L.) growth and yield in a Mediterranean agroecosystem. *J. Saudi Soc. Agric. Sci.* **2021**, *20*, 454–458. [CrossRef]
381. Arpanahi, A.A.; Feizian, M.; Mehdipourian, G.; Khojasteh, D.N. Arbuscular mycorrhizal fungi inoculation improve essential oil and physiological parameters and nutritional values of *Thymus daenensis* Celak and *Thymus vulgaris* L. under normal and drought stress conditions. *Eur. J. Soil Biol.* **2020**, *100*, 103217. [CrossRef]
382. Li, J.-F.; He, X.-H.; Li, H.; Zheng, W.-J.; Liu, J.-F.; Wang, M.-Y. Arbuscular mycorrhizal fungi increase growth and phenolics synthesis in *Poncirus trifoliata* under iron deficiency. *Sci. Hortic.* **2015**, *183*, 87–92. [CrossRef]
383. Amanifar, S.; Toghranegar, Z. The efficiency of arbuscular mycorrhizal for improving tolerance of *Valeriana officinalis* L. and enhancing valerenic acid accumulation under salinity stress. *Ind. Crops Prod.* **2020**, *147*, 112234. [CrossRef]
384. Li, J.; Awasthi, M.K.; Xing, W.; Liu, R.; Bao, H.; Wang, X.; Wang, J.; Wu, F. Arbuscular mycorrhizal fungi increase the bioavailability and wheat (*Triticum aestivum* L.) uptake of selenium in soil. *Ind. Crops Prod.* **2020**, *150*, 112383. [CrossRef]
385. Li, J.; Liu, R.; Zhang, C.; Yang, J.; Lyu, L.; Shi, Z.; Man, Y.B.; Wu, F. Selenium uptake and accumulation in winter wheat as affected by level of phosphate application and arbuscular mycorrhizal fungi. *J. Hazard. Mat.* **2022**, *433*, 128762. [CrossRef] [PubMed]

386. Ma, X.; Luo, W.; Li, J.; Wu, F. Arbuscular mycorrhizal fungi increase both concentrations and bioavailability of Zn in wheat (*Triticum aestivum* L.) grain on Zn-spiked soils. *Appl. Soil Ecol.* **2019**, *135*, 91–97. [CrossRef]
387. Tchabi, A.; Coyne, D.; Hountondji, F.; Lawouin, L.; Wiemken, A.; Oehl, F. Efficacy of indigenous arbuscular mycorrhizal fungi for promoting white yam (*Dioscorea rotundata*) growth in West Africa. *Appl. Soil Ecol.* **2010**, *45*, 92–100. [CrossRef]
388. Shahrajabian, M.H.; Sun, W. The important nutritional benefits and wonderful health benefits of cashew (*Anacardium occidentale* L.). *Nat. Prod. J.* **2023**, *13*, 2–10. [CrossRef]
389. Eke, P.; Adamou, S.; Fokom, R.; Nya, V.D.; Fokou, P.V.T.; Wakam, L.N.; Nwaga, D.; Boyom, F.F. Arbuscular mycorrhizal fungi alter antifungal potential of lemongrass essential oil against *Fusarium solani*, causing root rot in common bean (*Phaseolus vulgaris* L.). *Heliyon* **2020**, *6*, e05737. [CrossRef]
390. Baum, C.; El-Tohamy, W.; Gruda, N. Increasing the productivity and product quality of vegetable crops using arbuscular mycorrhizal fungi: A review. *Sci. Hortic.* **2015**, *187*, 131–141. [CrossRef]
391. Siddiky, M.R.; Schaller, J.; Caruso, T.; Rillig, M.C. Arbuscular mycorrhizal fungi and collembola non-additively increase soil aggregation. *Soil Biol. Biochem.* **2012**, *47*, 93–99. [CrossRef]
392. Kara, Z.; Arslan, D.; Guller, M.; Guler, S. Inoculation of arbuscular mycorrhizal fungi and application of micronized calcite to olive plant: Effects on some biochemical constituents of olive fruit and oil. *Sci. Hortic.* **2015**, *185*, 219–227. [CrossRef]
393. Yu, Z.; Zhao, X.; Su, L.; Yan, K.; Li, B.; He, Y.; Zhan, F. Effect of an arbuscular mycorrhizal fungus on maize growth and cadmium migration in a sand column. *Ecotoxicol. Environ. Saf.* **2021**, *225*, 112782. [CrossRef] [PubMed]
394. Saleh, A.M.; El-Soud, W.M.A.; Alotaibi, M.O.; Beemster, G.T.S.; Mohammed, A.E.; AbdElgawad, H. Chitosan nanoparticles support the impact of arbuscular mycorrhizae fungi on growth and sugar metabolism of wheat crop. *Int. J. Biol. Macromol.* **2023**, *235*, 123806. [CrossRef] [PubMed]
395. Leventis, G.; Tsiknia, M.; Feka, M.; Ladikou, E.V.; Papadakis, I.E.; Chatzipavlidis, I.; Papadopoulou, K.; Ehaliotis, C. Arbuscular mycorrhizal fungi enhance growth of tomato under normal and drought conditions, via different water regulation mechanisms. *Rhizosphere* **2021**, *19*, 100394. [CrossRef]
396. Tekaya, M.; Dahmen, S.; Mansour, M.B.; Ferhout, H.; Chehab, H.; Hammami, M.; Attia, F.; Mechri, B. Foliar application of fertilizers and biostimulant has a strong impact on the olive (*Olea europaea*) rhizosphere microbial community profile and the abundance of arbuscular mycorrhizal fungi. *Rhizosphere* **2021**, *19*, 100402. [CrossRef]
397. Sarkar, A.; Asaeda, T.; Wang, Q.; Kaneko, Y.; Rashid, M.H. Response of *Miscanthus saccharidlorus* to zinc stress mediated by arbuscular mycorrhizal fungi. *Flora* **2017**, *234*, 60–68. [CrossRef]
398. You, Y.; Wang, L.; Ju, C.; Wang, G.; Ma, F.; Wang, Y.; Yang, D. Effects of arbuscular mycorrhizal fungi on the growth and toxic element uptake of *Phragmites australis* (Cav.) Trin. Ex Steud under zinc/cadmium stress. *Ecotoxicol. Environ. Saf.* **2021**, *213*, 112023. [CrossRef]
399. Ruscitti, M.; Arango, M.; Beltrano, J. Improvement of copper stress tolerance in pepper plants (*Capsicum annuum* L.) by inoculation with arbuscular mycorrhizal fungi. *Theor. Exp. Plant Physiol.* **2017**, *29*, 37–49. [CrossRef]
400. Jiang, Q.Y.; Zhuo, F.; Long, S.H.; Zhao, H.D.; Yang, D.J.; Ye, Z.H.; Li, S.S.; Jing, Y.X. Can arbuscular mycorrhizal fungi reduce Cd uptake and alleviate Cd toxicity of *Lonicera japonica* grown in Cd-added soils? *Sci. Rep.* **2016**, *6*, 21805. [CrossRef]
401. Dagher, D.J.; Pitre, F.E.; Hijri, M. Ectomycorrhizal fungi inoculation of sphaerosporella brunnea significantly increased stem biomass of *Salix miyabeana* and decreased lead, tin, and zinc, soil concentrations during the phytoremediation of an industrial landfill. *J. Fungi* **2020**, *6*, 87. [CrossRef]
402. Alam, M.Z.; Hoque, M.A.; Ahammed, G.J.; Carpenter-Boggs, L. Effects of arbuscular mycorrhizal fungi, biochar, selenium, silica gel, and sulfur on arsenic uptake and biomass growth in *Pisum sativum* L. *Emerg. Contam.* **2020**, *6*, 312–322. [CrossRef]
403. Baghaie, A.H.; Aghili, F.; Jafarinia, R. Soil-indigenous arbuscular mycorrhizal fungi and zeolite addition to soil synergistically increase grain yield and reduce cadmium uptake of bread wheat (through improved nitrogen and phosphorus nutrition and immobilization of Cd in roots). *Environ. Sci. Pollut. Res.* **2019**, *26*, 30794–30807. [CrossRef] [PubMed]
404. Sbrana, C.; Avio, L.; Giovannetti, M. Beneficial mycorrhizal symbionts affecting the production of health-promoting phytochemicals. *Electrophoresis* **2014**, *35*, 1535–1546. [CrossRef] [PubMed]
405. Nguyen, T.D.; Cavagnaro, T.R.; Watts-Williams, D.J. The effects of soil phosphorus and zinc availability on plant responses to mycorrhizal fungi: A physiological and molecular assessment. *Sci. Rep.* **2019**, *9*, 14880. [CrossRef] [PubMed]
406. Watts-Williams, S.J.; Cavagnaro, T.R. Arbuscular mycorrhizal fungi increase grain zinc concentration and modify the expression of root *ZIP* transporter genes in a modern barley (*Hordeum vulgare*) cultivar. *Plant Sci.* **2018**, *274*, 163–170. [CrossRef]
407. Chen, S.; Zhao, H.; Zou, C.; Li, Y.; Chen, Y.; Wang, Z.; Jiang, Y.; Liu, A.; Zhao, P.; Wang, M.; et al. Combined inoculation with multiple arbuscular mycorrhizal fungi improves growth, nutrient uptake and photosynthesis in cucumber seedlings. *Front. Microbiol.* **2017**, *8*, 2516. [CrossRef]
408. Abdel-Fttah, G.M.; El-Haddad, S.A.; Hafez, E.E.; Rashad, Y.M. Induction of defense responses in common bean plants by arbuscular mycorrhizal fungi. *Microbiol. Res.* **2011**, *166*, 268–281. [CrossRef]
409. Li, H.; Zhang, L.; Wu, B.; Li, Y.; Wang, H.; Teng, H.; Wei, D.; Yuan, Z.; Yuan, Z. Physiological and proteomic analyses reveal the important role of arbuscular mycorrhizal fungi on enhancing photosynthesis in wheat under cadmium stress. *Ecotoxicol. Environ. Saf.* **2023**, *261*, 115105. [CrossRef]
410. Tofighi, C.; Khavari-Nejad, R.A.; Najafi, F.; Razavi, K.; Rejali, F. Brassinosteroid (BR) and arbuscular mycorrhizal (AM) fungi alleviate salinity in wheat. *J. Plant Nutr.* **2017**, *40*, 1091–1098. [CrossRef]

411. Sallaku, G.; Sanden, H.; Babaj, I.; Kaciu, S.; Balliu, A.; Rewad, B. Specific nutrient absorption rates of transplanted cucumber seedlings are highly related to RGR and influenced by grafting method, AMF inoculation and salinity. *Sci. Hortic.* **2019**, *243*, 177–188. [CrossRef]
412. Pandey, R.; Garg, N. High effectiveness of *Rhizophagus irregularis* is linked to superior modulation of antioxidant defence mechanisms in *Cajanus cajan* (L.) Millsp. genotypes grown under salinity stress. *Mycorrhiza* **2017**, *27*, 669–682. [CrossRef]
413. Yu, H.; Liu, X.; Yang, C.; Peng, Y.; Yu, X.; Gu, H.; Zheng, X.; Wang, C.; Xiao, F.; Shu, L.; et al. Co-symbiosis of arbuscular mycorrhizal fungi (AMF) and diazotrophs promote biological nitrogen fixation in mangrove ecosystems. *Soil Biol. Biochem.* **2021**, *161*, 108382. [CrossRef]
414. Selvaraj, A.; Thangavel, K.; Uthandi, S. Arbuscular mycorrhizal fungi (*Glomus intraradices*) and diazotrophic bacterium (*Rhizobium* BMBS) primed defense in blackgram against herbivorous insect (*Spodoptera litura*) infestation. *Microbiol. Res.* **2020**, *231*, 126355. [CrossRef] [PubMed]
415. Porcel, R.; Ruiz-Lozano, J.M. Arbuscular mycorrhizal influence on leaf water potential, solute accumulation, and oxidative stress in soybean plants subjected to drought stress. *J. Exp. Bot.* **2004**, *55*, 1743–1750. [CrossRef]
416. Habibzadeh, Y.; Pirzad, A.; Zardashti, M.R.; Jalilian, J.; Eini, O. Effects of arbuscular mycorrhizal fungi on seed and protein yield under water deficit stress in mungbean. *Agron. J.* **2013**, *105*, 79–84. [CrossRef]
417. Barzana, G.; Aroca, R.; Bienert, G.P.; Chaumont, F.; Ruiz-Lozano, J.M. New insights into the regulation of aquporins by the arbuscular mycorrhizal symbiosis in maize plants under drought stress and possible implications for plant performance. *Mol. Plant Microbe Interact.* **2014**, *27*, 349–363. [CrossRef] [PubMed]
418. Jumrani, K.; Bhatia, V.S.; Kataria, S.; Alamri, S.A.; Siddiqui, M.H.; Rastogi, A. Inoculation with arbuscular mycorrhizal fungi alleviate the adverse effects of high temperature in soybean. *Plants* **2022**, *11*, 2210. [CrossRef] [PubMed]
419. Wu, Q.S.; Zou, Y.N.; Liu, W.; Ye, X.F.; Zai, H.F.; Zhao, L.J. Alleviation of salt stress in citrus seedlings inoculated with mycorrhiza: Changes in leaf antioxidant defense systems. *Plant Soil Environ.* **2010**, *56*, 470–475. [CrossRef]
420. Farghaly, F.A.; Nafady, N.A.; Abdel-Wahab, D.A. The efficiency of arbuscular mycorrhiza in increasing tolerance of *Triticum aestivum* L. to alkaline stress. *BMC Plant Biol.* **2022**, *22*, 490. [CrossRef]
421. Mayo, K.; Davis, R.E.; Motta, J. Stimulation of germination spores of *Glomus versiforme* by spore-associated bacteria. *Mycologia* **1986**, *78*, 426–431. [CrossRef]
422. Xavier, L. Response of lentil under controlled conditions to co-inoculation with arbuscular mycorrhizal fungi and rhizobia varying in efficacy. *Soil Biol. Biochem.* **2002**, *34*, 181–188. [CrossRef]
423. Cruz, A.F.; Horii, S.; Ochiai, S.; Yasuda, A.; Ishii, T. Isolation and analysis of bacteria associated with spores of *Gigaspora margarita*. *J. Appl. Microbiol.* **2008**, *104*, 1711–1717. [CrossRef] [PubMed]
424. Hasani, H.; Aminpanah, H. Effect of *Pseudomonas fluorescens* inoculation on yield and yield components of rice (*Oryza sativa* L.) under different levels of phosphorus fertilizer. *J. Agric. Sci.* **2015**, *48*, 157–163.
425. Lally, R.D.; Galbally, P.; Moreira, A.S.; Spink, J.; Ryan, D.; Germaine, K.J.; Dowling, D.N. Application of endophytic *Pseudomonas fluorescens* and a bacterial consortium to *Brassica napus* can increase plant height and biomass under greenhouse and field conditions. *Front. Plant Sci.* **2017**, *8*, 2193. [CrossRef] [PubMed]
426. Santana-Fernandez, A.; Beovides-Garcia, Y.; Simo-Gonzalez, J.E.; Perez-Penaranda, M.C.; Lopez-Torres, J.; Rayas-Cabrera, A.; Santos-Pino, A.; Basail-Perez, M. Effect of a *Pseudomonas fluorescens*-based biofertilizer on sweet potato yield components. *Asian J. Appl. Sci.* **2021**, *9*, 105–113. [CrossRef]
427. Rungin, S.; Indananda, C.; Suttiviriya, P.; Kruasuwan, W.; Jaemsaeng, R.; Thamchaipenet, A. Plant growth enhancing effects by a siderophore-producing endophytic streptomycete isolated from a Thai jasmine rice plant (*Oryza sativa* L. cv. KDML105). *Antonie Van Leeuwenhoek* **2012**, *102*, 463–472. [CrossRef]
428. Verma, V.C.; Singh, S.K.; Prakash, S. Bio-control and plant growth promotion potential of siderophore producing endophytic *Streptomyces* from *Azadirachat indica*. A. *Juss. J. Basic Microbiol.* **2011**, *51*, 550–556. [CrossRef]
429. Mehnaz, S.; Mirza, M.S.; Haurat, J.; Bally, R.; Normand, P.; Bano, A.; Malik, K.A. Isolation and 16S rRNA sequence analysis of the beneficial bacteria from the rhizosphere of rice. *Can. J. Microbiol.* **2001**, *472*, 110–117. [CrossRef]
430. Arkhipova, T.N.; Veselov, S.U.; Melentiev, A.I.; Martynenko, E.V.; Kudoyarova, G.R. Ability of bacterium *Bacillus subtilis* to produce cytokinins and to influence the growth and endogenous hormone content of lettuce plants. *Plant Soil* **2005**, *272*, 201–209. [CrossRef]
431. Barazani, O.; Friedman, J. Is IAA the major root growth factor secreted from plant-growth-mediating bacteria? *J. Chem. Ecol.* **1999**, *25*, 2397–2406. [CrossRef]
432. Hussain, A.; Hasnain, S. Cytokinin production by some bacteria: Its impacts on cell division in cucumber cotyledons. *Afr. J. Microbiol. Res.* **2009**, *3*, 704–712.
433. Fulchieri, M.; Lucangeli, C.; Bottini, R. Inoculation with *Azospirillum lipoferum* affects growth and gibberellin status of corn seedling roots. *Plant Cell Physiol.* **1993**, *34*, 1305–1309.
434. Bacilio, M.; Rodriguez, H.; Moreno, M.; Hernandez, J.P.; Bashan, Y. Mitigation of salt stress in wheat seedlings by a gfp-tagged *Azospirillum liopferum*. *Biol. Fert. Soils* **2004**, *40*, 188–193. [CrossRef]
435. Marulanda, A.; Barea, J.-M.; Azcon, R. Stimulation of plant growth and drought tolerance by native microorganisms (AM fungi and bacteria) from dry environments: Mechanisms related to bacterial effectiveness. *J. Plant Growth Regul.* **2009**, *28*, 115–124. [CrossRef]

436. Bae, H.; Sicher, R.C.; Kim, M.S.; Kim, S.-H.; Strem, M.D.; Melnick, R.L.; Bailey, B.A. The beneficial endophyte *Trichoderma hamatum* isolate DIS 219b promotes growth and delays the onset of the drought response in *Theobroma cacao*. *J. Exp. Bot.* **2009**, *60*, 3279–3295 [CrossRef] [PubMed]
437. Singh, L.P.; Gill, S.S.; Tuteja, N. Unraveling the role of fungal symbionts in plant abiotic stress tolerance. *Plant Signal. Behav.* **2011**, *6*, 175–191. [CrossRef]
438. Dal Cortivo, C.; Ferrari, M.; Visioli, G.; Lauro, M.; Fornasier, F.; Barion, G.; Panozzo, A.; Vamerali, T. Effects of seed-applied biofertilizers on rhizosphere biodiversity and growth common wheat (*Triticum aestivum* L.) in the field. *Front. Plant Sci.* **2020**, *11*, 1–14. [CrossRef]
439. Das, K.; Prasanna, R.; Saxena, A.K. Rhizobia: A potential biocontrol agent for soilborne fungal pathogens. *Folia Microbiol.* **2017**, *62*, 425–435. [CrossRef]
440. Mondal, M.; Skalicky, M.; Garai, S.; Hossain, A.; Sarkar, S.; Banerjee, H.; Kundu, R.; Brestic, M.; Barutcular, C.; Erman, M. Supplementing nitrogen in combination with rhizobium inoculation and soil mulch in peanut (*Arachis hypogaea* L.) production system: Part II. Effect on phenology, growth, yield attributes, pod quality, profitability and nitrogen use efficiency. *Agronomy* **2020**, *10*, 1513. [CrossRef]
441. Igiehon, N.O.; Babalola, O.O.; Aremua, B.R. Genomic insights into plant growth promoting rhizobia capable of enhancing soybean germination under drought stress. *BMC Microbiol.* **2019**, *19*, 159. [CrossRef]
442. Jain, D.; Sharma, J.; Kaur, G.; Bhojiya, A.A.; Chauhan, S.; Sharma, V.; Suman, A.; Mohanty, S.R.; Maharjan, E. Phenetic and molecular diversity of nitrogen fixation plant growth promoting Azotobacter isolated from semiarid regions of India. *BioMed Res. Int.* **2021**, *2021*, 6686283. [CrossRef]
443. Fahsi, N.; Mahdi, I.; Mesfioui, A.; Biskri, L.; Allaoui, A. Plant growth-promoting rhizobacteria isolated from the jujube (*Ziziphus lotus*) plant enhance wheat growth, Zn uptake, and heavy metal tolerance. *Agriculture* **2021**, *11*, 316. [CrossRef]
444. Anli, M.; Baslam, M.; Tahiri, A.; Raklami, A.; Symanczik, S.; Boutasknit, A.; Ait-El-Mokhtar, M.; Ben-Laouane, R.; Toubali, S.; Ait Rahou, Y.; et al. Biofertilizers as strategies to improve photosynthetic apparatus, growth, and drought stress tolerance in date palm. *Front. Plant Sci.* **2020**, *11*, 516818. [CrossRef] [PubMed]
445. Ghadbane, M.; Medjekal, S.; Benderradji, L.; Belhadj, H.; Daoud, H. Assessment of arbuscular mycorrhizal fungi status and rhizobium on date palm (*Phoenix dactylifera* L.) cultivated in a Pb contaminated soil. In *Recent Advances in Environmental Science from the Euro-Mediterranean and Surrounding Regions*, 2nd ed.; Springer: Cham, Switzerland, 2021; pp. 703–707. [CrossRef]
446. Karthikeyan, B.; Abitha, B.; Henry, A.J.; Sa, T.; Joe, M.M. Interaction of rhizobacteria with arbuscular mycorrhizal fungi (AMF) and their role in stress abetment in agriculture. In *Fungal Biology*; Springer Science and Business Media LLC: Berlin, Germany, 2016; pp. 117–142. [CrossRef]
447. Vafadar, F.; Amooaghaie, R.; Otroshy, M. Effect of plant-growth-promoting rhizobacteria and arbuscular mycorrhizal fungus on plant growth, stevioside, NPK, and chlorophyll content of *Stevia rebaudiana*. *J. Plant. Interact.* **2013**, *9*, 128–136. [CrossRef]
448. Hashem, A.; Abd-Allah, E.F.; Alqarawi, A.A.; Al-Huqail, A.A.; Shah, M.A. Induction of osmoregulation and modulation of salt stress in *Acacia gerrardii* Benth. by arbuscular mycorrhizal fungi and *Bacillus subtilis* (BERA 71). *Biomed Res. Int.* **2016**, *2016*, 6294098. [CrossRef] [PubMed]
449. Zamljen, T.; Medic, A.; Veberic, R.; Hudina, M.; Grohar, M.C.; Slatnar, A. Influence of hydrolyzed animal protein-based biostimulant on primary, soluble and volatile secondary metabolism of Genovese and Greek-type basil grown under salt stress. *Sci. Hortic.* **2023**, *319*, 112178. [CrossRef]
450. Anton-Herrero, R.; Garcia-Delgado, C.; Anton-Herrero, G.; Mayans, B.; Delgado-Moreno, L.; Eymar, E. Design of a hydroponic test to evaluate the biostimulant potential of new organic and organomineral products. *Sci. Hortic.* **2023**, *310*, 111753. [CrossRef]
451. Distefano, M.; Steingass, C.B.; Leonardi, C.; Giuffrida, F.; Schweiggert, R.; Mauro, R.P. Effects of a plant-derived biostimulant application on quality and functional traits of greenhouse cherry tomato cultivars. *Food Res. Int.* **2022**, *157*, 111218. [CrossRef]
452. Ahammed, G.J.; Shamsy, R.; Liu, A.; Chen, S. Arbuscular mycorrhizal fungi-induced tolerance to chromium stress in plants. *Environ. Pollut.* **2023**, *327*, 121597. [CrossRef]
453. Bagu, H.M.M.K.; Hassan, E.A.; Nafady, N.A.; Dawood, M.F.A. Efficacy of arbuscular mycorrhizal fungi and endophytic strain *Epicoccum nigrum* ASU11 as biocontrol agents against blackleg disease of potato caused by bacterial strain *Pectobacterium carotovora* subsp. *Biol. Control.* **2019**, *134*, 103–113. [CrossRef]
454. Juila, D.; Ghosh, S. Aspects, problems and utilization of arbuscular mycorrhizal (AM) application as bio-fertilizer in sustainable agriculture. *Curr. Res. Microb. Sci.* **2022**, *3*, 100107. [CrossRef]
455. Larimer, A.L.; Clay, K.; Bever, J.D. Synergism and context dependency of interactions between arbuscular mycorrhizal fungi and rhizobia with a prairie legume. *Ecology* **2014**, *95*, 1045–1054. [CrossRef]
456. Liao, X.; Zhao, J.; Xu, L.; Tang, L.; Li, J.; Zhang, W.; Xiao, J.; Xiao, D.; Hu, P.; Nie, Y.; et al. Arbuscular mycorrhizal fungi increase the interspecific competition between two forage plant species and stabilize the soil microbial network during a drought event: Evidence from the field. *Appl. Soil Ecol.* **2023**, *185*, 104805. [CrossRef]
457. Lin, Y.; Jones, M.L. Evaluating the growth-promoting effects of microbial biostimulants on greenhouse floriculture crops. *HortScience* **2022**, *57*, 97–109. [CrossRef]
458. Lu, X.; Koide, R.T. The effects of mycorrhizal infection on components of plant growth and reproduction. *N. Phytol.* **1994**, *128*, 211–218. [CrossRef] [PubMed]

459. Pan, J.; Huang, C.-H.; Peng, F.; Wang, T.; Liao, J.; Ma, S.-X.; You, Q.-G.; Xue, X. Synergistic combination of arbuscular mycorrhizal fungi and plant growth-promoting rhizobacteria modulates morpho-physiological characteristics and soil structure in *Nitraria tangutorum* bobr. under saline soil conditions. *Res. Cold Arid Reg.* **2022**, *14*, 393–402. [CrossRef]
460. Sanjuan, J.; Napoles, M.C.; Perez-Mendoza, D.; Lorite, M.J.; Rodriguez-Navarro, D.N. Microbials for agriculture: Why do they call them biostimulants when they mean probiotics? *Microorganisms* **2023**, *11*, 153. [CrossRef]
461. Ven, A.; Verlinden, M.S.; Verbruggen, E.; Vicca, S. Experimental evidence that phosphorus fertilization and arbuscular mycorrhizal symbiosis can reduce the carbon cost of phosphorus uptake. *Funct. Ecol.* **2019**, *33*, 2215–2225. [CrossRef]
462. Xavier, L.J.C.; Germida, J.J. Bacteria associated with *Glomus clarum* spores influence mycorrhizal activity. *Soil Biol. Biochem.* **2003**, *35*, 471–478. [CrossRef]
463. Zou, Y.N.; Wu, Q.S.; Kuca, K. Unravelling the role of arbuscular mycorrhizal fungi in mitigating the oxidative burst of plants under drought stress. *Plant Biol.* **2021**, *23* (Suppl. S1), 50–57. [CrossRef]

**Disclaimer/Publisher's Note:** The statements, opinions and data contained in all publications are solely those of the individual author(s) and contributor(s) and not of MDPI and/or the editor(s). MDPI and/or the editor(s) disclaim responsibility for any injury to people or property resulting from any ideas, methods, instructions or products referred to in the content.

*plants*

MDPI

*Review*

# Developing Sustainable Agriculture Systems in Medicinal and Aromatic Plant Production by Using Chitosan and Chitin-Based Biostimulants

**Wenli Sun [1,*,†], Mohamad Hesam Shahrajabian [1,†], Spyridon A. Petropoulos [2] and Nazanin Shahrajabian [3]**

1 Biotechnology Research Institute, Chinese Academy of Agricultural Sciences, Beijing 100081, China; hesamshahrajabian@gmail.com
2 Department of Agriculture, Crop Production and Rural Environment, University of Thessaly, 38446 Volos, Greece; spetropoulos@uth.gr
3 Department of Economics, Isfahan (Khorasgan) Branch, Islamic Azad University, Isfahan 81595-158, Iran; nazanin.rajabian@gmail.com
* Correspondence: sunwenli@caas.cn; Tel.: +86-134-2608-3836
† These authors contributed equally to this work.

**Abstract:** Chitosan is illustrated in research as a stimulant of plant tolerance and resistance that promotes natural defense mechanisms against biotic and abiotic stressors, and its use may lessen the amount of agrochemicals utilized in agriculture. Recent literature reports indicate the high efficacy of soil or foliar usage of chitin and chitosan in the promotion of plant growth and the induction of secondary metabolites biosynthesis in various species, such as *Artemisia annua*, *Curcuma longa*, *Dracocephalum kotschyi*, *Catharanthus roseus*, *Fragaria × ananassa*, *Ginkgo biloba*, *Iberis amara*, *Isatis tinctoria*, *Melissa officinalis*, *Mentha piperita*, *Ocimum basilicum*, *Origanum vulgare* ssp. *Hirtum*, *Psammosilene tunicoides*, *Salvia officinalis*, *Satureja isophylla*, *Stevia rebaudiana*, and *Sylibum marianum*, among others. This work focuses on the outstanding scientific contributions to the field of the production and quality of aromatic and medicinal plants, based on the different functions of chitosan and chitin in sustainable crop production. The application of chitosan can lead to increased medicinal plant production and protects plants against harmful microorganisms. The effectiveness of chitin and chitosan is also due to the low concentration required, low cost, and environmental safety. On the basis of showing such considerable characteristics, there is increasing attention on the application of chitin and chitosan biopolymers in horticulture and agriculture productions.

**Keywords:** natural products; chitin; chitosan; medicinal and aromatic plants; biostimulant; stress resistance

**Citation:** Sun, W.; Shahrajabian, M.H.; Petropoulos, S.A.; Shahrajabian, N. Developing Sustainable Agriculture Systems in Medicinal and Aromatic Plant Production by Using Chitosan and Chitin-Based Biostimulants. *Plants* **2023**, *12*, 2469. https://doi.org/10.3390/plants12132469

Academic Editor: Daniela Businelli

Received: 5 June 2023
Revised: 26 June 2023
Accepted: 27 June 2023
Published: 28 June 2023

**Copyright:** © 2023 by the authors. Licensee MDPI, Basel, Switzerland. This article is an open access article distributed under the terms and conditions of the Creative Commons Attribution (CC BY) license (https://creativecommons.org/licenses/by/4.0/).

## 1. Introduction

Chitin is a linear cationic heteropolymer consisting of $\beta$-(1-4)-linked 2-acetamde-2-deoxy-D-glucopyranose (N-acetyl-D-glucosamine, GlcNAc) and randomly distributed units of 2-amino-2-deoxy-D-glucopyranose (D-glucosamine, GlcN) [1]. Chitin can be converted to chitosan by both biological and chemical or enzymatic procedures [1]. Chitin is generally available as a structural factor in the exoskeletons of an extensive range of eukaryotic organisms, such as insects; in the cell walls of fungi; in crustaceans, such as crabs, shrimps, and lobsters; and it is the second most plentiful biopolymer on the planet, alongside cellulose [2–8]. As a natural amino polysaccharide, it indicates different practical applications, owing to its functional characteristics [9,10]. Both chitooligosaccharides and chitosan are obtained from the depolymerization and deacetylation of chitin, respectively [11,12]. Chitin exists naturally in three polymorphic structures: $\alpha$-, $\beta$-, and $\gamma$-chitin [13]. A-chitin has anti-parallel chains, whereas $\beta$-chitin has parallel chains forming monoclinic crystals with intramolecular H-bonds and intermolecular H-bonds [14]. Li et al. [15] reported that a chitin/cellulose nanofiber complex (CCNFC), which was extracted from the spent mushroom substrate, indicated significant plant growth promotion activity and plant disease resistance activity. It has been reported that chitin-based nanofibers (CNFs) were outstanding in decreasing pathogenic fungal spread in pepper [16,17]. Parada et al. [18] noted that chitin nanofiber may cause a decrease in the number of spots caused by *Alternaria brassicicola* and the size of lesions induced by *Colletotrichum fructicola*, while it can promote disease resistance in strawberry and cabbage plants. Chitin and chitosan are derived from different sources: (1) aquatic: crabs (*Chionoecetes opilio, Podophthalmus vigil, Paralithodes camtschaticus, Carcinus mediterraneus*), water lobster, prawn, krill (*Anax imperator, Daphnia longispina, Notonecta glauca, Hydrophilus piceus, Asellus aquaticus, Agabus bipustulatus*), and squid pens; (2) terrestrial: spiders (*Geolycosa vultuosa, Hogna radiata, Nephila edulis*), scorpions (*Mesobuthus gibbosus*), beetles (*Bombyx mori, Holotrichia parallela, Leptinotarsa decemlineata*), cockroaches, and brachiopods; and (3) microorganisms.

Chitosan and its oligosaccharides may increase photosynthesis by increasing the activities of different enzymes of nitrogen and carbon metabolism, as well as dark and light responses of photosynthetic apparatus. They also have an important function in photosynthetic machinery stimulation by modulating principal photochemistry; they can also manage the restraints of stomata, escalate the carbon fixation effectiveness in dark reactions, and improve the synthesis of metabolites. Chitosan and its oligosaccharides may also promote the biosynthesis of secondary metabolites and the activity of enzymes [19]. Chitosan derivatives have superior antifungal activity compared to chitosan [20]. Chitosan can be used for different purposes based on the specificity of its physicochemical properties [21,22], as well as its biodegradability [23] and biocompatibility [24–26]. The potential utilization of chitosan and its derivatives include their use as antimicrobial and antioxidant agents, a drug delivery system, gene therapy, food technology, bio-nanotechnology, regenerative technology, effluent treatments, and electrolytes, as well as in various industries, including paper manufacturing, cosmetics, as absorption enhancers, photography, wood industries, environmental protection, agriculture, obesity treatment, catalyst research, immune therapy, as permeation enhancers, energy production, and in the immobilization of cells. Chitosan leads to various defensive genes in plants, such as genes related to pathogenesis, namely, *chitinase* and *glucanase* [27]. It may also trigger several enzymes in the reactive oxygen species scavenging system, such as superoxide dismutase, peroxidase, and catalase [27–30]. Chitosan also leads to the accumulation of auxin in the apex of roots, thus promoting root elongation [31]. Moreover, chitosan and its oligosaccharides act as the suited ligand for the initiation of accessible receptors and, therefore, elicit different signaling pathways in the plant, namely, GPCR and PLC/PK, MAPK, $H_2O_2$ burst, and the promotion of transcription parameters, thus generating plant responses to environmental stimuli [32]. Chitosan and its oligosaccharides also show antimicrobial properties and can be used as biopesticides, preventing the proliferation of pathogens and improving crop quality and yield [32–35]. For example, chitosan has innate antibacterial characteristics that prevent the development of fungi and germs [36]; its nanoparticles were utilized to control the release of citronella essential oil for use against cotton leafworm (*Spodoptera littoralis* (Boisd.) [37]. Moreover, it can noticeably decrease plant oxidative stress and improve crop yield [38,39], while root exposure to chitosan increased both antioxidant and defense enzyme activity [38]. Chitosan can decrease plant diseases via two major procedures: direct antimicrobial function against pathogens, or through the induction of plant defense reactions [39–43]. The effects of chitosan on plant defense mechanisms are connected to its application method and dose [44]. The potential applications of chitosan as eco-friendly tools in agricultural production are shown in Figure 1.

Oligochitosan, obtained from chitosan, is a potential plant immunity regulator [45,46]. It has shown an extensive range of biological applications, including as a plant growth stimulator, antimicrobial agent, and feed additive, among others [47,48]. Oligochitosan can activate a plant's innate immunity via signal transduction, signal perception, and the oligochitosan response to proteins and genes, thus leading to accumulation of defense-related secondary metabolites [49,50]. It can elicit different defense responses, including the scavenging of reactive oxygen species, the synthesis of jasmonate, the increase of cell walls' lignification, the production of phytoalexins, and the accumulation of defense-related proteinases inhibitors, as well as the induction of pathogenesis-related (PR) proteins and salicylic acid [51–54]. Zhang et al. [55] reported that oligochitosan could decrease black spots on tomato fruits, while Deng et al. [56] suggested that a preharvest application of oligochitosan can be an important substitute to conventional control practices for the prevention of post-harvest anthracnose in navel oranges. Oligochitosan showed a significant impact on controlling fruit brown rot; it could boost antioxidant and defense-related enzymes activities, accelerate the corresponding gene transcript expression, and also delay fruit senescence and softening [57]. Oligochitosan remarkably activated phenylpropanoid pathway metabolism and induced defense responses against green and blue mold in citrus fruit [58,59]. Wang et al. [60] proposed that primary metabolites induced by oligochitosan behave as the source of energy and signaling or as substrate molecules that have indirect or direct roles in defense reactions against *Geotrichum candidum*, also known as sour rot, in citrus fruit.

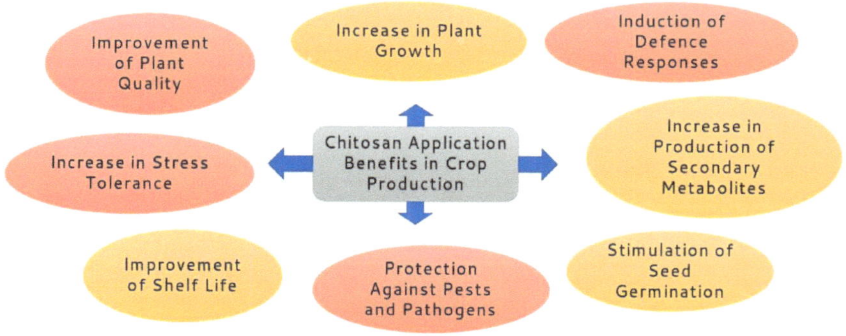


**Figure 1.** Utilization of chitosan and its derivatives in crop production.

Relevant literature was obtained using the keywords "Biostimulants", "Chitin", "Chitosan", "Oligochitosan", "Medicinal plants", "Aromatic plants", "Natural products", "Biotic stress", and "Abiotic stress" from scientific literature databases, such as "Web of Science", "SciFinder", "PubMed", and "Elsevier". The current review gives a literature overview with regards to the impacts of chitosan, chitin, and its derivatives on medicinal and aromatic plants, focusing on their noticeable functions in modern sustainable production. We also highlight the major restrictions and illustrate the future possibilities of applications of this form of biostimulant.

## 2. The Effects of Biostimulants on Medicinal and Aromatic Plants

A plant biostimulant is defined as any substance or microorganism used on plants that may lead to increased nutrition efficiency and improvement of abiotic stress tolerance and crop quality traits. Biostimulants consist of algae or plant extracts, mushrooms, amino acids, humic substances, microorganisms, biomolecules, natural ingredients, and metabolites of fermentation, among others [61–64]. Utilization of Moringa leaf extract led to greater growth, lower weed infestation, and an increased final yield of basil (*Ocimum basilicum* L.) [65,66], while the application of salicylic increased the content of different metabolite components in leaves [67] and improved the values of shoot fresh weight, plant height, and dry weight of basil seedlings under water shortage treatments [68]. The combination of *Azotobacter chroococcum* and *Azospirillum lipoferum* significantly enhanced the fresh and dry yield of coriander [69], while foliar spraying of a biostimulant (Asahi SL or Goemar Goto Arysta Life Science) induced the photochemical effectiveness of photosystem II, stomatal conductance, and transpiration rate, but decreased the intercellular carbon dioxide concentration [70]; moreover, it promoted the total phenolic compounds' concentration,

L-ascorbic acid, and the total antioxidant activity of coriander seedlings [70]. The utilization of different biostimulants as a seed treatment during germination boosted the germination percentage of garden cress (*Lepidium sativum* L.) [71]. Humic acids increased the bulb yield and clover and bulb diameter of garlic (*Allium sativum* L.) [72], and demonstrated a positive effect on the mineral nutrition content [73]. *Bacillus subtilis* and humic acids boosted the level of protein and minerals and increased the antioxidant activities of garlic [74]. Salicylic acid improved the secondary metabolites in a callus culture of Iranian sodab (*Ruta graveolens* L.) [75]. The application of salicylic acid led to a significant decrease of the toxic impact of mercury on the growth and final yield of lemon balm (*Melissa officinalis* L.) [76], while seed priming with *Pseudomonas* spp. accelerated the germination and inoculation of seedlings and positively influenced lemon balm growth [77]. Potassium silicate ($K_2SiO_3$) application increased the pigments, leucine, isoleucine, phenylalanine, ascorbic acid, valine, protein, Ca, Mg, and K content and the uptake of P, K, and N of horseradish tree (*Moringa oleifera* Lam.) [78]. Some other biostimulants, such as benzoic acid, phenylalanine, *N*-benzoyl glycine, glycine, and serine, significantly improved the taxol accumulation in a callus of Japanese yew [79]. The application of humic acids containing L-ascorbic acid and thiamine, both alone (ROOTS) and together with fertilizer (ROOT PLUS), increased the production of marigold (*Calendula officinalis* L.) plants; it also induced earlier flowering [80]. For the same species, bioactive amino acid constituents, namely Kadostim and Humiforte, promoted the total flavonoids and carbohydrate constituents of leaves, P, K, and N [81]. The application of Radifarm® boosted the nutrient status of French marigold (*Tagetes erecta* and *Tagetes patula* L.) plants and flowering [82]. The utilization of chitosan reduced the adverse effects of salt stress, and stimulated the enzyme activities and antioxidants of milk thistle (*Silybum marianum* L.) [83]. Salicylic acid decreased the osmotic adjustment under drought stress, increased the antioxidant compounds' content in seeds, and boosted the vegetative growth and yield of milk thistle by escalating the relative water content in regions under the influence of a water shortage [84]. *Trichoderma ovalisporum* (NFCCI2689) and *T. harzianum* (NFCCI 2241) improved the menthol content and oil yield of mint (*Mentha arvensis* L.) [85], while the application of CRADLE™, Mobilizer™, and Nanozim NXT™ biostimulants promoted metabolic and physiological responses, such as the leaf water potential, gas exchange, proline accumulation, and relative water content of stressed seedlings of mint [86]. The application of ethanol and humic acids improved the photosynthetic pigments and yield potential of Moldavian dragonhead (*Dracocephalum moldavica* L.) [87]. The application of salicylic acid stimulated the oil components of peppermint (*Mentha piperita* L.) compared to control plants, particularly menthol, menthone, methyl acetate, and 1,8-cineole [88]. Applications of chitosan, citric acid, and humic acid on the same species boosted the stem and leaves dry weight [89,90]. A combined application of arbuscular mycorrhizal fungi with chitosan positively affected the secondary metabolites and dry herbal weight of peppermint [91,92]. Moreover, *Pseudomonas* sp. P17 strain was used as a potential plant growth promoter for better plant growth and higher nutrient uptake in pigeon pea (*Cajanus cajan* L.) [93]. A chitosan and salicylic acid application stimulated the production of amarogentin, mangiferin, and swertiamarin from shoot cultures of swertia (*Swertia paniculata* Wallich), which have significant bioactive properties that render them very important in the pharmaceutical industries [94].

## 3. Practical Usage of Chitosan on Medicinal and Aromatic Plants

The multiple functions of chitosan's chemically and naturally synthesized components could act as potent elicitors and biostimulants and change the physicochemical status of medicinal crops [95–97]. They participate in the activation of defensive genes via the overexpression of pathogenesis-related genes; they may also provide photoprotection by increasing secondary metabolite biosynthesis. They play an important function in oxidative damage reversal by managing hydrogen peroxide and nitric oxide signaling. They also stimulate activity at the cellular level of antioxidant enzymes, such as peroxidase, dismutase superoxide, and catalase; retain ion homeostasis; improve the photosynthesis rate by interacting with abscisic acid-induced stomatal closure; improve plant vigor and environmental stress tolerance; and generally improve crop performance [98–100]. Aristolochic acid I (AA I) is one of the substituted 1-phaenanthrene carboxylic acids usually discovered in *Asarum* and *Aristolochia* plants, and chitosan-changed carbon microcoils were used as suitable adsorbents for the selective extraction of aristolochic acid I from Aristolochiaceae plants [101]. *Psammosilene tunicoides* W. C. Wu et C. Y. Wu, a monotypic species of Caryophyllaceae that is endemic to China, has been used as folk medicine for centuries, and it is one of the major ingredients in the traditional Chinese medicine "Yunnan Baiyao". According to the literature, hairy roots of the species elicited with 200 mg $L^{-1}$ of chitosan increased the total saponin content by 4.55-fold [102]. High-performance liquid chromatography (HPLC) indicated that the yields of gypsogenin, quillaic acid, and gypsogenin-3-O-$\beta$-D-glucuronopyranoside were significantly escalated after a chitosan application [102]. Furthermore, exogenous chitosan usage significantly triggered the reactive oxygen species (ROS), scavenging enzyme activities, and nitric oxide (NO) content in the hairy roots of *Psammosilene tunicoides* [102].

In basil seedlings, the level of rosmarinic acid (RA) increased after the application of 100 mg/L of chitosan lactate [103]. The mixture of common chili (*Capsicum annuum*) fruit extracts and chitosan can be effective against pathogenic bacteria, especially against *Staphylococcus aureus,* which is a Gram-positive, spherically shaped bacterium [104]. A chitosan application increased the content of the EO components, the positive impacts of chitosan utilization under drought stress conditions [105]. The usage of water-soluble carboxymethyl chitosan-grafted daphnetin nanoparticles (DA@CMCS-NPs) improved the activities of defense enzymes in tobacco, and effectively inhibited the development of tobacco bacterial wilt (*Ralstonia solanacearum*) [106]. The effects of chitosan on different medicinal and aromatic plants are shown in Table 1.

**Table 1.** The effects of chitosan on medicinal and aromatic plants.

| Plant | Scientific Name | Plant Family | Key Point | References |
|---|---|---|---|---|
| Basil | *Ocimum basilicum; Ocimum ciliatum* | Lamiaceae | Various amounts of chitosan had considerable impacts on the antioxidant activity and total phenol content of the extracts of two species. | [107,108] |
| | | | Chitosan could be a promising material used to reduce the adverse impacts of water stress on the growth factors of basil seedlings. | [108] |
| Bitter melon | *Momordica charantia* L. | Cucurbitaceae | An application of chitosan and their conjugated forms as nanoparticles (Se-CS NPs) improved the biochemical and morphophysiological characteristics of bitter melon plants under moderate and severe salinity stress condition. | [109] |
| | | | Exogenous utilization of Se-CS NPs leads to the expression of multiple defense- and secondary metabolism-related transcripts in bitter melon plants. | [109] |
| *Dendrobium* orchids | *Dendrobium* | Orchidaceae | Chitosan showed the ability to improve floral production of the *Dendrobium*. | [110] |

**Table 1.** *Cont.*

| Plant | Scientific Name | Plant Family | Key Point | References |
|---|---|---|---|---|
| Dragonhead | *Dracocephalum kotschyi* Boiss | Lamiaceae | It also stimulated the number of vascular bundles in both young and old leaves. | [110] |
| | | | It was an effectual elicitor for increasing rosmarinic acid and quercetin content. | [111] |
| | | | Chitosan increased the content of apigenin noticeably. | [111] |
| Fenugreek | *Trigonella foenum-graecum* L. | Leguminosae | A chitosan spray had a significant influence on the principle essential oil components, such as *p*-cymene and thymol. | [112] |
| | | | The activity of nitrate reductase and the contents of photosynthetic pigments and carbonic anhydrase enzymes were significantly increased after an application of Co-60 gamma-irradiated chitosan. | [113] |
| | | | Co-60 gamma-irradiated chitosan significantly boosted the total alkaloid content, seed yield, and trigonelline constituent. | [113] |
| Galega | *Galega officinalis* L. | Leguminosae | An application of chitosan had a more optimal impact on the morphological characteristics. | [114] |
| German chamomile | *Matricaria recutita* L. | Asteraceae | Its foliar spray in the flowering stage increased the chamomile flower number and weight per plant. | [115] |
| | | | Its usage at the rate of 40 ppm is suggested for German chamomile cultivation. | [115] |
| Ginger | *Zingiber officinale* L. | Zingiberaceae | Its application induced the HR marker gene *HSR203J* in ginger and inhibited the expression of cell death markers *AIF2*, *HIN1*, and *AIF1*. | [116] |
| | | | Chitosan has the potential to prime standing ginger plants against soft rot disease and facilitate appropriate sustainable management of ginger soft rot (*Pythium myriotylum*). | [116] |
| Hyssop | *Hyssopus officinalis* L. | Lamiaceae | The results show that spraying chitosan at 2.5 g/L significantly increased the measurements of the canopy diameter, inflorescence height, plant height, numbers of auxiliary and flowering branches, dry herbal weight, and the components of photosynthesis pigments under various levels of irrigation frequencies. | [117] |

**Table 1.** *Cont.*

| Plant | Scientific Name | Plant Family | Key Point | References |
|---|---|---|---|---|
| | | | The maximum values of the volatile oil yield were related to the chitosan-spraying in a reduced irrigation condition. | [117] |
| | | | Chitosan had noticeable effects on the percentage of (*cis*- and *trans*-) pinocamphone, as the predominant components of hyssop volatile oil, under a reduced irrigation condition. | [117] |
| Indian borage (Indian mint) | *Coleus aromaticus* Benth (L.) | Lamiaceae | A mixture of Chitosan and cytokinin showed a positive impact on multiple shoot induction than cytokinin alone. | [118] |
| | | | Its application also increased the percentage of alkaloids, flavonoids, terpenoids, saponins, tannins, and total phenolic content. | [118] |
| Lavender-cotton | *Santolina chamaecyparissus* L. | Asteraceae | Its application improved antifungal capacities and increased some physical characteristics. | [119] |
| Lemon balm | *Melissa officinalis* L. | Lamiaceae | It increased plant regeneration in a callus culture of a medicinal herb. | [120] |
| | | | Its consumption consists of the trigger of defense-related enzymes; stimulated expression of rosmarinic acid synthase (RAS) genes and tyrosine aminotransferase (TAT); and stimulation of methyl jasmonate biosynthesis. | [121] |
| Milk thistle | *Silybum marianum* (L.) Gaertn. | Asteraceae | Significant reduction of the negative impact of salinity and increased plant growth and improved physiological characteristics occurred after its application. | [83] |
| | | | Its usage at 0.01% increased the chlorophyll and total chlorophyll content. | [83] |
| | | | Its application promoted the enzymatic activity and reduced $H_2O_2$ components in the leaves. | [83] |
| | | | Chitosan may protect plants from salt stress damage by regulation of the intracellular ion component and through increasing the capability of antioxidant enzyme activities. | [83] |
| Moringa | *Moringa oleifera* L. | Moringaceae | Chitosan improved the plant growth parameters: root length, root fresh weight, shoot length, shoot fresh weight, shoot dry weight, root dry weight, and contents of chlorophyll-b and chlorophyll-a. | [122] |

**Table 1.** *Cont.*

| Plant | Scientific Name | Plant Family | Key Point | References |
|---|---|---|---|---|
| Peppermint | *Mentha piperita* L. | Lamiaceae | Under drought stress, the usage of chitosan-coated iron oxide nanoparticles (Fe-CTs NPS) increased essential oils production in the mint plant. | [123] |
| | | | Under foliar spray of chitosan, the maximum constituents of the essential oil yield, menthol, and the balance of menthol/menthone of the essential oil from peppermint were achieved from the peppermint plants. | [92] |
| Roselle | *Hibiscus sabdariffa* L. | Malvaceae | Chitosan nanofiber (CNF) treatments affected the plant growth regulators' impact on most of the traits evaluated. | [124] |
| | | | It could be applied for enhancement of the number of calyxes, plant height, total chlorophyll, β-carotene, antioxidant activity, and flavonoids. | [124] |
| Savory | *Satureja hortensis* L. | Lamiaceae | Under severe water stress, its application improved the total soluble sugar, antioxidant activity, total phenolic and proline contents, and essential oil content of the seedlings. | [125] |
| | | | Under water stress conditions, it may have positive influences on the essential oil quality and quantity, osmotic adjustment, antioxidant activity, and growth of summer savory. | [125] |
| Spearmint | *Mentha spicata* L. | Lamiaceae | Chitosan–melatonin nanoparticles could be used as an innovative protective agent to reduce the impact of salinity in spearmint plants. | [126] |
| St. John's wort | *Hypericum perforatum* L. | Clusiaceae | It showed a significant rise in xanthone production and a simultaneous decline in flavonoid production. | [127] |
| | | | Chitosan also led to the production of 1,7-dihydroxyxanthone (euxanthone). | [127] |
| Sweet marjoram | *Majorana hortensis* Moench | Lamiaceae | Its application improved the marjoram biomass and secondary metabolites' assimilation. | [128] |
| Tashenehdari | *Scrophularia striata* Boiss. | Scrophulariaceae | The increase in amino acid content and phenylalanine ammonia-lyase (PAL) activity was related to rises in the phenolic components after application of chitosan treatments. | [129] |

**Table 1.** *Cont.*

| Plant | Scientific Name | Plant Family | Key Point | References |
|---|---|---|---|---|
| | | | Chitosan, by up-regulating the *PAL* gene, increases the production of the phenylpropanoid content. | [129] |
| Thyme | *Thymus daenensis* Celak | Lamiaceae | Under the mild stress, the maximum essential oil yield was related to the utilization of 400 $\mu L^{-1}$ chitosan. | [130] |
| Mediterranean thyme | *Thymbra spicata* L. | Lamiaceae | The carvacrol level in the essential oil escalated in the plants after application of chitosan. | [131] |
| | | | The spray of chitosan could reduce the harmful impacts of water shortage on the oil yield and the carvacrol percentage. | [131] |
| Russian sage | *Salvia abrotanoides* (Kar.) Sytsma | Lamiaceae | A chitosan foliar application can increase drought tolerance. | [132] |
| | | | The leaves showed significant antioxidant enzyme activity under drought stress using chitosan nanoparticles. | [132] |
| Turmeric | *Curcuma longa* L. | Zingiberaceae | Its application improved the production of curcumin. | [133] |
| | | | The chitosan could boost the activity or protease inhibitor in rhizomes and leaves. | [133] |
| Wild mint | *Mentha arvensis* L. | Lamiaceae | The 0.125% content of chitosan provides a higher accumulation of roots, shoots, and total dry weight. The maximum menthol content is obtained after application of 0.06% chitosan. | [134] |
| Winged-fruited marigold | *Calendula tripterocarpa* Rupr. | Asteraceae | The chitosan application also increased the growth parameters, carotenoids, and levels of chlorophyll *a*, *b* under both Ni stress and normal conditions. | [135] |
| | | | Chitosan decreased the level of malondialdehyde and the activities of catalase (CAT) and superoxide dismutase (SOD) in the shoot and roots under Ni stress. | [135] |

## 4. Activities and Applications of Oligochitosan

Chitooligosaccharides or oligochitosan are plant elicitors that have similar impacts to chitosan on herbs and plants against biotic stress [136]. Moreover, oligochitosan, with a higher degree of polymerization, indicated a stronger elicitation impact via the overexpression of defensive genes [137]. Oligochitosan may promote protection against pathogens through the activation of salicylic acid (SA) or jasmonic acid/ethylene (JA/ET)-dependent pathways; boost growth by escalating gibberellin, auxin, photosynthesis content, and N and C assimilation; induce protection against abiotic stress via the abscisic acid (ABA)-dependent pathway; and increase the synthesis of secondary metabolites with medicinal and antipathogenic activities [138]. The application of chitosan oligosaccharide at both 60 g $L^{-1}$ and 120 g $L^{-1}$ levels increased fennel growth and yield, while it increased estragole and reduced the anethole content, thus demonstrating adverse effects on the essential oil quality of fennel [139]. Liu et al. [140] reported that oligochitosan sprayed at 5 g/L noticeably inhibited rhizome rot caused by *Fusarium oxysporum*, and also decreased weight loss; however, it did not influence the amount of soluble solids or titratable acidity of the rhizomes of ginger (*Zingiber officinale* Roscoe). The potential of oligochitosan to decrease rot in stored rhizomes could be related to their ability to induce defense responses in ginger [140]. Li et al. [141] also indicated that oligochitosan caused the accumulation of reactive oxygen species, mitochondrial dysfunction, metacaspase activation, and $Ca^{2+}$ homeostasis dysregulation, together with hallmarks of apoptosis consisting of DNA fragmentation, phosphatidylserine externalization, and nuclear condensation to control pathogens in *Ceratocystis fimbriata*. Lu et al. [142] observed that oligochitosan led to tobacco resistance to tobacco mosaic virus (TMV) by the $Ca^{2+}$ signaling pathway. Ahmad et al. [143] indicated that a foliar spray of irradiated chitosan (80 $mgL^{-1}$) significantly increased the physiology and growth of peppermint. On the basis of LC-MS analysis, the chitosan oligosaccharide application slightly increased artemisinin production in the leaves, while the impacts of chitosan oligosaccharide and salicylic acid on artemisinin production varied, indicating that these elicitors may not be an appropriate method for increasing the artemisinin yield or that the regulation takes longer to be effective in sweet wormwood (*Artemisia annua* L.) [144].

## 5. The Application of Chitin as Biostimulant

Chitin can induce both a local immune reaction and a systemic disease resistance when provided as a supplement in soils [145]. The most important chitin derivatives are chitosan, alkyl chitin, N and O sulfated chitin, dibutyryl chitin, carboxymethyl chitin (CMCH), chito-oligosaccharides (COS), chitin nanofibers (CNF), chitin nano-whiskers (CNW), chitin nanoparticles (CNP), chitin nanocomposites (CNC), and chitin hydrogels (CHG). Unlike chitosan, the application of chitin had no remarkable effects on defense mechanisms, such as enzymatic activities, gene expression levels, hormone quantities, and root secreted proteins of medicinal cannabis (*Cannabis sativa* L.) [146–149]. Cretoiu et al. [150] added chitinous matter obtained from shrimps to the soil's top layer to assess the suppressiveness of the oil toward *Verticillium dahliae*, as well as against plant pathogenic nematodes. They reported a change in both structures and an abundance of soil microbial communities, both total soil fungi and bacteria; they also found that the structures and abundance of soil actinobacteria and the Oxalobacteraceae were influenced by chitin. Liopa-Tsakalidi et al. [151] reported that a chitin application in the peat substrate did not influence the weight and length of lemon balm. Chitin influenced the tarragon leaves, leading to an increase of the total chlorophyll content [151]. The seed germination percentage of chervil (*Anthriscus cerefolium* L.) was improved in the substrate with 1% and 2% chitin and 200, 500, and 1000 ppm $GA_3$ compared to the control treatments ($H_2O$) [152]. With the increasing rate of chitin and $GA_3$, the seed germination of the chervil was decreased (Liopa-Tsakalidi and Barouchas, 2011). In the combinations of 1% chitin + 80 mM NaCl, 1% chitin + 120 NaCl, 120 NaCl + $GA_3$, 2% chitin + 180 NaCl, and 200 $GA_3$ +180 NaCl, germination was boosted and the Timson index of germination velocity was greater than the corresponding velocity in $H_2O$ [152]. Kanawi et al. [153] found that the percentage of germination of basil seeds increased with the increasing concentration of chitin, while the percentage of inhibition of pathogenic fungi also escalated with the rise of the concentration of chitin. The application of chitin/nano-$TiO_2$ and chitin/nano-$SiO_2$ resulted in a notable increase of the preservation of seed quality of Ginkgo (*Ginkgo biloba* L.) [154].

## 6. Conclusions—Future Remarks

Biostimulants are known and described as molecules of biological origin with positive effects on the growth and development of different crops, increasing resistance to both biotic and abiotic stress parameters and improving crop quality and yield. The most important biostimulant impacts on crops are improved stress tolerance, the production of deeper roots, increased cation exchange, accelerated crop establishment, improved seed germination rates, reduced leaching, increased nutrient uptake, detoxification of heavy metals, increased stomata opening and plant transpiration, improved drought tolerance, increased natural plant toxins, repelling of pests, stimulation of immune system of plants, and improved yields and crop performance. The accurate application of chitin may show a beneficial function in the growth, development, and bioactive phytoconstituents accumulation of medicinal plants, which, in turn, could upscale the possibility of industrial exploitation of medicinal plants in the most sustainable way possible. Chitosan is the principal tool in the agricultural field because it is biocompatible, nontoxic, and biodegradable, and also has adsorption capabilities, since it can absorb different heavy metals, insecticides from water or pollutants, soil, and pesticides. Chitosan can be formed in several ways, as nanoparticles, nanomaterials, and nanocomposites, which are appropriate for plant protection, while its characteristics can be improved by combining it with other substances. The importance of chitosan in the formation of photosynthetic metabolites and pigments; the absorption of elements; assimilation; and the expression of genes as micro-, macro-, and nano-structures and in the form of composites or pure components with metal elements; and, lastly, the qualitative and quantitative performance of different medicinal plants and crops influenced by chitosan components has been investigated thus far. Chitosan and chitosan-based nanoparticles have shown promising findings in medicinal plants as fertilizers, plant disease control and insecticidal agents, soil-conditioning agents, ripening retardants, and fruit and seed coatings. These components are able to increase the assimilation of secondary metabolites through the stimulation of biochemical modification, defensive responses, and the biosynthesis of phytoalexins. Using chitosan as a foliar spray significantly increased plant growth; ions; chlorophyll; antioxidant capacity; and phytopharmaceutical components, such as anthocyanins, total soluble phenols and flavonoids, and alkaloids; it also reduced the number of oxidative biomarkers. Biocompatible and biodegradable chitosan and chitosan-based nanomaterials are becoming important in agriculture because of their unique characteristics, such as eliciting, biostimulating, antimicrobial activity, and stimulation of plant growth and tolerance to environmental stresses. Currently, these substances are not being commonly utilized in agriculture, as the procedures of their biological activity in plants and action against pathogenic microorganisms have not been completely elucidated to date. Chitosan successfully increases the physiological functions of plants; additionally, chitosan treatment modulates various genes, especially through the activation of plant defense signaling pathways. Furthermore, it is useful in decreasing fertilizer losses because of its coating ability, which is essential in maintaining control of environmental pollution. Chitin and its derivatives are highly recognized as growth stimulants, plant growth regulators, and elicitors for the production of secondary metabolites. In summary, despite the open challenges, molecules of chitosan and chitin have been proven practical in different aspects of plant biology, from the increase in crop yield to the protection against attacks from pathogens. Hence, more research is needed to survey and fine tune the application of chitosan and chitin to the root zone to prevent soil-borne diseases and to decrease the overuse of synthetic fertilizers in medicinal and aromatic crops in the future.

**Author Contributions:** W.S., writing—original draft preparation; M.H.S., writing—original draft preparation and editing; S.A.P., writing and editing; N.S., writing and editing. All authors have read and agreed to the published version of the manuscript.

**Funding:** This research was supported by the National Key R&D Program of China (research grant 2019YFA0904700). This research was funded by the Natural Science Foundation of Beijing, China (Grant No. M21026). This research was also supported by the European Regional Development Fund of the European Union and Greek national funds through the Operational Program for Competitiveness, Entrepreneurship, and Innovation under the call RESEARCH—CREATE—INNOVATE (project code: T2EDK-05281). The article processing charges (APC) were also covered by the European Regional Development Fund of the European Union and Greek national funds through the Operational Program for Competitiveness, Entrepreneurship, and Innovation under the call RE-SEARCH—CREATE—INNOVATE (project code: T2EDK-05281).

**Institutional Review Board Statement:** Not applicable.

**Informed Consent Statement:** Not applicable.

**Data Availability Statement:** Not applicable.

**Conflicts of Interest:** The authors declare no conflict of interest.

## References

1. Chakravarty, J.; Edwards, T.A. Innovation from waste with biomass-derived chitin and chitosan as green and sustainable polymer: A review. *Energy Nexus* **2022**, *8*, 100149. [CrossRef]
2. Joseph, S.M.; Krishnamoorthy, S.; Paranthaman, R.; Moses, J.A.; Anandharamakrishnan, C. A review on source-specific chemistry, functionality, and applications of chitin and chitosan. *Carbohydr. Polym. Technol. Appl.* **2021**, *2*, 100036. [CrossRef]
3. Shahrajabian, M.H. Medicinal herbs with anti-inflammatory activities for natural and organic healing. *Curr. Org. Chem.* **2021**, *25*, 2885–2901. [CrossRef]
4. Shahrajabian, M.H.; Chaski, C.; Polyzos, N.; Petropoulos, S.A. Biostimulants application: A low input cropping management tool for sustainable faring of vegetables. *Biomolecules* **2021**, *11*, 698. [CrossRef] [PubMed]
5. Al Shaqsi, N.H.K.; Al Hoqani, H.A.S.; Hossain, M.A.; Al Sibani, M.A. Optimization of the demineralization process for the extraction of chitin from Omani *Portunidae segnis*. *Biochem. Biophys. Rep.* **2020**, *23*, 100779. [CrossRef]
6. Liao, J.; Wang, Y.; Hou, B.; Zhang, J.; Huang, H. Nano-chitin reinforced agarose hydrogels: Effects of nano-chitin addition and acidic gas-phase coagulation. *Carbohydr. Polym.* **2023**, *313*, 120902. [CrossRef]
7. Hou, F.; He, L.; Ma, X.; Wang, D.; Ding, T.; Ye, X.; Liu, D. Ultrasound enhanced the binding ability of chitinase onto chitin: From an AFM insight. *Ultrason. Sonochem.* **2020**, *67*, 105117. [CrossRef]
8. Vaikuntapu, P.R.; Rambabu, S.; Madhuprakash, J.; Podile, A.R. A new chitinase-D from a plant growth promoting *Serratia marcescens* GPS5 for enzymatic conversion of chitin. *Bioresour. Technol.* **2016**, *220*, 200–207. [CrossRef]
9. Kertmen, A.; Ehrlich, H. Patentology of chitinous biomaterials. Part I: Chitin. *Carbohydr. Polym.* **2022**, *282*, 119102. [CrossRef]
10. Lv, J.; Lv, X.; Ma, M.; Oh, D.-H.; Jiang, Z.; Fu, X. Chitin and chitin-based biomaterials: A review of advances in processing and food applications. *Carbohydr. Polym.* **2023**, *299*, 120142. [CrossRef]
11. Curto, M.A.; Butassi, E.; Ribas, J.C.; Svetaz, L.A.; Cortes, J.C.G. Natural products targeting the synthesis of β(1,3)-D-glucan and chitin of the fungal cell wall. Existing drugs and recent findings. *Phytomedicine* **2021**, *88*, 153556. [CrossRef] [PubMed]
12. Mohan, K.; Ganesan, A.R.; Ezhilarasi, P.N.; Kondamareddy, K.K.; Rajan, D.K.; Sathishkumar, P.; Rajarajeswaran, J.; Conterno, L. Green and eco-friendly approaches for the extraction of chitin and chitosan: A review. *Carbohydr. Polym.* **2022**, *287*, 119349. [CrossRef] [PubMed]
13. Sultan, R.; Reghunadhan, A.; Sambhudevan, S. A new era of chitin synthesis and dissolution using deep eutectic solvents-comparison with ionic liquids. *J. Mol. Liq.* **2023**, *380*, 121794. [CrossRef]
14. Singh, S.K. Solubility of lignin and chitin in ionic liquids and their biomedical applications. *Int. J. Biol. Macromol.* **2019**, *132*, 265–277. [CrossRef] [PubMed]
15. Li, J.; Liu, R.; Zhang, C.; Yang, J.; Lyu, L.; Shi, Z.; Man, Y.B.; Wu, F. Selenium uptake and accumulation in winter wheat as affected by level of phosphate application and arbuscular mycorrhizal fungi. *J. Hazard. Mater.* **2022**, *433*, 128762. [CrossRef]
16. Aiman, U.; Nisar, N.; Tsuzuki, T.; Lowe, A.; Rossiter, J.T.; Javaid, A.; Powell, G.; Waseem, R.; Al-Mijalli, S.H.; Iqbal, M. Chitin nanofibers trigger membrane bound defense signaling and induce elicitor activity in plants. *Int. J. Biol. Macromol.* **2021**, *178*, 253–262. [CrossRef]
17. Singh, R.; Upadhyay, S.K.; Singh, M.; Sharma, I.; Shara, P.; Kamboj, P.; Saini, A.; Voraha, R.; Sharma, A.K.; Upadhyay, T.K.; et al. Chitin, chitinases and chitin derivatives in biopharmaceutical, agricultural and environmental perspective. *Biointerface Res. Appl. Chem.* **2021**, *11*, 9985–10005. [CrossRef]
18. Parada, R.Y.; Egusa, M.; Aklog, Y.F.; Miura, C.; Ifuku, S.; Kaminaka, H. Optimization of nanofibrillation degree of chitin for induction of plant disease resistance: Elicitor activity and systemic resistance induced by chitin nanofiber in cabbage and strawberry. *Int. J. Biol. Macromol.* **2018**, *118 Pt B*, 2185–2192. [CrossRef]
19. Alamdari, S.; Mirzaee, O.; Jahroodi, F.N.; Tafreshi, M.J.; Ghamsari, M.S.; Shik, S.S.; Ara, M.H.M.; Lee, K.-Y.; Park, H.-H. Green synthesis of multifunctional ZnO/chitosan nanocomposite film using wild *Mentha pulegium* extract for packaging applications. *Surf. Interfaces* **2022**, *34*, 102349. [CrossRef]
20. Wei, L.; Zhang, J.; Tan, W.; Wang, G.; Li, Q.; Dong, F.; Guo, Z. Antifungal activity of double Schiff bases of chitosan derivatives bearing active halogeno-benzenes. *Int. J. Biol. Macromol.* **2021**, *179*, 292–298. [CrossRef]
21. Desti, A.; Kavetsou, E.; Kostopoulou, I.; Pitterou, I.; Pontillo, A.R.N.; Tzani, A.; Christodoulou, P.; Siliachli, A.; Zoumpoulakis, P. Nanosystems for the encapsulation of natural products: The case of chitosan biopolymer as a matrix. *Pharmaceutics* **2020**, *12*, 669. [CrossRef]
22. Wahba, M.I. Enhancement of the mechanical properties of chitosan. *J. Biomater. Sci. Poly. Ed.* **2020**, *31*, 30–375. [CrossRef]
23. Landrischina, A.; Rosen, J.; Friedman, A.J. Biodegradable chitosan nanoparticles in drug delivery for infectious disease. *Nanomedicine* **2015**, *10*, 1609–1619. [CrossRef] [PubMed]
24. Islam, N.; Dmour, I.; Taha, M.O. Degradability of chitosan micro/nanoparticles for pulmonary drug delivery. *Heliyon* **2019**, *5*, e01684. [CrossRef]
25. Shahrajaian, M.H.; Sun, W.; Soleymani, A.; Cheng, Q. Traditional herbal medicines to overcome stress, anxiety and improve mental health in outbreaks of human coronaviruses. *Phytother. Res.* **2020**, *35*, 1237–1247. [CrossRef] [PubMed]

26. Shahrajabian, M.H.; Sun, W.; Cheng, Q. Exploring *Artemisia annua* L., artemisinin and its derivatives, from traditional Chinese wonder medicinal science. *Not. Bot. Horti Agrobot. Cluj-Napoca* **2020**, *48*, 1719–1741. [CrossRef]
27. Pichyangkura, R.; Chadchawan, S. Biostimulant activity of chitosan in horticulture. *Sci. Hortic.* **2015**, *196*, 49–65. [CrossRef]
28. Hemmati, F.; Salehi, R.; Ghotaslou, R.; Kafil, H.S.; Hasani, A.; Gholizadeh, P.; Rezaee, M.A. The assessment of antibiofilm activity of chitosan-zinc oxide-gentamicin nanocomposite on *Pseudomonas aeruginosa* and *Staphylococcus aureus*. *Int. J. Biol. Macromol.* **2020**, *163*, 2248–2258. [CrossRef]
29. Shahrajabian, M.H.; Sun, W.; Cheng, Q. The importance of flavonoids and phytochemicals of medicinal plants with antiviral activities. *Mini Rev. Org. Chem.* **2022**, *19*, 293–318. [CrossRef]
30. Shahrajabian, M.H.; Sun, W.; Cheng, Q. Foliar application of nutrients on medicinal and aromatic plants, the sustainable approaches for higher and better production. *Beni-Suef Univ. J. Basic. Appl. Sci.* **2022**, *11*, 26. [CrossRef]
31. Suarez-Fernandez, M.; Marhuenda-Egea, F.C.; Lopez-Moya, F.; Arnao, M.B.; Cabrera-Escribano, F.; Nueda, M.J.; Gunse, B.; Lopez-Llorca, L.V. Chitosan induces plant hormones and defenses in tomato root exudates. *Front. Plant Sci.* **2020**, *11*, 572087. [CrossRef] [PubMed]
32. Ahmad, B.; Khan, M.M.A.; Jahan, A.; Shabbir, A.; Jaleel, H. Increased production of valuable secondary products in plants by leaf applied radiation-processed polysaccharides. *Int. J. Biol. Macromol.* **2020**, *164*, 286–294. [CrossRef] [PubMed]
33. Do, N.H.N.; Truong, Q.T.; Le, P.K.; Ha, A.C. Recent developments in chitosan hydrogels carrying natural bioactive compounds. *Carbohydr. Polym.* **2022**, *294*, 119726. [CrossRef] [PubMed]
34. Roshni, A.; Thambidurai, S. Enhanced photocatalytic and antibacterial activity of ZnO with rice field crab chitosan and *Plectranthus amboinicus* extract. *Mater. Chem. Phys.* **2022**, *291*, 126739. [CrossRef]
35. Sultan, M.; Abdelhakim, A.A.; Nassar, M.; Hassan, Y.R. Active packaging of chitosan film modified with basil oil encapsulated *in silica* nanoparticles as an alternate for plastic packaging materials. *Food Biosci.* **2023**, *51*, 102298. [CrossRef]
36. Kabiriyel, J.; Jeyanthi, R.; Jayakumar, K.; Amalraj, A.; Arjun, P.; Shanmugarathinam, S.; Vignesh, G.; Raja Mohan, C. Green synthesis of carboxyl methyl chitosan based curcumin nanoparticles and its biological activity: Influence of size and conductivity. *Carbohydr. Polym. Technol. Appl.* **2023**, *5*, 100260. [CrossRef]
37. Ibrahim, S.S.; Abou-Elseoud, W.S.; Elbehery, H.H.; Hassan, M.L. Chitosan-cellulose nanoencapsulation systems for enhancing the insecticidal activity of citronella essential oil against the cotton leafworm *Spodoptera littoralis*. *Ind. Crops Prod.* **2022**, *184*, 115089. [CrossRef]
38. Balusamy, S.R.; Rahimi, S.; Sukweenadhi, J.; Sunderraj, S.; Shanmugam, R.; Thangavelu, L.; Mijakovic, I.; Perumalsamy, H. Chitosan, chitosan nanoparticles and modified chitosan biomaterials, a potential tool to combat salinity stress in plants. *Carbohydr. Polym.* **2022**, *284*, 119189. [CrossRef]
39. Ji, H.; Wang, J.; Chen, F.; Fan, N.; Wang, X.; Xiao, Z.; Wang, Z. Meta-analysis of chitosan-mediated effects on plant defense against oxidative stress. *Sci. Total Environ.* **2022**, *851 Pt 1*, 158212. [CrossRef]
40. Khairy, A.M.; Tohamy, M.R.A.; Zayed, M.A.; Mahmoud, S.F.; El-Tahan, A.M.; El-Saadony, M.T.; Mesiha, P.K. Eco-friendly application of nano-chitosan for controlling potato and tomato bacterial wilt. *Saudi J. Biol. Sci.* **2022**, *29*, 2199–2209. [CrossRef]
41. Du, M.; Yi, Y.; Yin, Y.; Cai, Z.; Cai, W.; Li, J.; He, G.; Zhang, J. Bacterial-triggered photodynamic nano-system based on hematoporphyrin-modified chitosan for sustainable plant disease control. *Eur. Polym. J.* **2023**, *191*, 112035. [CrossRef]
42. Riseh, R.S.; Hassanisaadi, M.; Vatankhah, M.; Babaki, S.A.; Barka, E.A. Chitosan as a potential natural compound to manage plant diseases. *Int. J. Biol. Macromol.* **2022**, *220*, 998–1009. [CrossRef] [PubMed]
43. Sathiyabama, M.; Gandhi, M.; Indhumathi, M. Suppression of dry root rot disease caused by *Rhizoctonia bataticola* (Taub.) Butler in chickpea plants by application of thiamine loaded chitosan nanoparticles. *Microb. Pathog.* **2022**, *173 Pt B*, 105893. [CrossRef]
44. Fan, Z.; Wang, L.; Qin, Y.; Li, P. Activity of chitin/chitosan/chitosan oligosaccharide against plant pathogenic nematodes and potential modes of application in agriculture: A review. *Carbohydr. Polym.* **2023**, *306*, 120592. [CrossRef] [PubMed]
45. Zhao, X.; She, X.; Du, Y.; Liang, X. Induction of antiviral resistance and stimulary effect by oligochitosan in tobacco. *Pestic. Biochem. Physiol.* **2007**, *87*, 78–84. [CrossRef]
46. Yin, H.; Zhao, X.; Du, Y. Oligochitosan: A plant diseases vaccine—A review. *Carbohydr. Polym.* **2010**, *82*, 1–8. [CrossRef]
47. Yang, R.; Zuo, P.; Zhang, M.; Meng, D.; Wang, B.; Zhen, T. Transglutaminase induced oligochitosan glycosylation of ferritin as a novel nanocarrier for food bioactive molecules. *Food Hydrocoll.* **2019**, *94*, 500–509. [CrossRef]
48. Lee, T.; Chang, Y.H. Structural, physicochemical, and in vitro release properties of hydrogel beads produced by oligochitosan and de-esterified pectin from yuzu (*Citrus junos*) peel as a quercetin delivery system for colon target. *Food Hydrocoll.* **2020**, *108*, 106086. [CrossRef]
49. Yin, H.; Li, S.; Zhao, X.; Du, Y.; Ma, X. cDNA microarray analysis of gene expression in *Brassica napus* treated with oligochitosan elicitor. *Plant Physiol. Biochem.* **2006**, *44*, 910–916. [CrossRef]
50. Svirshchevskaya, E.V.; Alekseeva, L.G.; Reshetov, P.D.; Phomicheva, N.N.; Parphenyuk, S.A.; Ilyina, A.V.; Zueva, V.S.; Lopatin, S.A.; Levov, A.N.; Varlamov, V.P. Mucoadjuvant properties of lipo- and glycoconjugated derivatives of oligochitosans. *Eur. J. Med. Chem.* **2009**, *44*, 2030–2037. [CrossRef]
51. Feng, J.; Zhao, L.; Yu, Q. Receptor-mediated stimulatory effect of oligochitosan in macrophages. *Biochem. Biophys. Res. Commun.* **2004**, *317*, 414–420. [CrossRef] [PubMed]
52. Han, Y.; Zhao, L.; Yu, Z.; Yu, Q. Role of mannose receptor in oligochitosan-mediated stimulation of macrophage function. *Int. Immunopharmacol.* **2005**, *5*, 1533–1542. [CrossRef] [PubMed]

53. Li, Y.; Yin, H.; Wang, Q.; Zhao, X.; Du, Y.; Li, F. Oligochitosan induced *Brassica napus* L. production of NO and $H_2O_2$ and their physiological function. *Carbohydr. Polym.* **2009**, *75*, 612–617. [CrossRef]
54. Lu, H.; Zhao, X.; Bai, X.; Du, Y. Involvement of *N*-mediated defense in oligochitosan inducing resistance to tobacco mosaic virus. *J. Biotechnol.* **2008**, *136*, S581–S582. [CrossRef]
55. Zhang, X.; Zhou, H.; Han, Z.; Huang, W.; Gu, X.; Li, B.; Zhao, L.; Zhou, S.; Zhang, H. *Pichia caribbica* combined with oligochitosan controlling black spot of tomatoes and the regulation on ROS metabolism of the fruits. *Biol. Control* **2022**, *176*, 105109. [CrossRef]
56. Deng, L.; Zhou, Y.; Zeng, K. Pre-harvest spray of oligochitosan induced the resistance of harvested navel oranges to anthracnose during ambient temperature storage. *Crop. Prot.* **2015**, *70*, 70–76. [CrossRef]
57. Ma, Z.; Yang, L.; Yan, H.; Kennedy, J.F.; Meng, X. Chitosan and oligochitosan enhance the resistance of peach fruit to brown rot. *Carbohydr. Polym.* **2013**, *94*, 272–277. [CrossRef]
58. Zhou, Y.; Ma, J.; Xie, J.; Deng, L.; Yao, X.; Zeng, K. Transcriptomic and biochemical analysis of highlighted induction of phenylpropanoid pathway metabolism of citrus fruit in response to salicylic acid, *Pichia membranaefaciens* and oligochitosan. *Postharvest Biol. Technol.* **2018**, *142*, 81–92. [CrossRef]
59. Yang, L.-Y.; Zhang, J.-L.; Bassett, C.L.; Meng, X.-H. Difference between chitosan and oligochitosan in growth of *Monilinia fructicola* and control of brown rot in peach fruit. *LWT-Food. Sci. Technol.* **2012**, *46*, 254–259. [CrossRef]
60. Wang, S.; Zhou, Y.; Luo, W.; Deng, L.; Yao, S.; Zeng, K. Primary metabolites analysis of induced citrus fruit disease resistance upon treatment with oligochitosan, salicylic acid and *Pichia membranaefaciens*. *Biol. Control* **2020**, *148*, 104289. [CrossRef]
61. Shahrajabian, M.H.; Chaski, C.; Polyzos, N.; Tzortzakis, N.; Petropoulos, S.A. Sustainable agriculture systems in vegetable producing using chitin and chitosan as plant biostimulants. *Biomolecules* **2021**, *11*, 819. [CrossRef] [PubMed]
62. Shahrajabian, M.H.; Sun, W. Sustainable approaches to boost yield and chemical constituents of aromatic and medicinal plants by application of biostimulants. *Recent Adv. Food Nutr. Agric.* **2022**, *13*, 72–92. [CrossRef]
63. Shahrajabian, M.H.; Sun, W.; Cheng, Q. Using bacteria and fungi as plant biostimulants for sustainable agricultural production systems. *Recent Pat. Biotechnol.* **2022**, *16*, 206–244. [CrossRef]
64. Shahrajabian, M.H.; Petropoulos, S.A.; Sun, W. Survey of the influences of microbial biostimulants on horticultural crops: Case studies and successful paradigms. *Horticulturae* **2023**, *9*, 193. [CrossRef]
65. Prabhu, M.; Kumar, A.R.; Rajamani, L. Influence of different organic substances on growth and herb yield of sacred basil (*Ocimum sanctum* L.). *Indian J. Agric. Res.* **2010**, *44*, 48–52. [CrossRef]
66. Kwiatkowski, C.A.; Juszczak, J. The response of sweet basil (*Ocimum basilicum* L.) to the application of growth stimulators and forecrops. *Acta Agrobot.* **2012**, *64*, 69–76. [CrossRef]
67. Elhindi, K.M.; Al-Amri, S.M.; Abdel-Salam, E.M.; Al-Suhaibani, N.A. Effectiveness of salicylic acid in mitigating salt-induced adverse effects on different physico-biochemical attributes in sweet basil (*Ocimum basilicum* L.). *J. Plant Nutr.* **2017**, *40*, 908–919. [CrossRef]
68. Damalas, C.A. Improving drought tolerance in sweet basil (*Ocimum basilicum*) with salicylic acid. *Sci. Hortic* **2019**, *246*, 360–365. [CrossRef]
69. Shirkhodaei, M.; Darzi, M.T.; Haj Seyed Hadi, M.R. Influence of vermicompost and biostimulant on the growth and biomass of coriander (*Coriandrum sativum* L.). *Int. J. Adv. Biol. Biomed. Res.* **2014**, *2*, 706–714.
70. Pokluda, R.; Sekara, A.; Jezdinsky, A.; Kalisz, A.; Neugebauerova, J.; Grabowska, A. The physiological status and stress biomarker concentration of *Coriandrum sativum* L. plants subjected to chilling are modified by biostimulant application. *Biol. Agric. Hortic.* **2016**, *32*, 258–268. [CrossRef]
71. Lisjak, M.; Tomic, O.; Spoljarevic, M.; Teklic, T.; Stanisavljevic, A.; Balas, J. Garden cress germinability and seedling vigour after treatment with plant extracts. *Poljoprivreda (Osijek)* **2015**, *21*, 41–46. [CrossRef]
72. Abdel-Razz, H.S.; El-Sharkaw, G.A. Effect of biofertilizer and humic acid applications on growth, yield, quality and storability of two garlic (*Allium sativum* L.) cultivars. *Asian J. Crop Sci.* **2012**, *5*, 48–64. [CrossRef]
73. Manas, D.; Soumya, G.; Kheyali, S. Effect of humic acid application on accumulation of mineral nutrition and pungency in garlic (*Allium sativum* L.). *Int. J. Biotechnol. Mol. Biol. Res.* **2014**, *5*, 7–12. [CrossRef]
74. Jedrszczyk, E.; Kopec, A.; Bucki, P.; Ambroszczyk, A.M.; Skowera, B. The enhancing effect of plants growth biostimulants in garlic cultivation on the chemical composition and level of bioactive compounds in the garlic leaves, stems and bulbs. *Not. Bot. Horti Agrobot. Cluj-Napoca* **2018**, *47*, 81–91. [CrossRef]
75. Sharifi, Y.; Nematzadeh, G.A.; Ghasemi, O.V.; Tavabe, G.T.S.; Ebrahimzadeh, M.A. Effect of salicyic acid on phenols and flavonoids content in callus culture of Iranian sodab (*Ruta graveolens*): A threatened medicinal plant of North of Iran. *Tabari Biomed. Stud. Res. J.* **2019**, *1*, 32–36. [CrossRef]
76. Safari, F.; Akramian, M.; Salehi-Arjmand, H.; Khadivi, A. Physiological and molecular mechanisms underlying salicylic acid-mitigated mercury toxicity in lemon balm (*Melissa officinalis* L.). *Ecotoxicol. Environ. Saf.* **2019**, *183*, 109542. [CrossRef]
77. Hatami, M.; Khanizadeh, P.; Bovand, F.; Aghaee, A. Silicon nanoparticle-mediated seed priming and Pseudomonas spp. Inoculation augment growth, physiology and antioxidant metabolic status in *Melissa officinalis* L. plants. *Ind. Crops Prod.* **2021**, *162*, 113238. [CrossRef]
78. Moustafa, H.E.B.; Ahmed, S.S.; Shahin, S.M. Effect of foliar spray with potassium silicate on growth and active constituents of horseradish armiractetra (*Moringa oleifera* Lam.) plants grown in some soils of Egypt. *Middle East J. Agric. Res.* **2018**, *7*, 60–70. [CrossRef]

79. Fett-Neto, A.G.; Melanson, S.J.; Nicholson, S.A.; Pennington, J.J.; DiCosmo, F. Improved taxol yield by aromatic carboxylic acid and amino acid feeding to cell cultures of *Taxus cuspidata*. *Biotechnol. Bioeng.* **1994**, *44*, 967–971. [CrossRef]
80. Russo, R.; Poincelot, R.P.; Berlyn, G.P. The use of a commercial organic biostimulant for improved production of marigold cultivars. *J. Home. Consumer. Horticultr.* **1993**, *1*, 83–93. [CrossRef]
81. Rafiee, H.; Mehrafarin, A.; Qaderi, A.; Kalate Jari, S.; Naghdi Badi, H. Phytochemical, agronomical and morphological responses of pot Marigold (*Calendula officinalis* L.) to foliar application of bio-stimulators (Bioactive amino acid compounds). *Faslnamahi Giyahan-I Daruyi* **2013**, *12*, 48–61.
82. Zeljkovic, S.; Paradikovic, N.; Vinkovic, T.; Tkalec, M.; Maksimovic, I.; Haramija, J. Nutrient status, growth and proline concentration of French marigold (*Tagetes patula* L.) as affected by biostimulant treatment. *J. Food Agric. Environ.* **2013**, *11*, 2324–2327. [CrossRef]
83. Safikhan, S.; Khoshbakht, K.; Chaichi, M.R.; Amini, A.; Motesharezadeh, B. Role of chitosan on the growth, physiological parameters and enzymatic activity of milk thistle (*Silybum marianum* (L.) Gaertn.) in a pot experiment. *J. Appl. Red. Med. Aromat. Plants* **2018**, *10*, 49–58. [CrossRef]
84. Estaji, A.; Niknam, F. Foliar salicylic acid spraying effect on growth, seed oil content, and physiology of drought-stressed *Silybum marianum* L. plant. *Agric. Water Manag.* **2020**, *234*, 106116. [CrossRef]
85. Ratnakumari, R.; Nagamani, A.; Sarojini, C.K.; Adinarayana, G. Effect of *Trichoderma* species on yield of *Mentha arvensis* L. *Int. J. Adv. Res. (Indore)* **2014**, *2*, 864–867. [CrossRef]
86. Elansary, H.O.; Mahmoud, E.A.; El-Ansary, D.O.; Mattar, M.A. Effects of water stress and modern biostimulants on growth and quality characteristics of mint. *Agronomy* **2020**, *10*, 6. [CrossRef]
87. Samadimatin, A.; Hani, A. Effect of ethanol and humic acid foliar spraying on morphological traits, photosynthetic pigments, and quality and quantity of essential oil content of *Dracocephalum moldavica* L. *Iran. J. Plant Physiol.* **2017**, *8*, 2299–2306.
88. Saharkhiz, M.J.; Goudarzi, T. Folliar application of salicylic acid changes essential oil content and chemical compositions of peppermint (*Mentha piperita* L.). *J. Essent. Oil-Bear. Plants* **2014**, *17*, 435–440. [CrossRef]
89. Santoro, M.V.; Capperllari, L.R.; Giordano, W.; Banchio, E. Plant growth-promoting effects of native *Pseudomonas* strains on *Mentha piperita* (peppermint): An in vitro study. *Plant Biol.* **2015**, *17*, 1218–1226. [CrossRef]
90. Shahabivan, S.; Padash, A.; Aghaee, A.; Nasiri, Y.; Fathi Rezaei, P. Plant biostimulants (*Funnerliformis mosseae* and humic substances) rather than chemical fertilizer improved biochemical responses in peppermint. *Iran. J. Plant Physiol.* **2018**, *8*, 2333–2344.
91. Caruso, G.; De Pacale, S.; Cozzolino, R.; Giordano, M.; El-Nakhel, C.; Cuciniello, A.; Cenvinzo, V.; Colla, G.; Rouphael, Y. Protein hydrolysate or plant extract-based biostimulants enhanced yield and quality performances of greenhouse perennial wall rocket growth in different seasons. *Plants* **2019**, *8*, 208. [CrossRef] [PubMed]
92. Goudarzian, A.; Pirbalouti, A.G.; Hossaynzadeh, M. Menthol balance of menthol/menthone, and essential oil contents of *Mentha piperita* L. under foliar-applied chitosan and inoculation of arbuscular mycorrhizal fungi. *J. Essent. Oil-Bear. Plants* **2020**, *23*, 1012–1021. [CrossRef]
93. Praveen Kumar, G.; Desai, S.; Reddu, G.; Leo Daniel Amalraj, E.; Rasul, A.; Mir Hassan Ahmed, S.K. Seed bacterization with fluorescent *Pseudomonas* spp. Enhances nutrient uptake and growth of *Cajanus cajan* L. *Commun. Soil Sci. Plant Anal.* **2015**, *46*, 652–665. [CrossRef]
94. Kaur, P.; Gupta, R.C.; Dey, A.; Malik, T.; Pandey, D.K. Optimization of salicylic acid and chitosan treatment for bitter secoiridoid and xanthone glycosides production in shoot cultures of *Swertia paniculata* using response surface methodology an artificial neural network. *BMC Plant Biol.* **2020**, *20*, 225. [CrossRef] [PubMed]
95. Bektas, N.; Senel, B.; Yenilmez, E.; Ozatik, O.; Arslan, R. Evaluation of wound healing effect of chitosan-based gel formulation containing vitexin. *Saudi Pharm. J.* **2020**, *28*, 87–94. [CrossRef]
96. Kaboudi, Z.; Peighambardoust, S.H.; Nourbakhsh, H.; Soltanzadeh, M. Nanoencapsulation of Chavir (*Ferulago angulata*) essential oil in chitosan carrier: Investigating physicochemical, morphological, thermal, antimicrobial and release profile of obtained nanoparticles. *Int. J. Biol. Macromol.* **2023**, *237*, 123963. [CrossRef] [PubMed]
97. Yin, W.; Yan, R.; Zhou, X.; Li, X.; Sang, S.; McClements, D.J.; Chen, L.; Long, J.; Jiao, A.; Wang, J.; et al. Preparation of robust, water-resistant, antibacterial, and antioxidant chitosan-based films by incorporation of cinnamaldehyde-tannin acid-zinc acetate nanoparticles. *Food Chem.* **2023**, *419*, 136004. [CrossRef]
98. Yadav, N.; Mudgal, D.; Anand, R.; Jindal, S.; Mishra, V. Recent development in nanoencapsulation and delivery of natural bioactives through chitosan scaffolds for various biological applications. *Int. J. Biol. Macromol.* **2022**, *220*, 537–572. [CrossRef]
99. Mujtaba, M.; Ali, Q.; Yilmaz, B.A.; Kurubas, M.S.; Ustun, H.; Erkan, M.; Kaya, M.; Cicek, M.; Oner, E.T. Understanding the effects of chitosan, chia mucilage, levan based composite coatings on the shelf life of sweet cherry. *Food Chem.* **2023**, *416*, 135816. [CrossRef]
100. Das, S.; Chaudhari, A.K.; Singh, V.K.; Dwivedy, A.K.; Dubey, N.K. Chitosan based encapsulation of Valeriana officinalis essential oil as edible coating for inhibiting of fungi and aflatoxin $B_1$ contamination, nutritional quality improvement, and shelf life extension of *Citrus sinensis* fruits. *Int. J. Biol. Macromol.* **2023**, *233*, 123565. [CrossRef]
101. Shu, H.; Ge, Y.-H.; Xu, X.-Y.; Guo, P.-Q.; Luo, Z.-M.; Du, W.; Chang, C.; Liu, R.-L.; Fu, Q. Hybrid-type carbon microcoil-chitosan composite for selective extraction of aristolochic acid I from *Aristolochiaceae* medicinal plants. *J. Chromatogr. A* **2018**, *1561*, 13–19. [CrossRef]

102. Qiu, H.; Su, L.; Wang, H.; Zhang, Z. Chitosan elicitation of saponin accumulation in *Psammosilene tunicoides* hairy roots by modulating antioxidant activity oxide production and differential gene expression. *Plant Physiol. Biochem.* **2021**, *166*, 115–127. [CrossRef] [PubMed]
103. Hawrylak-Nowak, B.; Dresler, S.; Rubinowska, K.; Matraszek-Gawron, R. Eliciting effect of foliar application of chitosan lactate on the phytochemical properties of *Ocimum basilicum* L. and *Melissa officinalis* L. *Food Chem.* **2021**, *342*, 128358. [CrossRef] [PubMed]
104. Aldayel, M.F. The synergistic effect of capsicum aqueous extract (*Capsicum annuum*) and chitosan against multidrug-resistant bacteria. *J. King Saud Univ. Sci.* **2023**, *35*, 102438. [CrossRef]
105. Khodadadi, F.; Ahmadi, F.S.; Talebi, M.; Moshtaghi, N.; Matkowski, A.; Szumny, A.; Rahimmalek, M. Essential oil composition, physiological and morphological in *Salvia abrotanoides* and *S. yangii* under drought stress and chitosan treatments. *Ind. Crop. Prod.* **2022**, *187 Pt B*, 115429. [CrossRef]
106. Wang, Y.; Yang, L.; Zhou, X.; Wang, Y.; Liang, Y.; Luo, B.; Dai, Y.; Wei, Z.; Li, S.; He, R.; et al. Molecular mechanism of plant elicitor daphnetin-carboxymethyl chitosan nanoparticles against *Ralstonia solanacearum* by activating plant system resistance. *Int. J. Biol. Macromol.* **2023**, *241*, 124580. [CrossRef]
107. Mathew, R.; Sankar, P.D. Effect of methyl jasmonate and chitosan on growth characteristics of *Ocimum basilicum* L., *Ocimum santum* L. and *Ocimum gratissimum* L. cell suspension cultures. *Afr. J. Biotechnol.* **2012**, *11*, 4759–4766. [CrossRef]
108. Pirbalouti, A.G.; Malekpoor, F.; Salimi, A.; Golparvar, A. Exogenous application of chitosan on biochemical and physiological characteristics, phenolic content and antioxidant activity of two species of basil (*Ocimum ciliatum* and *Ocimum basilicum*) under reduced irrigation. *Sci. Hortic.* **2017**, *217*, 114–122. [CrossRef]
109. Sheikhalipour, M.; Mohammadi, S.A.; Esmaielpour, B.; Spanos, A.; Mahmoudi, R.; Mahdavinia, G.R.; Milani, M.H.; Kahnamoei, A.; Nouraein, M.; Antoniou, C.; et al. Seedling nanopriming with selenium-chitosan nanoparticles mitigates the adverse effects of salt stress by inducing multiple defence pathways in bitter melon plants. *Int. J. Biol. Macromol.* **2023**, *242*, 124923. [CrossRef]
110. Limpanavech, P.; Chaiyasuta, S.; Vongpromek, R.; Pichyangkura, R.; Khunwasi, C.; Chadchawan, S.; Lotrakul, P.; Bunjongrat, R.; Chaidee, A.; Bangyeekhun, T. Chitosan effects on floral production, gene expression, and anatomical changes in the *Dendrobium* orchid. *Sci. Hortic.* **2008**, *116*, 65–72. [CrossRef]
111. Kahromi, S.; Khara, J. Chitosan stimulates secondary metabolites production and nutrient uptake in medicinal plant *Dracocephalum kotschyi*. *J. Sci. Food Agric.* **2020**, *101*, 3898–3907. [CrossRef] [PubMed]
112. Razavizadeh, R.; Adabavazeh, F. In vitro application of chitosan effects on essential oil content and physiological characteristics of *Dracocephalum kotschyi* Boiss. *J. Plant Process. Funct.* **2019**, *9*, 23–30.
113. Dar, T.A.; Uddin, M.; Khan, M.M.; Ali, A.; Mir, S.R.; Varshney, L. Effect of Co-60 gamma irradiated chitosan and phosphorus fertilizer on growth, yield and trigonelline content of *Trigonella foenum-graecum* L. *J. Radiat. Res. Appl. Sci.* **2015**, *8*, 446–458. [CrossRef]
114. Angouti, F.; Nourafcan, H.; Saeedi Sar, S.; Asadi, A.; Ebrahimi, R. The effect of different levels of chitosan and salicylic acid on morphological traits of the medicinal plant Galega (*Galega officinalis* L.). *J. Crop. Improv.* **2022**, *24*, 1341–1358. [CrossRef]
115. Abdul-Hafeez, E.Y.; Ibrahim, O.H.M. Effects of chitosan and BABA foliar application on flowering and chemical characteristics of German chamomile Bode-gold. *S. Afr. J. Bot.* **2021**, *139*, 241–245. [CrossRef]
116. Varghese, L.; Thomas, G. Chitosan triggers tolerance to *Pythium myriotylum* infection in ginger (*Zingiber officinale*) by modulating multiple defense signaling pathways. *Physiol. Mol. Plant Pathol.* **2023**, *125*, 101983. [CrossRef]
117. Samany, S.M.A.; Pirbalouti, A.G.; Malekpoor, F. Phytochemical and morpho-physiological changes of hyssop in response to chitosan-spraying under different levels of irrigation. *Ind. Crops Prod.* **2022**, *176*, 114340. [CrossRef]
118. Govindaraju, S.; Arulselvi, P.I. Effect of cytokinin combined elicitors (L-phenylalanine, salicylic acid and chitosan) on in vitro propagation, secondary metabolites and molecular characterization of medicinal herb- *Coleus aromaticus Benth* (L.). *J. Saudi Soc. Agric. Sci.* **2018**, *17*, 435–444. [CrossRef]
119. Elguea-Culebras, G.O.D.; Bourbon, A.; Costa, M.J.; Munoz-Tebar, N.; Carmona, M.; Molina, A.; Sanchez-Vioque, R.; Berruga, M.I.; Vicente, A.A. Optimization of a chitosan solution as potential carrier for the incorporation of *Santolina chamaecyparissus* L. solid by product in an edible vegetal coating on *Manchego* cheese. *Food Hydrocoll.* **2019**, *89*, 272–282. [CrossRef]
120. Coskun, Y.; Duran, R.E.; Ozgul, Z.; Yavuz, N.; Turkmen, K. Chitosan improves plant regeneration in callus culture of a medicinal herb *Melissa officinalis* L. *J. Biotechnol.* **2015**, *208*, S107. [CrossRef]
121. Fooladi vanda, G.; Shabani, L.; Razavizadeh, R. Chitosan enhances rosmarinic acid production in shoot cultures of *Melissa officinalis* L. through the induction of methyl jasmonate. *Bot. Stud.* **2019**, *60*, 26. [CrossRef] [PubMed]
122. Zubair, M.; Ramzani, P.M.A.; Rasool, B.; Khan, M.A.; ur-Rahman, M.; Akhtar, I.; Turan, V.; Tauqeer, H.M.; Farhad, M.; Khan, S.A.; et al. Efficacy of chitosan-coated textile waste biochar applied to Cd-polluted soil for reducing Cd mobility in soil and its distribution in moringa (*Moringa oleifera* L.). *J. Environ. Manag.* **2021**, *284*, 112047. [CrossRef] [PubMed]
123. Giglou, M.T.; Giglou, R.H.; Esmaeilpour, B.; Azarmi, R.; Padash, A.; Falakian, M.; Sliwka, J.; Gohari, G.; Lajayer, H.M. A new method in mitigation of drought stress by chitosan-coated iron oxide nanoparticles and growth stimulant in peppermint. *Ind. Crops Prod.* **2022**, *187 Pt A*, 115286. [CrossRef]
124. Mirheidari, F.; Hatami, M.; Ghorbanpour, M. $GA_3$ and chitosan nano-fiber on physio-morphological characteristics and metabolite contents in roselle (*Hibiscus sabdariffa* L.). *S. Afr. J. Bot.* **2022**, *145*, 323–333. [CrossRef]
125. Alizadeh, A.; Moghaddam, M.; Asgharzade, A.; Sourestani, M.M. Phytochemical and physiological response of *Satureja hortensis* L. to different irrigation regimes and chitosan application. *Ind. Crops Prod.* **2020**, *158*, 112990. [CrossRef]

126. Gohari, G.; Farhadi, H.; Panahirad, S.; Zareei, E.; Labib, P.; Jafari, H.; Mahdavinia, G.; Hassanpouraghdam, M.B.; Ioannou, A.; Kulak, M.; et al. Mitigating of salinity impact in spearmint plants through the application of engineered chitosan-melatonin nanoparticles. *Int. J. Biol. Macromol.* **2023**, *224*, 893–907. [CrossRef]
127. Tocci, N.; Ferrari, F.; Santamaria, A.R.; Valletta, A.; Rovardi, I.; Pasqua, G. Chitosan enhances xanthone production in *Hypericum perforatum* subsp. *angustifolium* cell cultures. *Nat. Prod. Res.* **2010**, *24*, 286–293. [CrossRef]
128. Sayed, T.E.; Ahmed, E.-S.S. Elicitation promoability with gamma irradiation, chitosan and yeast to perform sustainable and inclusive development for Marjoram under organic agriculture. *Sustainability* **2022**, *14*, 9608. [CrossRef]
129. Kamalipourazad, M.; Sharifi, M.; Maivan, H.Z.; Behmanesh, M.; Chashimi, N.A. Induction of aromatic amino acids and phenylpropanoid compounds in *Scrophularia striata* Boiss. cell culture in response to chitosan-induced oxidative stress. *Plant Physiol. Biochem.* **2016**, *107*, 374–384. [CrossRef]
130. Bistgani, Z.E.; Siadat, S.A.; Bakhshandeh, A.; Pirbalouti, A.G.; Hashemi, M. Interactive effects of drought stress and chitosan application on physiological characteristics and essential oil yield of *Thymus daenensis* Celak. *Crop. J.* **2017**, *5*, 407–415. [CrossRef]
131. Momeni, M.; Pirbalouti, A.G.; Mousavi, A.; Naghdi Badi, H. Effect of foliar applications of salicylic acid and chitosan on the essential oil of *Thymbra spicata* L. under different soil moisture conditions. *J. Essent. Oil Bear. Plants* **2020**, *23*, 1142–1153. [CrossRef]
132. Dowom, S.A.; Karimian, Z.; Dehnavi, M.M.; Samiei, L. Chitosan nanoparticles improve physiological and biochemical responses of *Salvia abrotanoides* (Kar.) under drought stress. *BMC Plant Biol.* **2022**, *22*, 364. [CrossRef]
133. Sathiyabama, M.; Bernstein, N.; Anusuya, S. Chitosan elicitation for increased curcumin production and stimulation of defence response in turmeric (*Curcuma longa* L.). *Ind. Crop. Prod.* **2016**, *89*, 87–94. [CrossRef]
134. Silva, E.A.D.; Paula, A.C.C.F.F.D.; Silva, V.N.B.; Alvarenga, A.A.D.; Bertolucci, S.K.V. Biostimulating effect of chitosan and acetic acid on the growth and profile of the essential oil of *Mentha arvensis* L. *Ind. Crops Prod.* **2021**, *171*, 113987. [CrossRef]
135. Heidari, J.; Amooaghaie, R.; Kiani, S. Impact of chitosan on nickel bioavailability in soil, the accumulation and tolerance of nickel in *Calendula triptrocarpa*. *Int. J. Phytoremediat.* **2020**, *22*, 1175–1184. [CrossRef] [PubMed]
136. Darvill, A.; Augur, C.; Bergmann, C.; Carlson, R.W.; Cheong, J.-J.; Eberhard, S.; Hahn, M.G.; Lo, V.-M.; Marfa, V.; Meyer, B.; et al. Oligosaccharins-oligosaccharides that regulate growth, development and defence responses in plants. *Glycobiology* **1992**, *2*, 181–198. [CrossRef] [PubMed]
137. Zhang, B.; Ramonell, K.; Somerville, S.; Stacey, G. Characterization of early, chitin-induced gene expression in Arabidopsis. *Mol. Plant Microbe Interact.* **2002**, *15*, 963–970. [CrossRef]
138. Moenne, A.; Gonzalez, A. Chitosan-, alginate- carrageenan-derived oligosaccharides stimulate defense against biotic and abiotic stresses, and growth in plants: A historical perspective. *Carbohydr. Res.* **2021**, *503*, 108298. [CrossRef]
139. El-Serafy, R.S.; El-Sheshtawy, A.-N.; Dahab, A.A.; Al-Ashkar, I. Can yeast extract and chitosan-oligosaccharide improve fruit yield and modify the pharmaceutical active ingredients of organic fennel? *Ind. Crop. Prod.* **2021**, *173*, 114130. [CrossRef]
140. Liu, Y.; Wisniewski, M.; Kennedy, J.F.; Jiang, Y.; Tang, J.; Liu, J. Chitosan and oligochitosan enhance ginger (*Zingiber officinale* Roscoe) resistance to rhizome rot caused by *Fusarium oxysporum* in storage. *Carbohydr. Polym.* **2016**, *151*, 474–479. [CrossRef]
141. Li, T.; Zhang, Y.; Xu, M.; Liu, Y.; Zhang, C.; Zhang, Y.; Peng, X.; Li, Z.; Qin, S.; Xing, K. Novel antifungal mechanism of oligochitosan by triggering apoptosis through a metacaspase-dependent mitochondrial pathway in *Ceratocystis fimbriata*. *Carbohydr. Polym.* **2020**, *245*, 116574. [CrossRef] [PubMed]
142. Lu, H.; Zhao, X.; Wang, W.; Yin, H.; Xu, J.; Bai, X.; Du, Y. Inhibition effect on *tobacco mosaic virus* and regulation effect on calreticulin of oligochitosan in tobacco by induced $Ca^{2+}$ influx. *Carbohydr. Polym.* **2010**, *82*, 136–142. [CrossRef]
143. Ahmad, B.; Dar, T.A.; Khan, M.M.A.; Ahmad, A.; Rinklebe, J.; Chen, Y.; Ahmad, P. Oligochitosan fortifies antioxidative and photosynthetic metabolism and enhances secondary metabolite accumulation in arsenic-stressed peppermint. *Front. Plant. Sci.* **2022**, *13*, 987746. [CrossRef] [PubMed]
144. Yin, H.; Kjaer, A.; Frette, X.C.; Du, Y.; Christensen, L.P.; Jensen, M.; Grevsen, K. Chitosan oligosaccharide and salicylic acid up-regulate gene expression differently in relation to the biosynthesis of artemisinin in *Artemisia annua* L. *Process. Biochem.* **2012**, *47*, 1559–1562. [CrossRef]
145. Takagi, M.; Hotamori, K.; Naito, K.; Matsukawa, S.; Egusa, M.; Nishizawa, Y.; Kanno, Y.; Seo, M.; Ifuku, S.; Mine, A.; et al. Chitin-induced systemic disease resistance in rice requires both OsCERK1 and OsCEBiP and is mediated via perturbation of cell-wall biogenesis in leaves. *Front. Plant Sci.* **2022**, *13*, 1064628. [CrossRef] [PubMed]
146. Suwanchaikasen, P.; Nie, S.; Idnurm, A.; Selby-Pham, J.; Walker, R.; Boughton, B.A. Effects of chitin and chitosan on root growth, biochemical defense response and exudate proteome of *Cannabis sativa*. *Plant Environ. Interact.* **2023**, *4*, 115–133. [CrossRef]
147. Sun, W.; Shahrajabian, M.H.; Cheng, Q. Barberry (*Berberis vulgaris*), a medicinal fruit and food with traditional and modern pharmaceutical uses. *Isr. J. Plant Sci.* **2021**, *68*, 1–11. [CrossRef]
148. Sun, W.; Shahrajabian, M.H.; Cheng, Q. Natural dietary and medicinal plants with anti-obesity therapeutics activities for treatment and prevention of obesity during lock down and in post-Covid-19 era. *Appl. Sci.* **2021**, *11*, 7889. [CrossRef]
149. Sun, W.; Shahrajabian, M.H.; Lin, M. Research progress of fermented functional foods and protein factory-microbial fermentation technology. *Fermentation* **2022**, *8*, 688. [CrossRef]
150. Cretoiu, M.S.; Korthals, G.W.; Visser, J.H.M.; Elsas, D.V. Chitin amendment increases soil suppessiveness toward plan pathogens and modulates the actinobacterial and oxalobacteraceal communities in an experimental agricultural field. *Appl. Environ. Microbiol.* **2013**, *79*, 5291–5301. [CrossRef]

151. Liopa-Tsakalidi, A.; Chalikiopoulos, D.; Papasavvas, A. Effect of chitin on growth and chlcrophyll content of two medicinal plants. *J. Med. Plant. Res.* **2010**, *4*, 499–508.
52. Liopa-Tsakalidi, A.; Barouchas, P.E. Salinity, chitin and GA3 effects on seed germination of chervil (*Anthriscus cerefolium*). *Aust. J. Crop Sci.* **2011**, *5*, 973–978.
53. Kanawi, M.A.; Haydar, M.A.; Radhi, W.N. Effect of chitin and chitosan in improvement of plant growth and anti-fungal activity. *Egypt. J. Bot.* **2021**, *61*, 513–519. [CrossRef]
54. Tian, L.; Shi, S.; Ma, L.; Zhou, X.; Luo, S.; Zhang, J.; Lu, B.; Tian, C. The effect of *Glomus intraradices* on the physiological properties of *Panax ginseng* and on rhizospheric microbial diversity. *J. Ginseng. Res.* **2019**, *43*, 77–85. [CrossRef] [PubMed]

**Disclaimer/Publisher's Note:** The statements, opinions and data contained in all publications are solely those of the individual author(s) and contributor(s) and not of MDPI and/or the editor(s). MDPI and/or the editor(s) disclaim responsibility for any injury to people or property resulting from any ideas, methods, instructions or products referred to in the content.

MDPI AG
Grosspeteranlage 5
4052 Basel
Switzerland
Tel.: +41 61 683 77 34

*Plants* Editorial Office
E-mail: plants@mdpi.com
www.mdpi.com/journal/plants

Disclaimer/Publisher's Note: The statements, opinions and data contained in all publications are solely those of the individual author(s) and contributor(s) and not of MDPI and/or the editor(s). MDPI and/or the editor(s) disclaim responsibility for any injury to people or property resulting from any ideas, methods, instructions or products referred to in the content.

www.ingramcontent.com/pod-product-compliance
Lightning Source LLC
Chambersburg PA
CBHW041359210926
51850CB00013B/3130
* 9 7 8 3 7 2 5 8 2 3 6 5 9 *